I0084832

Women Feeding Cities

Praise for the book...

'One of the key findings of the recent International Assessment of Agricultural Knowledge, Science and Technology for Development was that strengthening and redirecting AKST to address gender issues will advance progress toward achievement of sustainability and development goals. *Women Feeding Cities* contributes significant new information and insight into practice about a relatively neglected opportunity to act on this finding. A climate-changing world demands that we seek every means possible to increase the resilience of food systems and livelihoods if urban civilization is to survive - this collection of experiences shows that there is much that can be done to increase resilience, even in what at first glance might seem unpromising circumstances.'

Janice Jiggins, Professor and guest researcher, Communication and Innovation Studies, Wageningen University Research, Netherlands.

Women Feeding Cities
Mainstreaming gender in urban agriculture and food security

Edited by
Alice Hovorka, Henk de Zeeuw, and Mary Njenga

PRACTICAL ACTION
Publishing

Practical Action Publishing Ltd
Schumacher Centre for Technology and Development
Bourton on Dunsmore, Rugby,
Warwickshire CV23 9QZ, UK
www.practicalactionpublishing.org

© Practical Action Publishing 2009

First published 2009

ISBN 978 1 85339 685 4

All rights reserved. No part of this publication may be reprinted or reproduced or utilized in any form or by any electronic, mechanical, or other means, now known or hereafter invented, including photocopying and recording, or in any information storage or retrieval system, without the written permission of the publishers.

A catalogue record for this book is available from the British Library.

The contributors have asserted their rights under the Copyright Designs and Patents Act 1988 to be identified as authors of their respective contributions.

Since 1974, Practical Action Publishing (formerly Intermediate Technology Publications and ITDG Publishing) has published and disseminated books and information in support of international development work throughout the world. Practical Action Publishing Ltd (Company Reg. No. 1159018) is the wholly owned publishing company of Practical Action Ltd. Practical Action Publishing trades only in support of its parent charity objectives and any profits are covenanted back to Practical Action (Charity Reg. No. 247257, Group VAT Registration No. 880 9924 76).

Indexed by Indexing Specialists (UK) Ltd
Typeset by SJI Services

Cover photo: Rob Small, Cape Town, South Africa

Technical Centre for Agricultural and Rural Cooperation (ACP-EU)

The Technical Centre for Agricultural and Rural Cooperation (CTA) was established in 1983 under the Lomé Convention between the ACP (African, Caribbean and Pacific) Group of States and the European Union Member States. Since 2000, it has operated within the framework of the ACP-EU Cotonou Agreement. CTA's tasks are to develop and provide services that improve access to information for agricultural and rural development, and to strengthen the capacity of ACP countries to produce, acquire, exchange and utilise information in this area. CTA's programmes are designed to: provide a wide range of information products and services and enhance awareness of relevant information sources; promote the integrated use of appropriate communication channels and intensify contacts and information exchange (particularly intra-ACP); and develop ACP capacity to generate and manage agricultural information and to formulate ICM strategies, including those relevant to science and technology. CTA's work incorporates new developments in methodologies and cross-cutting issues such as gender and social capital.
CTA, Postbus 380, 6700 AJ Wageningen, The Netherlands, www.cta.int

Contents

Part II Guidelines for Gender Mainstreaming in Urban Agriculture Research and Development Projects

Tables

Boxes

Figures

Preface

Certainly, 'home and mother' are written over every phase of neolithic agriculture and not least over the new village centers...Women's presence made itself felt in every part of the village: not least in its physical structures, with their protective enclosures...in the house and the oven, the byre and the bin, the cistern, the storage pit, the granary and from there pass on to the city, in the wall and the moat, and all inner spaces, from the atrium to the cloister. House and village, eventually the town itself, are woman writ large. (Lewis Mumford, *The City in History: Its Origins, Its Transformation and Its Prospects*, p. 12, 1961, Harcourt, Brace and World, New York)

A major motive for the commissioning and writing of this book was the recognition of two contradictory realities. On the one hand, we know from different kinds of study the key role that women have played in agriculture – from its emergence ten thousand years ago to current practice. We also recognize that agriculture has been intimately linked to the growth of urban settlements, and that women have been central to that growth, especially because of their primary contribution to the food security of urban families, either by virtue of their own food production or through trade. On the other hand, food production and food security have been given extremely limited attention in the history of urban development and planning, a fact which has contributed to the invisibility of women's role in provisioning cities with food.

Fortunately, urban food production and its contribution to the food security of low-income households within and around cities have begun to receive more serious research and development attention over the past 25 years. Studies have begun to quantify the contributions of women and men to various types of urban agriculture, and results confirm the centrality and diversity of women's roles, particularly in Africa. In many cases the studies suggest that gender roles typically associated with rural-based social structures are often transformed in the city, because urban agriculture is embedded in a wide range of complex social and economic processes in and around the city to which individuals and households have to adjust. They also highlight the fact that urban agriculture is embedded in ecological processes, producing both benefits and risks to human health, with important gender-related implications. These factors reveal the need for research and development tools that contribute to more detailed understanding of the roles of women and men in urban food production, so that research and development organizations can improve their support for this livelihood strategy in the future.

This was the common recognition shared by two leading actors in the field of urban agriculture research and development which led to the publication of this book. The Cities Farming for the Future Programme of the RUAF Foundation (International Network of Resource Centres on Urban Agriculture and Food Security) and Urban Harvest, the System-wide Initiative on Urban and Peri-urban Agriculture of the Consultative Group on International Agricultural Research (CGIAR), have sought to bring attention to gender issues in research, policy, and development interventions relating to urban food production and urban food security. Our efforts began in 2002 with the joint organization of an international workshop on appropriate methodologies for urban agriculture research and development programmes in Nairobi, Kenya. One of the specific conclusions of that workshop was the necessity of highlighting women's role in urban food production and distribution, and the importance of strengthening the capacity of different types of organization to give proper attention to gender issues in their research, policy, and development activities.

In the following years Urban Harvest and RUAF jointly and individually implemented a number of activities to address the issue of gender mainstreaming – ensuring that the goal of gender equality is central to all interventions in support of urban agriculture. A state-of-the-art review of gender in urban agriculture was commissioned by RUAF, an initial inventory of gender-sensitive methods and tools was compiled, and several case studies with a strong gender focus were undertaken. In 2004, Urban Harvest and RUAF jointly organized 'Women Feeding Cities', a workshop on gender mainstreaming in urban food production and food security, held in Accra, Ghana. Based on initial ideas of Diana Lee-Smith, a specialist in gender and urban agriculture with Urban Harvest, and guided by her keynote presentation, the workshop brought together urban-agriculture researchers involved in gender research and development to analyse a number of case studies from cities across the developing world, and to systematize lessons from those case studies.

Stimulated by that workshop, both organizations have subsequently devoted major attention to gender issues. They have supported staff capacity building and have sought and received funding for further exploration and development of guidelines, methods, and tools for integrating gender in the project cycle of Urban Harvest research projects and in RUAF multi-stakeholder strategic planning processes and pilot projects. (See Chapter 16 of this publication for more details of these activities.)

The production of this book has involved broad consultation and active participation by a global network of urban-agriculture specialists, co-ordinated by the book's editors, Alice Hovorka, Henk de Zeeuw, and Mary Njenga. These three specialists in gender and urban agriculture have been supported by an Editorial Committee, consisting of the three editors together with Diana Lee-Smith, Gordon Prain, and Joanna Wilbers (later succeeded by Femke Hoekstra).

This book seeks to do two main things. First it draws attention to women's crucial role in bringing food to the tables of urban families, and especially to the ways in which low-income women locally produce, carry, or trade food in multiple strategies to keep their families food-secure. A series of case studies enrich our understanding of these strategies. Second, the book aims to put into the hands of researchers, development practitioners, and local government officers the guidelines and tools that will ensure the centrality of gender concerns in future projects and initiatives related to urban agriculture and food security.

The book's Introduction, prepared by Alice Hovorka with important inputs from Joanna Wilbers and Diana Lee-Smith, provides a rationale for the centrality of gender concerns in relation to urban agriculture and offers an analytical approach to its study. Pulling together common threads linking the case studies, the Introduction illuminates key issues on the theme of women feeding cities. The chapter concludes with a discussion of the need for gender mainstreaming of urban agriculture research, planning, and implementation activities and the main ways to achieve it.

Part I consists of case studies on gender in urban agriculture in various cities of Africa, Asia, and Latin America; some are refined, updated versions of the papers presented during the Women Feeding Cities workshop in 2004, others are new work. The case studies, preparation of which was guided by Joanna Wilbers and Mary Njenga, analyse urban agriculture activities in varying political, socio-economic, and cultural situations, revealing the gender dynamics that underlie people's abilities to secure fresh, affordable, and accessible food and (often complementary) income. They serve as an entry point into considering differences between men and women in urban agriculture activities, highlighting especially the role and significance of women in this context, as well as the gender dynamics within which food production, processing, and marketing take place in urban areas.

Part II provides researchers and development practitioners with specific guidelines on how to include and incorporate gender-related aspects in each phase of urban agriculture research or development project cycles. It also includes a set of gender-sensitive tools that are referred to in the guidelines for each phase of the cycle. The tools are drawn from the experiences gained in the field by RUAF and Urban Harvest partners, but they draw also on earlier frameworks and 'tool boxes', which are included in the list of Resources in the final chapter of this publication for use by readers in their own projects.

The collaborative and interactive process involved in developing the guidelines and tools deserves some further explanation. Alice Hovorka took the lead in the preparation of draft guidelines and tools, with additional inputs and revisions provided by other members of the Editorial Committee. The draft drew on outputs from earlier activities of RUAF (especially two documents by Joanna Wilbers and Henk de Zeeuw) and Urban Harvest, and work by Hovorka and Lee-Smith, as well as adding new material.

The field-testing of the guidelines and tools took place between October 2007 and July 2008 in on-going urban agriculture projects around the world, and it was supported by Mary Njenga and Femke Hoekstra. Project personnel and external consultants knowledgeable about gender issues evaluated the usefulness of the guidelines and tools in particular phases of the project cycle, depending on where the project was in the cycle. The valuable experiences gained and conclusions reached about the draft guidelines and tools were reported back to the Editorial Committee through structured-format reports that were synthesized and discussed during a workshop in the Netherlands in August 2008. We wish to acknowledge with thanks the excellent contribution made by those involved in the field-testing. They are responsible for significant improvements in the final publication.

It is hoped that the guidelines and tools will aid efforts to understand the nature and extent of women's participation in urban agriculture activities, in order to facilitate appropriate, effective, and beneficial policy and planning interventions in urban centres. Of course, we fully recognize that this is not an exhaustive listing of guidelines or tools. It has not been possible to include all of the specialist research and development initiatives that have seriously addressed the relationship between gender and agriculture. Research on nutrition, for example, has yielded a growing literature in this field which is only partially reflected in the text and in the final chapter on Resources. Nevertheless, the book has concentrated on experiences that link gender with urban agriculture in its broad practice, and we hope that it will contribute to more gender-sensitive and responsive policies and projects on urban agriculture and food security in the future; to cities that are better and more equitably fed; and to women who are better recognized and supported in this crucial role.

Finally, we want to thank IDRC for making possible the preparation and publication of this book.

Gordon Prain, Programme Co-ordinator, CGIAR–Urban Harvest
Henk de Zeeuw, Programme Co-ordinator, RUAF–Cities Farming for the Future

Acronyms and abbreviations

CAUS	Cabinet d'Architecture et d'Urbanisme du Sénégal
CEDICAM	Centre for Integral Small Farmer Development in the Mixteca
CEPAR	Centro de Estudios de Producciones Agro-Ecológicas (Centre of Studies on Agro-ecological Production)
CGIAR	Consultative Group on International Agricultural Research
CIAT	International Center for Tropical Agriculture
CIP	Centro Internacional de Papa (International Potato Center)
CODIS	Community Organisational Development and Institutional Strengthening
DFID	Department for International Development
DWCD	Department of Women and Child Development
FAO	United Nations Food and Agriculture Organization
FGFC	Fodder Grass Farmers Committee
GWA	Gender and Water Alliance
IAGU	Institut Africain de Gestion Urbaine (African Institute for Urban Management)
ICRAF	World Agroforestry Centre
ICRISAT	International Crop Research Institute for the Semi-Arid Tropics
IDRC	International Development Research Centre
IFPRI	International Food Policy Research Institute
IGSNRR	Institute of Geographical Sciences and Natural Resources Research
ILRI	International Livestock Research Institute
INTA	National Institute of Agricultural Technologies
IPES	IPES Promoción del Desarrollo Sostenible (IPES Sustainable Development Promotion)
IRRI	International Rice Research Institute
IWMI	International Water Management Institute
KARI	Kenya Agricultural Research Institute
KGTPA	Kenya Green Towns Partnership Association
MDP	Municipal Development Partnership
NEFSALF	Nairobi and Environs Food Security Agriculture and Livestock Forum
PRGA	Programme on Participatory Research and Gender Analysis
PROVANIA	The Farmers' Association of the Niayes Valley
RUAF–CFF	RUAF Foundation–Cities Farming for the Future programme

SOYIA   Soweto Youth in Action
UNDP   United Nations Development Programme
UNEP   United Nations Environment Programme
UPWARD   Users' Perspectives with Agricultural Research and
            Development
WHO   World Health Organization

CHAPTER 1

Gender in urban agriculture: an introduction

Abstract

This chapter discusses the important role that women play in feeding urban populations (in particular the urban poor), and the need to mainstream gender in policies and programmes regarding urban food security and urban agriculture. It identifies and explains key issues related to gender and urban agriculture that require the attention of policy makers, planners, researchers, and practitioners. The chapter also provides an overview of urban agriculture, highlighting its importance in the light of increasing urban poverty and urban food insecurity.

Why *Women Feeding Cities*?

Women are in the majority among urban farmers in many cities around the world, but they tend to predominate in subsistence farming, whereas men play a greater role in urban food production for commercial purposes. This observation prompted the initial exploration of an area of scholarship termed 'Women Feeding Cities' which led to a similarly titled workshop in September 2004, and ultimately to the production of this book. These endeavours combined initiatives by two organizations, the RUAF Foundation (International Network of Resource Centres on Urban Agriculture and Food Security) and Urban Harvest (the System-Wide Initiative on Urban and Peri-urban Agriculture of the Consultative Group on International Agricultural Research – CGIAR). The result is a comprehensive framework for gender mainstreaming – ensuring that the goal of gender equality is central to all activities – in urban agriculture research and development

Frameworks on gender and agriculture already exist (see for example Hovorka, 1998; Feldstein and Jiggins, 1994); however, the context of gender and agriculture differs in urban settings compared with rural settings. This is largely owing to the fact that structures, institutions, and circumstances create specific gender dynamics in and around cities. There is a diversity of cultural values merging in urban areas such that traditional definitions of gender roles, responsibilities, characteristics, and behaviour are not necessarily appropriate, and often become hybridized with alternative perspectives. Thus there is potential for different political, economic, and social scenarios in urban settings, compared with rural contexts. Specifically, the urban setting often brings with it more diverse sources of family income; greater opportunities for women's schooling, wage labour, and financial credit; new configurations of mobility; heightened insecurity regarding land tenure and ownership (particularly for women); a greater incidence of theft; less involvement of children in food-production activities; new opportunities for co-operative efforts among community members; and flexibility in gender roles, as men and women adapt to urban life. In other words, it cannot be assumed that what is happening in rural agriculture is necessarily the case in urban agriculture within a particular context, for example within a particular country. Livelihood systems will differ, political/economic structures will differ, and social-network dynamics will differ. These key agricultural variables of difference need to be documented in urban areas in the context of agriculture, in order to avoid prescriptive and assumptive assertions for any potential research or development interventions.

Through family and social networks in cities of many poor countries, women have been making sure that their families get food. Sometimes this is achieved by means of kitchen gardens and urban agriculture plots, sometimes by remittances of food from rural or peri-urban farms. Women try their best to sustain their families in often difficult circumstances, and that is why in a very real sense they end up being the ones who are feeding the cities. The important distinction between subsistence and commercial production of food – with planners often prioritizing the latter – has frequently rendered invisible women's role in feeding cities. Similarly within households women are often marginalized and accorded lower status than men on account of the fact that they engage in subsistence farming rather than commercial agricultural ventures .

It was often taken for granted that women stay home while men go to work, and that still is the case in many places, including rural areas of poor countries. It is the case even though women work very hard, providing the majority of agricultural labour world-wide. The social changes that came with industrialization created divisions between work and home, as large numbers of people became wage labourers in production enterprises. Nineteenth-century laws passed to prevent the exploitation of women and children meant that men became the predominant wage labourers and employees. This gendered division of labour has stayed with us, so much so that the unpaid work of the home ceased to be seen as 'work' at all until recently, even if it meant producing food for the family.

This invisibility of women in the economy happened in parallel with the industrialization of food production, along with the production of other commodities. Women's role was to maintain the household, with men bringing in cash from their earnings. In poor countries with low wages and a rural economic base, it was mainly the women who continued producing food for home consumption on small farms or in kitchen gardens. To this day, women's work on small family farms, whether rural or urban, and their work in moving foodstuffs around on their backs or on public transport is scarcely regarded as of any significance and at worst is harassed or prohibited, as is the case with some of women's food-marketing activities.

However, the idealized economic plan whereby workers in cities paid for food from commercial farms and industries that produced it in rural areas and trucked it to them soon ran into difficulties. The urban food crises of the 1970s and 1980s in poor countries were attributed to the malfunctioning of official schemes that failed to match supply and demand. In fact the remittances of food from rural families and social food networks were invisible to food and agriculture policy makers, as were the women who operated them, either as small-scale producers or as petty traders or 'market women'. As official schemes collapsed, the explosion of informal-sector trade and enterprise – mainly involved in food supply – drew attention to the importance of the informal sector – but not enough to the role of women in food production and trade (Hovorka and Lee-Smith, 2006). Urban agriculture, compared with other informal-sector activities in cities, has been especially convenient for married women with children, given that it can be practised close to home, requiring little cash investment, and combining multiple roles of urban women. In urban agriculture production, women tend to focus on saving on family cash expenditures by growing their family's food (and eventually selling some surplus production), whereas women active in other informal urban-sector activities are focused more on generating a cash income.

Women's role in feeding cities, through formal or informal means, has become more challenging recently, given the mounting global food crisis. Higher food prices are rooted in increased energy costs, rising demand resulting from economic growth in emerging economies, the growth of bio fuels, and increasing climatic shocks from droughts and floods. Food reserves are at their lowest in 25 years. Associated commodity markets are volatile, and in a bid to protect their own populations many countries have imposed export bans or restrictions on certain foodstuffs, further driving up prices as food becomes less available (World Food Programme, 2007). Millions of people are being pushed deeper into poverty and hunger, and the urban poor are no exception: in many of the world's poorest cities, people can suddenly no longer afford the food available on store shelves. This is increasing in particular the time and energy that urban women expend on producing food and/or procuring monies for foodstuffs. For many women, this means that their families will suffer not only in terms of not having adequate (let alone fresh) supplies of food on hand, but also because they will have less money to pay for school fees or health costs. Beyond women's daily struggles to secure food for their families, their own potential for empowerment is limited in the longer term, given the need to focus on the here-and-now demands of urban food production and procurement.

The case studies in this book attempt to build up a picture of the complexities of men's and women's activities in urban agriculture in three developing regions of the world. This is a subject which has become increasingly important and timely in the context of the global food crisis. The analytical framework, tools, and guidelines presented are designed to identify solutions to some of the problems of marginalization and inequality experienced by women. The book's approach to gender mainstreaming tries to contribute to the important social goal of gender equality, not through the exclusion and marginalization of men but through careful analysis of gender circumstances and experiences, and engagement of men in the process. Ultimately, women's goals, both practical and strategic, may be addressed through increased and systematic support for urban agriculture activities around the world.

Urban agriculture

What is urban agriculture?

Urban agriculture (often differentiated as intra-urban and peri-urban agriculture) can be defined as the production of food (for example, vegetables, fruits, meat, eggs, milk, fish) and non-food items (for example, fuel, herbs, ornamental plants, tree seedlings, flowers) within the urban area and its periphery, for home consumption and/or for the urban market, and related small-scale processing and marketing activities (including street vending of fresh or prepared food and other products). In many places urban agriculture is also closely linked with recycling and use of urban organic wastes and wastewater.

Urban agriculture takes place on private, leased, or rented land in peri-urban areas, in backyards, on roof tops, on vacant public lands (such as vacant industrial or residential lots, roadsides), or on semi-public land such as school grounds, in prisons and other institutions, as well as in ponds, lakes, and

Woman watering crops in Rosario
By Hans Peter Reinders

rivers. In 1996 the United Nations Development Programme estimated that eight hundred million people were practising urban agriculture, 200 million of them market producers employing 150 million people full time (UNDP, 1996). Since then the numbers have increased.

For a long time the importance of urban agriculture was overlooked or dismissed as merely the result of traditional habits brought by rural migrants to the city, expected to fade away over time when these people integrated into the city economy. There was opposition to urban agriculture from public health and urban planning circles, which perceived urban agriculture either as a threat to public health that should be abandoned, or as a low-rent land use that would not be able to compete with other urban land uses. Such perceptions were institutionalized in restrictive by-laws and regulations at national and city levels, although these have remained largely ineffective.

During the past 15 years, studies have shown that urban agriculture should be recognized as an integral and permanent element of the urban socio-economic and ecological system (Van Veenhuizen and Danso, 2007; Mougeot, 2006). It forms an important part of the livelihood strategies of large numbers of urban poor. In many countries, rapid urbanization is accompanied by increasing urban poverty, food insecurity, and malnutrition. As a result, in many cities the number of people involved in urban agriculture tends to increase with ongoing urbanization, rather than decreasing, as had been previously assumed. Another factor is the growing urban demand for perishable products, including vegetables, meat, milk, and eggs, coupled with the comparative advantages of production close to the markets, and the availability of productive resources, including urban organic wastes, wastewater, and vacant public land.

The increasing importance of urban agriculture

About 50 per cent of the world's population now lives in cities; 77 per cent of Latin Americans live in cities, while in Asia and Africa the proportion is currently 39 per cent, climbing at a rate of 3 and 4 per cent per year respectively (UN Habitat, 2003), and the numbers of urban poor are rapidly increasing.

It is hard for most cities in developing countries to create sufficient employment for their rapidly increasing population. Meanwhile, transmissible diseases such as HIV/AIDS have eroded the income- earning capacity and assets of millions of urban households. As a consequence, the urbanization process goes hand in hand with an increase in urban poverty, dubbed the 'urbanization of poverty' (Haddad et al., 1999). According to UN-HABITAT, slum populations in urban areas of developing countries were estimated at 870 million in 2001 and are expected to increase by an average of 29 million per year up to 2020. Forty per cent of the population of Mexico City, for instance, and a third of Sao Paulo's population are subsisting at or below the poverty line.

A lack of jobs and income is leading to increasing urban poverty, as well as to growing food insecurity among the urban poor. A substantial proportion of urban household expenditures is dedicated to food – for poor households as much as 60–80 per cent – and in the city context the lack of cash income translates more directly into food shortages and malnutrition than in the rural areas (Mougeot, 2006). On average, urban consumers spend at least 30 per cent more on food than rural consumers spend, but despite this their average calorie intake is lower and in many cases insufficient (Argenti, 2000).

Increasing food insecurity among the urban poor and increasing problems in accessing fresh nutritious food at affordable prices largely went unnoticed by municipal authorities until some years ago. This was due among other things to a middle-class bias in urban planning, a lack of attention to urban food issues, and an exclusive focus on food imports to the city; at the same time, planners paid little attention to problems of access to food and the actual and potential roles of urban food production. According to Dahlberg (1998), although during the entire history of humankind cities accorded high priority to ensuring their food supply, few cities nowadays show great concern about this or perceive that their future food safety is linked to the local food system and to the agricultural areas surrounding it. In this context, urban agriculture has belatedly been recognized as making a 'significant contribution to food security of urban households and generation of jobs and income, self-esteem and environmental improvement' (Report of the Ministers' Conference on Urban and Peri-Urban Agriculture; Prospects for Food Security and Growth in Eastern and Southern Africa, MDP Harare, Zimbabwe, August 2003).

Cities are fast becoming the principal territories for intervention and planning of strategies that aim to eradicate hunger and poverty and improve livelihoods, requiring innovative ways to enhance the food security and nutrition of the urban poor and vulnerable households. Urban agriculture is one such strategy. In 2000 the UN Food and Agriculture Organization (FAO) included urban agriculture in its programme and created the Priority Area for Interdisciplinary Action 'Food for the Cities'. The FAO is now supporting the development of national and local policies and programmes on urban agriculture. More and more cities (for example, Rosario in Argentina, Bulawayo in Zimbabwe) and countries (such as Brazil, Botswana, and China) worldwide are now promoting urban agriculture to enhance food security, stimulate local economic development, and facilitate social inclusion and poverty alleviation (Brazil Government, 2008; Hovorka and Keboneilwe, 2004).

Benefits and risks associated with urban agriculture

While urban agriculture has important positive effects – on poverty alleviation, local economic development, food security, nutrition and health of the urban poor, social inclusion, and urban ecology – it can also lead to some undesirable outcomes if certain associated risks are not taken into account and preventive measures are not taken.

Income and employment creation

Available research indicates that urban agriculture can be a profitable undertaking, especially in the case of products that are in high demand and have a comparative advantage over rural production. These include perishable products such as green leafy vegetables, eggs, milk, mushrooms, medicinal herbs, flowers, and ornamental plants (see Moustier and Danso, 2006 or Van Veenhuizen and Danso, 2007 for an overview).

Market-oriented urban agriculture generates net incomes that in most cases are equivalent to or better than the minimum urban wage. In cases where by-law barriers have been removed, urban agriculture has proved to be a highly dynamic sector, with features that include irrigated year-round production, production under cover, small-scale processing of fruits, vegetables, herbs, and mushrooms, development of certification, green or organic labelling, and shorter marketing chains for consumers and institutions (IBRD/World Bank, 2008).

Subsistence-oriented urban agriculture leads to important cash savings, since food is by far the largest component of household expenditures; as a result, a significant portion of family income becomes available for non-food expenditures. Urban farming also provides a source of employment, not only for urban farmers themselves but for hired labourers and workers in related micro-enterprises such as production of compost, herding, collection and selling of grass or manure, processing of agricultural produce, and street vending of food. In Cuba, urban agriculture generated 25,000 new jobs in the years 1994–8, with a smaller investment per job than in other sectors (Gonzalez Novo and Murphy, 2000).

Urban food security and resilience

During economic or political crises – which in some cities are more often the norm than stability and economic progress – urban agriculture tends to increase rapidly, since it provides a safety net for the poor and for other households seeking to augment their dwindling incomes. It enhances their access to fresh and nutritious food by making fresh food available at prices that are lower than imported food, due to savings on transport, storage, refrigeration, and middlemen.

While cities will in future remain largely dependent upon food, especially staple crops, brought in from the rural areas and from international sources, cities can and should pursue greater food self-reliance in order to enhance their resilience and reduce vulnerability to shocks and food insecurity. This has been made clear by food riots in various cities, in response to recent sharp increase in food prices, caused in part by climate change and higher incidences of natural and human-made disasters. All of these have led to problems with food-supply chains that are dependent on imports or long-distance transport from rural areas.

To illustrate these trends, Box 1.1 presents findings taken (with permission) from a brief on urban food security and urban agriculture prepared by the FAO Regional Office for Latin America (FAO, 2008).

Box 1.1 Urban food consumption in Latin America and the Caribbean

For the first time in history more than half the world's population lives in cities. In 2007, 78 per cent of Latin America and the Caribbean population was defined as urban, a figure that is growing at the rate of 1.3 per cent per year.

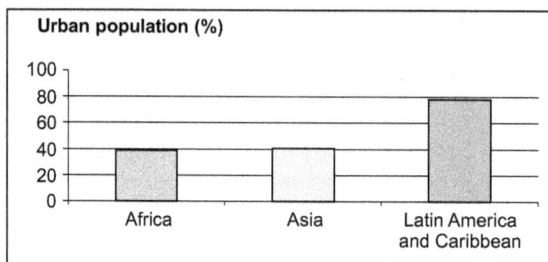

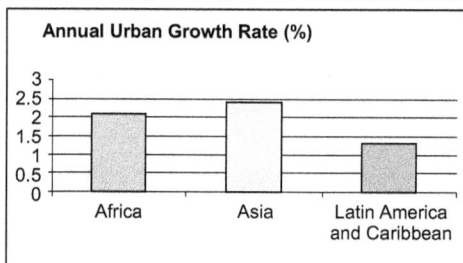

Source: UNFPA, 2007

The average daily vegetable and fruit consumption is 157 g/capita/day, well below the WHO-recommended level of 400 g/capita/day. It is interesting that the country with the highest fruit and vegetable consumption (Cuba) is also the country with greater advances in urban agriculture.

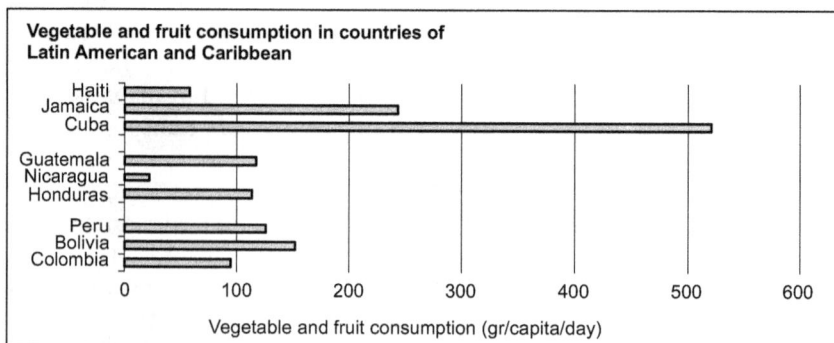

Source: FAO, 2005

Urban dwellers in poor neighbourhoods in Latin America are at risk of malnutrition and food insecurity, further aggravated by rising food prices as families have to allocate a higher share of their income to food. For example, a family in Bolivia earning $5 (Bolivian peso) a day typically spends $3 on food. Recent increase in prices of basic foodstuffs – some of which have risen 50 per cent or more – takes $1.50 of their purchasing power away, affecting the quantity and types of foods purchased as well as non-food expenses such as health care. The logical consequence is increased under-nutrition.

Consumer price index variation (CPV) in metropolitan areas of Latin America

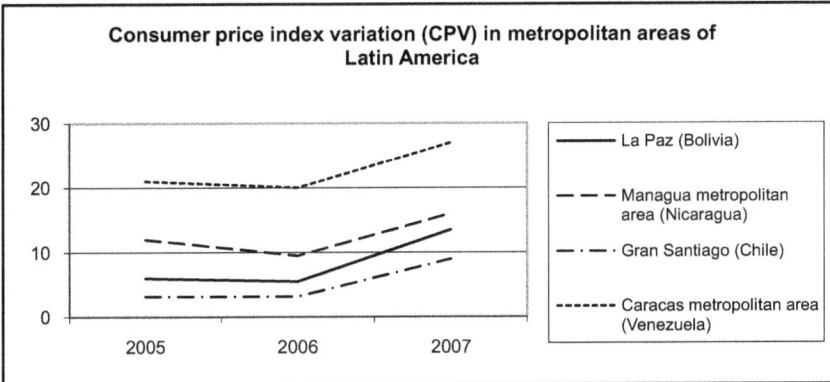

La Paz (Bolivia)

Managua metropolitan area (Nicaragua)

Gran Santiago (Chile)

Caracas metropolitan area (Venezuela)

Source: ECLAC, 2005

Nutrition and health

Some studies suggest that urban farming households have a better nutritional status (as shown by calorific and protein intake, and measures of stunting and wasting) compared with non-farming households. Further, creation of better conditions for poor urban families to produce and market items such as vegetables, livestock products, and fish would increase the access of other poor households to fresh and nutritious food at affordable prices. Medicinal plant production is increasingly explored, given that poor urban families may spend 10–20 per cent of their income on health care. Local production of medicines and medicinal herbs could contribute to improved health conditions for these people. Urban agriculture also now receives attention as part of HIV/AIDS mitigation programmes, being a strategy to enhance the nutritional condition of patients as well as reducing the negative effects of reduced working capacity or loss of adult members on household food security.

However, food produced in and around cities may be detrimental to human health if there is pathogenic contamination, potentially causing infectious disease. This is especially the case if contaminated water or fresh solid organic wastes are used to fertilize crops of foodstuffs that are eaten raw, or if hygiene is lacking in the production, processing, and marketing of food, as can happen if market produce is 'refreshed' with contaminated water, or street vendors do not observe hygiene precautions. Cultivated areas in cities may attract or provide breeding grounds for rodents and flies, which can contribute to the spread of infectious diseases, while certain diseases can also be transmitted to humans by livestock kept in close proximity to them, if proper precautions are not taken.

Food produced in and around cities can also be detrimental to human health if there is chemical contamination which might cause chronic disease. Heavy metals and complex organic compounds released by industry and traffic in particular may pollute urban crops through deposition and absorption via air, water, and soils.

The World Health Organization (WHO, 2001) has published an 'Action Plan on Urban Food Production and Consumption' as part of its strategy to stimulate the local production and consumption of fresh nutritious food and to improve the nutrition and health of disadvantaged urban groups.

Urban environmental management

Urban agriculture has a high potential for improving the urban environment by using organic wastes – solid wastes and wastewater – as inputs, by improving the micro-climate, and by preventing erosion and flooding through replanting bare lands. It also conserves energy and food, because there are fewer food losses during transport and handling, and greater energy savings due to the smaller need for storage, processing, and packaging.

The current highly capitalized and energy-consuming 'super market' model, based on the external supply of foodstuffs, increases the urban 'ecological food print'. Urban agriculture is one aspect of a different urban food system based on local or regional fresh products, with shorter market chains from producers to consumers, offering a policy alternative to the long-distance transport of food from elsewhere.

However, urban agriculture activities can create problems. They may contaminate local water sources if high inputs of fertilizers and pesticides are used. Neighbours may complain of dust, smells, and noise. Urban farmers may use high-cost treated drinking water for irrigation purposes. Further problems can be caused by poorly managed farm wastes clogging storm drains and piling up in the streets, while farming on steep slopes or along sensitive parts of river banks may lead to erosion and siltation.

The use of fully or partly treated urban wastewater for urban agriculture is investigated and promoted by the Sustainable Cities Programme of UN-HABITAT and the UN Environmental Programme (UNEP), and WHO and FAO recently updated their guidelines for the safe use of wastewater in agriculture (WHO/ UNEP, 2006).

Enhancing civic participation in urban management

Finally, urban agriculture has proved to be an effective strategy to enhance the participation of urban communities in the management of municipal resources, including land, water, and urban wastes. The planning and implementation of urban agriculture and related projects for recycling and reuse of urban organic wastes and wastewater can have direct positive effects on people's living conditions while generating feelings of self-reliance and creating links between the urban poor and other actors. The latter include NGOs providing technical support and training, and local authorities providing access to municipal land and services.

Gender in urban agriculture – an analytical approach

The concept of gender was introduced into international thinking during the 1980s as a counterpoint to ongoing work on 'Women in Development'. It was recognized that the underlying structures of the unequal relationships between men and women required analysis. Because different societies have so many varied customs governing men's and women's roles and behaviours, the word 'gender' was adopted to distinguish things about men and women that are socially ascribed – part of culture – as opposed to physical and biological difference, which is still described as difference in 'sex'. It is important to note that gender differences (wearing a headscarf, for example) do not necessarily mean that there is lack of equality. It is discrimination and denial of equal rights and opportunities that militate against gender equality.

Facts that are differentiated according to whether they relate to men or women are referred to as 'gender-disaggregated data', which are often called for in programmes or projects that are supposed to incorporate gender considerations. Another common term used is 'gender balance', denoting the fact that a certain ratio of women to men is noted (or called for). Getting a better gender balance is a target or indicator of better gender performance in a programme or organization. This kind of measurement is a useful indicator, but it does not substitute for 'gender analysis', which provides systematic documentation and understanding of the roles of men and women in a given situation. Achieving gender balance does not necessarily bring about gender equality or the empowerment of women. Thinking about the reasons why there are more women or more men doing urban agriculture (or any other activity) can be as important for a gender analysis as merely obtaining the numbers.

For example, a study in Nepal found that women and men peri-urban farmers shared many types of agricultural task, unlike rural farmers, who exercised a stricter gender division of labour. Some taboos of gender behaviour in agriculture were thus changing in the case of peri-urban farming as it became increasingly commercial, and both women and men in these farming households were benefiting economically and in terms of social status. But gender analysis showed that men had complete control of the marketing of the produce and controlled all commercial transactions; they also owned and controlled all land and property, while women had the additional burden of household work (Sapkota, 2004).

In another example, while almost equal numbers of women and men poultry entrepreneurs were found in Gaborone, Botswana in 2000, gender analysis revealed that the men had larger enterprises and were supplying more of the market. Even deeper investigation revealed that the women had less education and were often heads of household. The men who headed households generally had wives assisting them with extra income as well as labour – which put them at an advantage over women-headed households. Moreover, the men had larger plots, with more secure tenure. It was suggested that the women entrepreneurs were likely to be forced out of business as the market became more competitive, and that the reasons for this were mainly the structural factors of discrimination which disadvantaged them: women had less education, less land, less access to capital, and fewer property rights (Hovorka, 2005).

Several Latin American studies have shown that men may be more involved in agriculture than women, although, as in Africa, women are responsible for gardens near the house that produce food for home consumption. One of these studies demonstrated that, despite men's customary restriction of women to the home, an urban agriculture intervention designed to empower women was able to bring about some positive change. The young women with small children were also responsible for the families' food. Women's production of vegetables near their homes was at first resisted by the men, because they thought it was taking up the women's time and taking them out of the home, but as they saw the benefits they changed to supporting their wives' work (Olarte, 2004).

These examples show that only a thorough disentangling of the reasons behind some statistics can reveal the sources of gender inequalities, and thus enable us to find solutions to them.

Gender analysis helps to diagnose the issues that need to be addressed in order to bring about gender equality and women's empowerment. Another concept that is important to the process of gender analysis in order to identify strategies for change is the difference between practical and strategic gender needs. Women and men often express their need for change and improvement in terms of access to basic resources, infrastructure, or income. These are practical needs that usually bear no obvious relation to gender needs. Strategic needs are related to the gender division of labour, control over resources, legal rights, income equality, and decision-making power. While a gender analysis usually aims to understand the underlying strategic needs, dealing with practical needs can often have a transformative effect.

Gender equality is a basic societal value. Most countries have constitutions that assert people's equality, with no discrimination based on sex or other factors. The equality of men and women is also enshrined in the Universal

Declaration of Human Rights of 1948, and was developed in detail in the international Convention on the Elimination of All Forms of Discrimination Against Women (CEDAW) of 1979. But what is the situation in practice, and what does it mean in the context of real life? Some argue that since women and men have obvious biological differences – resulting in different needs and experiences – there is a need for a concept like 'equity' instead of 'equality', calling for fairness instead of sameness. But this discrimination contravenes the Universal Declaration and CEDAW. Women and men should be allowed the same opportunities, and efforts need to be made to ensure that they get them even if they differ biologically.

The extent of inequalities between men and women – with women clearly disadvantaged – is widely known and documented, in terms of their access to education, employment, and property ownership; inequalities in their incomes; and their presence in positions of leadership in just about every type of organization and institution, including governments, businesses, and community-based organizations. The correction of these inequalities necessarily involves the empowerment of women. However, to achieve this entails a long and complicated process of social learning and change that is ongoing all over the world. It can be particularly difficult for men – especially those who are themselves disempowered and marginalized – if losing their privileges and advantages over women touches on deep-rooted feelings of personal identity. Involving men in promoting gender awareness, analysis, programmes, and activities is seen as essential to meeting the social goal of gender equality.

Women feeding cities – key gender issues

This section focuses on illuminating and assessing the gender issues within urban agriculture activities that give rise to differences between men and women, and in many cases stifle women's opportunities for self-empowerment. Such gender analysis within site-specific contexts, stemming from the case studies featured in this book, is an important step in gender mainstreaming at a broader scale. Within this analytical process, it is important to recognize that gender is not a fixed category – indeed it is continually contested, such that individuals participate in (re-)defining gender relations on a daily basis. Further, gender dynamics may differ according to the ages, the size, and the life-cycle stage of the household, and according to marital status, religion, and caste. As a consequence, gender dynamics change over time and vary widely within and between cultures.

Nevertheless, the case studies presented in this book highlight the general consistencies in women's circumstances and experiences of urban agriculture that warrant attention, given that they reveal substantial gender differences and in some cases inequalities within cities around the world. In particular, the case studies illuminate the predominance of women in urban agriculture, numerous gender-based benefits and challenges of food production, and differential divisions of labour, knowledge/preferences, access to and control of resources, and decision-making power. The patterns of key gender issues that emerge from the case studies are overwhelmingly similar, revealing that in most urban agriculture contexts men own land, have access to greater resources, make decisions, and reap more benefits than women.

The remainder of this section expands on these common threads, detailing the ways in which gender matters in urban agriculture activities, and what these issues specifically mean for women's participation in this urban sector.

Women's predominance in urban agriculture

There are many women involved in urban agriculture worldwide, and data from several countries show them to be in the majority among urban farmers. This is the case in East Africa in general, with 80 per cent of farming households using only female labour in Kampala in the 1980s – although in neighbouring Kenya 56 per cent of the labour was female, with more men engaged in farming in the smaller towns, and more women (62 per cent) growing food in the capital. In West Africa more men than women were sometimes found in urban agriculture, the difference generally being attributed to the fact that much of the food produced was for sale (Obuobie et al., 2004). In Yaoundé, Cameroun, however, 87 per cent of urban farmers growing vegetables were women, 95 per cent of whom were growing mainly for subsistence and 79 per cent were growing commercially (Bopda et.al. forthcoming). Similarly in Kampala, women growing crops on contaminated sites were more likely to be doing it in order to feed their families than were the men, who tended to sell more of the produce (Nabulo et al., 2004). Poland, Thailand, Senegal, and Zimbabwe are other countries where women urban farmers have been noted to outnumber men (Smit et al., 1996). Statistical data on women in urban agriculture in Asian and Latin American cities are rare or even non-existent.

The predominance of women (illustrated in Box 1.2) in many contexts can be ascribed to two factors: first, women bear responsibility for household sustenance and well-being; second, women tend to have lower educational status than men, and therefore more difficulties in finding formal wage employment (Hovorka, 2001). In some contexts, men predominate in urban agriculture activities because of their access to land and resources, as well as the socio-economic status created by this activity. Box 1.3 illustrates this gender variability in urban agriculture participation.

Benefits and challenges of urban agriculture for women

Generally speaking, urban farming provides particular benefits to women as producers and/or procurers of foodstuffs in cities. Urban agriculture is a viable alternative to wage labour for women who lack access to formal employment because of their limited education and training, or socio-cultural factors that limit their freedom of movement. As a largely informal-sector

---

Box 1.2 Female predominance in urban agriculture in Harare, Zimbabwe

The Musikavanhu urban farmers' movement, which was originally concentrated in Budiriro and Glen View and has now spread to the other low-income suburbs of Harare, started with seven families meeting and agreeing to form a group that would work together and engage in urban farming. Currently, the movement has more than 5,000 members, of whom more than 90 per cent are women. Several factors explain the dominance of women in the group. First, generally speaking most of the urban producers in Zimbabwe's cities are women, given that putting food on the family table remains a responsibility for women. Second, until the mid-1980s in Zimbabwe, access to formal employment was a preserve of men, while women were supposed to concentrate on looking after the household, with some time available for farming activities. The third reason, and specific to Musikavanhu, was that the group emerged as a mechanism to resolve a conflict over the use of the land with people who wanted to construct houses there. The women producers felt that they needed to organize themselves in order to defend their use rights, while the male producers felt that they did not require the group to defend their land. Fourth, until the late 1990s men felt that urban agriculture was not a high-income earning activity and thus did not support their spouses when they started practising it. Women's freedom to attend training courses is still limited by their husbands, because it takes them away from home for more than a day. This severely impedes women's capacity-building opportunities and participation in leadership positions. Men finally joined in urban agriculture activities after successful harvests yielded useful profits. Further, massive retrenchments that were exercised in the late 1990s left many of the men with very little option but to engage in informal-sector employment, including urban agriculture. Currently, the engagement of men in the activity is causing conflicts, given increased demand for agricultural land, sometimes resulting in some men invading land belonging to women. (Toriro, Chapter 6 this volume)

Box 1.3 Male predominance in urban agriculture in Accra, Ghana

Men dominate urban farming in Accra as a result of the arduous nature of the farming tasks, including land preparation that is mainly manual, and vegetable production that requires regular watering, planting and transplanting, shading, tilling, and weeding. Land clearance, land preparation, and watering are considered the most difficult tasks and are designated as male activities. While men can supplement their effort by providing paid labour, half of independent women cultivators depend on paid male labourers to carry out land clearing and preparation. Women with limited financial resources cultivate relatively small plots that can easily be managed. This is illustrated in the following comments made by a typical woman producer who had been cultivating in Accra for 11 years: 'I started with five other women, but they have all left because of the difficulty of the tasks involved. Talking about land clearing and preparation, forking of beds, spraying of chemicals etc., it takes much determination to continue cultivating. I mostly use men hired labour for land clearing and preparation. When I have not got enough money to hire labour, I do the land preparation myself, but then I'm able to cultivate only part of my plot.' (Hope et al., Chapter 4 this volume)

---

activity, urban agriculture is in many cases especially effective and efficient for married women with children, or women heads-of-households, because it is often (but not always) performed close to the home and combines well with their household responsibilities. Urban agriculture requires little cash, given that it can be undertaken with relatively low capital, technology, and inputs.

It is thus attainable and affordable for women with limited education and resources, and often stimulates the use of indigenous practices.

By reducing household food expenditures through crop production and/ or livestock keeping, women in cities can redirect such finances to non-food items. Surplus production can turn into income-generation activities, either through direct sales of foodstuffs at the marketplace or through diversification into small-scale food processing. It is not unusual to find women in urban households earning more from food production than their husbands earn from formal jobs. The ownership of animals and/or independent cash income may strengthen a woman's social position within the household and the community. Animal rearing can also fulfil an important role as an economic safety net, and plays an important part in certain socio-cultural practices such as the payment of marriage dowries. Urban agriculture not only allows women to secure their daily household needs but provides a potential stepping stone for increased independence, confidence, and opportunity to improve their quality of life. The case of María del Triunfo, Lima in Box 1.4 illustrates the advantages and potential benefits of urban agriculture for women.

At the same time, the challenges facing women who aspire to participate in urban agriculture are numerous. For example, women face severe constraints in accessing, using, and/or controlling land in cities, compared with their male counterparts. Men tend to have the first choice of any available vacant land, leaving women with low-quality or less secure plots of land, often located at a considerable distance from home. Even within households with adequate land resources, wives may be at a disadvantage in terms of access to these plots (Hovorka, 1998). Distance is a related challenge: women are often left to travel extensive distances to marginal lands, their journeys requiring considerable time, physical effort, and financial expense for transportation, if it is available.

Women also face constraints in terms of urban agriculture production itself. They often lack inputs and working capital, as well as access to knowledge and information on the use of modern inputs and technologies. The latter is partly due to women's limited exposure to commercial urban agriculture or to their limited access to training courses offered by institutions or non-government organizations. Women are less likely to benefit from research or extension services that fail to consider gender-specific differences when

---

Box 1.4 Benefits of urban agriculture for women in Villa María del Triunfo, Lima, Peru

Urban agriculture in María del Triunfo, Lima does not reinforce gender inequities. On the contrary, this activity is contributing to women's empowerment, improved self-esteem, leadership, capacity building, and increased independence and freedom. It is important to note that for women urban agriculture is not an overload of activities but rather a means of building their personal development and their capacity for social interaction and organization, helping them to overcome many conditions including devaluation, subordination, and exclusion. (Soto et al., Chapter 8 this volume)

Box 1.5 Gender constraints on urban agriculture in Dakar, Senegal

Within the coastal fringe commonly known as the Niayes Valley, which runs from Saint-Louis to Dakar, Senegal, a strip about 350 km wide is often referred to as the 'green lung of the region'. In Pikine, located in this valley close to the capital city of Dakar, several urban agriculture activities are carried out on an area of 60 ha, including vegetable gardening, floriculture, fishing, and fruit and vegetable processing. A gender study conducted in the area determined the constraints encountered in the practice of urban agriculture by both men and women. Most constraints were identified or prioritized differently according to gender. For male urban farmers, land insecurity was deemed the most important constraint on their activities. They believed that their efforts were being threatened because of the rapid and uncontrolled growth of urbanization, resulting in the construction of collective housing and infrastructure. Another important constraint was access to water; watering was time-consuming and physically demanding, given that most farmers used watering cans to water their farms. This proved particularly challenging for women.

Another key problem for women was access to other inputs. For example, they had limited access to operational space, so that they could not rent out a room in which to carry out processing activities. Further, women are trained in processing techniques but they rarely receive the necessary assistance (in the form of equipment, functional premises, working capital, etc.) to carry out their activities effectively. Women were also constrained by the lack of follow-up after training courses and by the difficulty of mobilizing available labour within the family, especially among the children during the academic year. This latter difficulty illustrates (in some cases) the insignificant influence that women have on their husbands' decision to allow the children to pursue their schooling. As a result, the women have additional chores to carry out in the household. (Gaye and Touré, Chapter 14 this volume)

---

selecting technologies and working methodologies. Further, one must take into account the limited labour time available to urban women, and the local dynamics within which their daily activities take place. Women's response to opportunities to grow more food or better-earning crops will depend on the extent to which they can influence the decisions in the household about cultivation, the use or sale of produce, and the distribution of tasks and benefits within the household. The case of Pikine, close to Dakar, Senegal in Box 1.5 illustrates the differential challenges facing men and women involved in urban agriculture.

Division of labour

Within the household, various tasks and responsibilities are divided between male and female members. This division of labour is subject to context-specific circumstances such that within different cities, even within different households, the tasks between men and women relating to urban cultivation and livestock keeping differ according to the cultural group to which they belong, the socio-economic status of the household, the crop type or livestock type, and the location of the household in the city.

Division of labour in urban agriculture is also influenced by the reasons why urban households engage in the activity. On the one hand, households engage in urban agriculture because they have moved from rural to urban areas, bringing their agricultural practices with them. In this situation it is often the woman's task to provide food for the family through farming and gathering, but this may prove challenging, given that in the city family labour is hardly supplemented by casual labour, which increases the burden on the women in the households (Rakodi, 1988). On the other hand, households may have an urban background and are involved in agriculture by choice or by need. In this situation, there may be no recognition of traditional gender divisions of labour if the social norms brought from the countryside have lost their influence (Lee-Smith, 1994). Box 1.6 describes how changing cultural traditions in the context of urban agriculture can bring about changing gender roles within households.

As previously discussed, men's and women's involvement and predominance in urban agriculture activities may be different from one context to another. Beyond this, a number of other differences in the roles of men and women in urban agriculture can be observed. First, there is the difference in division of responsibility for certain crops. In most urban agriculture household systems, men are responsible for a few cash crops and larger livestock, and for generating cash income for the family, whereas women are responsible for a variety of food crops and small animals, and for household food security and nutrition (Hovorka, 1998). In research by Ofei-Aboagye in Ghana (1997), it was found that women are mainly responsible for crops with lower maintenance requirements, which leaves them with more time to spend on their household tasks.

Second, Ofei-Aboagye (1997) witnessed the difference between men and women in dry- and wet-season farming in Ghana. Usually, men are more actively engaged in irrigated dry-season agriculture, while women are more involved in wet-season farming. Women often lack the physical strength to clear the dry-season farmland, and their access to hired labour, oxen, or a tractor is limited. Fewer producers engage in dry-season farming, and so more money is made as a result of the relatively limited supply of foodstuffs

---

Box 1.6 Changing division of labour in Kisumu, Kenya

The city of Kisumu is situated on the shores of Lake Victoria. It has an area of 395 km², of which 35.5 per cent is covered by water. The cultural traditions of the Luo community that prevail in the rural areas apply differently here. For example, cases of women owning property are found, which would be impossible if the Luo tradition was completely adhered to. In Kisumu, control over property is largely determined by the identity of the household head; hence female heads of household hold absolute control over the household property. This is especially true for widows who control land, houses, and other property, including livestock. Women make independent decisions about finances, consumption, and production (even when adult sons and their families are living in the same compound). Sometimes the sons are consulted, but never the daughters. Another example of the influence of urbanization on gender roles is that women in Kisumu are inheriting livestock, even though tradition prescribes that wives and daughters do not inherit. (Ishani, Chapter 7 this volume)

Box 1.7 Gender division of labour in Kampala, Uganda

Urban agriculture in Kampala takes place predominantly on private land, in back yards, and on undeveloped public land. Due to rapid urbanization and population growth, people are increasingly utilizing hazardous places that are unsuitable for growing crops. Such places include road verges, banks of drainage channels, wetlands, and contaminated sites such as scrap yards and dump sites for solid and liquid waste. Most of the farmers in these hazardous locations produce and sell their food, with a higher proportion of women selling food directly to consumers. This may be attributed to the nature of crops grown: specifically, men grow crops on a larger scale and sell them on a wholesale basis to retailers, while women sell directly to consumers in the neighbourhood. In general, a higher proportion of the men sell some of the food that they produce from farming activities; women use the food crops to feed their families. The percentage of farmers who sell all of the food grown on contaminated sites to consumers is higher among women, who consequently use the funds to buy other foodstuffs from the market. (Nabulo et al., Chapter 5 this volume)

---

and unchanged levels of demand. Third, the gender division of labour at organizational and community levels is such that in some instances women participate in producer organizations, but it is quite common that they do so only as members or in supporting functions, not in key leadership roles with decision-making authority. Box 1.7 describes the different tasks of men and women involved in urban agriculture in Kampala, Uganda. Traditional divisions of labour continue to exist in urban households, such that women are responsible for reproductive (subsistence-oriented) tasks, while men are primary breadwinners, taking on formal jobs in the economy.

Differences in knowledge and preferences

Another key issue within the field of gender and urban agriculture is the differences in knowledge that exist between men and women in terms of the cultivation of certain crops and animals; the application of certain cultural practices (for example, women in the Andes know more than men about seed selection and storage, herding, processing of wool and natural medicines); the use of certain technologies (men generally have more knowledge of irrigation techniques, chemical inputs, and castration of bulls, for example); and certain social domains (men may know much more about formal marketing channels, whereas women may know more about informal barter relations). Men and women normally also differ strongly in their preferences and priorities in relation to their main roles and responsibilities (regarding, for instance, commercial or subsistence-oriented production goals); location of plots (women with young children often preferring to work close to the home, for example); mode of production (such as single versus multiple cropping); and use of the benefits (for household consumption or for sales, as an example). Differences between men's and women's preferences, priorities, and perceptions are illustrated in Box 1.8. In the urban context, men and women also come to acquire specific types of new knowledge, given their exposure and access

---

Box 1.8 Gendered differences in knowledge and preferences in Nakuru, Kenya

Information about the knowledge, opportunities, and constraints of men and women in respect to livelihoods and nutrition was obtained from a diagnostic study in which 85 male-headed households and 70 female-headed households were interviewed. The participation of men and women in the project has helped in tapping and exchanging their knowledge and skills in vegetable production and dairy-goat rearing. Women had a lot of experience and care in tending vegetables, including the production of traditional African vegetable seeds, while men knew more about the milking of goats, their reproductive cycle, and health issues. The knowledge and skills of both men and women in vegetable production and dairy-goat rearing have, however, been improved through training. (Njenga et al., Chapter 11 this volume)

---

to new crops or technologies that are conducive to urban cultivation; they negotiate new social-network patterns, given the need to establish production chains or marketing linkages that are specific to city life.

Access to and control over resources

A central issue in men's and women's differential circumstances and experiences of urban agriculture is that of access to and control over resources. This refers both to productive resources such as land, water, inputs, credit, information, and technology and to contacts, interpersonal networks, and organizations. It also refers to access to and control over one's own labour and the benefits of production, which include cash income, food, and other products for home consumption, sales, or exchange. Gendered access to and control over specifically natural resources often means that women have rights of renewable use (such as harvesting leaves from trees), while men have rights of consumption (harvesting the tree itself, for example). For female heads of household, their access to resources is often limited to those of poorer quality, and the consequence is lower agricultural production levels compared with male heads of household. Box 1.9 and Box 1.10 detail urban

---

Box 1.9 Accessing credit in Accra, Ghana

In general, urban farmers do not have access to formal credit schemes in Ghana. This is mainly because farmers, particularly women, cannot meet the collateral demands of the financial institutions. In addition, most of the urban female farmers have limited space for cultivation and do not own land. In spite of these problems, some have managed to create a win–win situation with the vegetable sellers in terms of access to informal credit. Sellers pre-finance farming activities by providing seeds, fertilizers, pesticides, or cash in order to obtain the vegetables subsequently produced. Sometimes sellers order the products before cultivation, through verbal agreements based on trust and confidence. The final sum of money received by the farmer may differ from the initial sum agreed on, as demand and supply might have changed during the growing period. Similar situations have been observed in Lomé, Togo, and Cotonou, Benin, in West Africa. (Hope et al., Chapter 4 this volume)

Box 1.10 Accessing education in Kampala, Uganda

In Kampala, most of the women involved in urban farming have only primary education, or none at all. Only a few have received a secondary education. This determines what kind of work they do, and it explains why poverty is a great problem among women: few of them participate in the formal sector, and many either work at home as housewives, or farm in their backyards, or trade foodstuffs at evening candlelight markets by the roadsides. (Nabulo, Chapter 5 this volume)

---

farmers' circumstances in relation to credit in Accra, Ghana, and education in Kampala, Uganda, respectively, in order to emphasize the implications for participation and potential success in the urban agriculture sector.

Access and control are highly influenced by structures and processes at the macro level, where socio-cultural ideas determine which roles men and women play, what responsibilities they each have, and the value placed on these roles. According to Moser (1993), external factors such as ideology, culture, and economics underlie intra-household resource allocation. Both *de facto* traditions and formal laws may prevent women from inheriting and controlling land on an equal basis with men. Traditions of patrilineal property inheritance limit women's access to a secure place to live, their ability to produce food for their families, and their ability to generate income. An example of the influence of culture and traditions on women's access to land is described in Box 1.11.

In contrast, urban areas facilitate a culture of individualist political and economic circumstances such that a 'survival of the fittest' scenario prevails. Box 1.12 illustrates this scenario in terms of land access in Ghana that is not consonant with gender traditions.

Box 1.11 Women's access to land in Hyderabad, India

The city of Hyderabad, India, is one of the fastest-growing cities in the world. Spread out over an area of 500 km², it has a population of six million. Various crops irrigated with wastewater are cultivated in the urban and peri-urban areas of the city along the Musi River, which flows through the centre of the city. Here, land is considered to be a resource for men, with legal inheritance of land equally distributed between sons and daughters. Land title, however, is usually in the name of the male head of the household, and after he dies it is inherited by the male members of the family (his sons). Indians still adhere to the dowry system, whereby a bride's father has to pay the family of the bridegroom before or during the wedding. Parents of the bride give cash and jewellery to the bridegroom and retain their land for their son, as he is the one who will support them in their old age. Women usually do not file cases against their fathers or brothers, even if they do not get their share of the land. The main reason is that a father pays a dowry to the bridegroom for the wedding of his daughter, which is supposed to compensate for the land that will go to his son. Culturally, women are taught that land is a man's property. Women get land titles only if their husbands die and their sons are too young (less than 18 years old) to work the land. Divorce is not a common phenomenon in the study area; even in the rare event of a divorce, the land remains with the husband. (Devi and Buechler, Chapter 2 this volume)

Box 1.12 Accessing land for urban agriculture in Accra, Ghana

Even though some communities disallow women from owning land, this restriction pertains mainly to communal land in peri-urban and rural areas and has little or no effect on access to land for farming in the open spaces within the cities in Ghana. Seventy per cent of the land cultivated in urban areas belongs to the government, and access to these lands is not based on gender differences. This is very interesting, as it would mean that established culture and traditions did not prevail in the urban situation, or at least that they would be less important. In a recent study, 87 per cent of the farmers in Accra indicated that men and women have equal access to government lands in urban open spaces. In essence, access to government land is based on availability and the lobbying strategies of individuals. In most cases, access is achieved via direct contact with the owner or caretaker, or through a third party working with the government institutions in the area. In some peri-urban areas of Accra, where share-cropping is used as payment for cultivating land owned by individuals, landowners or traditional leaders (such as chiefs) prefer men rather than women to cultivate larger plots, because they believe that men are likely to produce higher yields than women. (Hope et al., Chapter 4 this volume)

In many instances, men and women must compete with each other for scarce resources such as land and water for urban agriculture, given that these structures undermine traditionally established inheritance rules. In many of these scenarios, men out-compete women, given their generally elevated socio-economic status in cities. Women producers who are not landowners may demand their share of revenue derived from production, because they are the ones who are primarily responsible for the care of children (Hovorka, 1998). If they are not successful in persuading their husbands to share the earnings, women may retain part of the money from their vegetable produce sales without the knowledge or consent of their husbands (Maxwell, 1994). Box 1.13 illustrates access to and control of the benefits of urban agricultural production.

Additionally, female producers cannot always get access to transportation when distances from available plots of land may be considerable. The physical time and effort involved in such journeys is high and therefore proves to be a significant constraint for women, especially the elderly or those with young children, who aspire to become involved in food production. Lack of inputs

---

Box 1.13 Benefiting from urban agriculture in Kampala, Uganda

In Uganda it is the men who control the major source of household income and determine how to use it. The men purchase the farm inputs and equipment such as hoes and pangas. They have a strong hold on the household budget and allocate a certain amount of money to women, who in turn decide on household-expenditure priorities. One woman explained (focus-group discussion, Kigobe zone, Rubaga Division, Kampala, 2003): 'You grow the crops, but when it comes to selling it is your husband or male relative who sells and decides on how to spend the money. If you complain, he asks you if you are the one who owns the land. He then goes to spend the money on local brew.' (Nabulo et al., Chapter 5 this volume)

Women fetching water in Tamale
By IWMI Ghana

and working capital, as well as lack of information on the use of modern inputs and technologies, also limits women's ability to participate in urban agriculture activities. Women tend to have limited exposure to commercial urban agriculture or training courses offered by institutions or non-government organizations; they may also benefit less from research or extension services that fail to consider gender-specific differences regarding methods of plant production, crop species, and use of composts, manure, and fertilizer.

Decision-making power

There is a close relationship between access to resources and control over their use, and the power to make decisions; but they are distinguishable issues. Within the household, decisions are taken on the sale of products, land, or animals, the production process itself (what to produce, when, where, why, how), development of the infrastructure, whether to save or invest, and whether some members of the household should work on the farm or take other jobs outside the household. Productive activities can help to strengthen the position of women in the decision-making process within the household. For example, in Kampala, farming activities represent a means to economic self-reliance, as was found in the research of Maxwell (1994). For married women in particular, urban farming offers more than the opportunity to augment their family's food supply: while still within the margins of what is culturally expected of these women, participation in urban agriculture gives them access to their own source of income and thereby strengthens their position in intra-household conflicts. Culturally, urban agriculture is seen as a marginal economic activity, and the women may have good reason to maintain this image (Hetterschijt 2001; Maxwell 1994). Box 1.14 illustrates how men's views on urban agriculture can change once it has proven to be a profitable activity.

Within the community, contacts and participation in local networks and organizations often facilitate access to and control over productive resources. Women's groups play a pivotal role in this context, such that their activities are often co-operative mechanisms through which individual women successfully pool resources, skills, information, time, and energy. Box 1.15 provides an example of this scenario. In some societies, however, women's groups depend on a male chairman to represent their interests to the rest of the community, which may not be the best possible arrangement for addressing women's strategic needs (Peters, 1998).

Given the gender-related differences in the division of labour, knowledge and preferences, resource access, and decision making, urban agriculture research and development projects can have quite different impacts on men and women. If such gender aspects are not taken into account, projects may for example have positive effects on family income and thus reduce poverty, but they may also increase the workload of women, negatively affect the nutrition

---

Box 1.14 Men's views on urban agriculture in Lima

Of the total number of productive family units (PFUs) in Villa María del Triunfo, a municipality in the southern part of Lima, Peru, 76 per cent are controlled by women and 24 per cent by men. Of the total number of PFUs, 82 per cent practise urban agriculture recreationally and consume what they produce, while 3 per cent (all headed by women) practise urban agriculture with the goal of supplementing their family income. Fifteen per cent (all headed by women) see urban agriculture as a strategy for the potential generation of supplementary family income. Fewer men than women participate in urban agriculture, because men generally do not see this activity as a viable strategy for the generation of direct income. They therefore dedicate little time to it and give priority to other income-generating activities. However, they are interested in taking the next step and using the products of urban agriculture to generate income, particularly through processing activities. The current purposes (recreation and self-consumption) of urban agriculture in Villa María del Triunfo avoid conflicts within families about access to and control over resources and benefits of home gardens. Women make decisions without intervention from men, since this activity does not at present generate visible economic income and is therefore not relevant to men. However, when the possibility of generating visible income through commercialization arises, men want to take part in decision making. When striving to make urban agriculture an income-generating activity, it is necessary to identify strategies to avoid conflicts and inequalities in control over the benefits arising from home gardens. (Soto et al., Chapter 8 this volume)

Box 1.15 The Kachi Women's Association in Hyderabad, India

Kachiguda is an urban neighbourhood located almost in the centre of Hyderabad. Many of its inhabitants farm along the Musi river. There are four community associations in the neighbourhood: the Hyderabad Farmers' Association, Kachi Association, Kachi Women's Association, and Yadava Sangham. The Kachi Association and Yadava Sangham are caste-based associations, and only those belonging to the Kachi and Yadava castes respectively can become members; membership is completely male. The Kachi Women's Association is exclusively a women's association, formed in 2004 to help the women belonging to the Kachi caste to solve their domestic problems. According to the Secretary of the Women's Association, Ms Madhumathi Bai: 'The Kachi Association is entirely a men's association, and women cannot talk freely about their problems in front of the men. So the chairman of the Kachi Association himself encouraged us women to form a separate women's association, where we can freely discuss our problems such as domestic violence, access to water, blocked sewage drains, lack of electricity, disputes with neighbours, etc. If the problem cannot be solved, then we take it to the men. We still do not have a savings group, but plan to start one soon. As for agriculture, it is the only source of livelihood for some of the Kachi women, as they do not have any other skills or courage to go out and search for other jobs.' (Devi and Buechler, Chapter 2 this volume)

---

and health of the women and children, or leave women's strategic interests unaddressed. Literature suggests that urban agriculture projects that integrate gender issues to a high degree tend to have more positive effects, not only on the position of women but also on poverty alleviation, household food security, and health. For example, according to Talukder et al. (2001), home gardening activities in Bangladesh increase the income-earning capacity of the women and thus contribute to their empowerment; such activities provide important socio-economic returns through lower health and welfare costs, lower fertility rates, and lower maternal and infant mortality rates. Maxwell and Armar-Klemesu (1998) show that female-headed households in Accra have lower mean incomes than male-headed households, but their food-budget shares and calorie availability are significantly higher than those of the male-headed households. Female-headed households spend 60 per cent of their household budgets on food, compared with 50 per cent in male-headed households.

Gender analysis, as featured in this section, helps to diagnose issues that need to be addressed in order to bring about gender equality and women's empowerment. Going one step further, the concept of 'gender mainstreaming', as featured in the following section, proposes a comprehensive approach to this aspect of social change. Gender mainstreaming can provide a means of establishing urban agriculture research and project-planning methods that facilitate appropriate, effective, and beneficial policy and planning interventions in urban centres. Ultimately the goal is to ensure that urban agriculture activities help women to feed cities through their daily activities, as well as facilitating women's self-empowerment such that it allows them to change their inequitable circumstances relative to men and determine their own paths of development.

Mainstreaming gender in urban agriculture efforts

At the Social Development Summit organized by the United Nations in Copenhagen, national governments, NGOs, and international organizations agreed that the principle of equality of all rights for all people forms the basis for social inclusion. Specifically, all human beings are born free and equal in dignity and rights, and everyone is entitled to all human rights without distinction of any kind, such as race, colour, sex, language, religion, political or other opinion, national or social origin, property, birth, or other (United Nations, 1995).

A society where certain population groups are not part of the decision-making system is an unjust society. Male representatives normally do not automatically represent women's interests. Women's active participation in decision making is essential in order to ensure that women can promote and defend their specific needs and interests (for example, land rights and reproductive rights, their right not be subjected to violence, and their need for child-care services). Women can be prime actors in promoting gender-sensitive governance that addresses their needs as well as men's, thus enhancing access to and control over local resources for both. In some countries, women have succeeded in putting women's issues on the political agenda. But recognizing the need to enable the participation of all population groups in society is not enough. One also needs to know what strategies can be implemented to achieve a just and equal society, and how it can be done.

Gender mainstreaming is not only thorough in its documentation and analysis, but prescriptive in its method and comprehensive in its operation at all levels of a process or organization. It entails collecting gender-disaggregated data, identifying gaps, identifying strategies to close those gaps, investing resources in implementing the strategies, monitoring the implementation, and holding individuals and institutions accountable for the results (UNDP, 2003). For urban agriculture, it means that gender is considered in every aspect and at every step along the way in any initiative.

It is important to stress that, in empowering women and making them visible, gender mainstreaming is not aimed at excluding men, and certainly not at disempowering them. Rather it aims for the inclusion of men in the process. Set within the context of men's and women's circumstances and

---

Box 1.16 UN definition of gender mainstreaming

The United Nations (ECOSOC, 1997) defines gender mainstreaming as follows. 'Mainstreaming a gender perspective is the process of assessing the implications for women and men of any planned action, including legislation, policies or programmes, in all areas and at all levels. It is a strategy for making women's as well as men's concerns and experiences an integral dimension of the design, implementation, monitoring and evaluation of policies and programmes in all political, economic and societal spheres so that women and men benefit equally and inequality is not perpetuated. The ultimate goal is to achieve gender equality.'

experiences, gender mainstreaming aims to intervene in the gender dynamics within which food production, processing, and marketing take place in urban areas, for the purpose of creating greater social equality.

Gender mainstreaming must be taken into account in each of the various phases of research and development processes, including situational diagnosis and identification of problems, potentials, and actors; in-depth analytical research; intervention action planning, implementation, and monitoring/ evaluation; and policy development. To this end, the following measures are important:

- Recognize that women and men have different needs.
- Identify the mechanisms that keep women in a disadvantageous position.
- Identify the possible impacts of urban agriculture on the reproductive, productive, community, political, and cultural roles and areas of social interaction of women and men.
- Identify the practical and strategic needs of men and women and respond to both, paying particular attention to the strategic needs, given that meeting them tends to create more of a balance in gender relations.
- Define equality policies and affirmative actions that make gender equality a reality in the urban agriculture process.

Mainstreaming gender issues into development policy, planning, and decision-making mechanisms has proved challenging at all scales. As noted by UNDP (2003: 3), nowhere is the gap between stated intentions and operational reality as wide as it has been in the promotion of equality between men and women. Gender inequality is still continuously recreated by the 'mainstream', given that existing policies and practices are based on dominant and stereotyped views of the respective roles and status of men and women. While organizations and institutions continue to grapple with the incorporation of gender dynamics, the needs and priorities of one half of humankind have yet to feature on the development agenda.

As an emerging development strategy, the realm of urban agriculture is well poised to accommodate just and equitable guidelines for addressing the needs and interests of both men and women. Practitioners, planners, and policy makers can make a concerted effort to fully acknowledge and incorporate gender into the promotion and support of urban agriculture globally (Hovorka and Lee-Smith, 2006). Urban agriculture could and should be based on practices that generate more equitable social relations (Palacios, 2002: 1-2). If the data in a diagnosis are not disaggregated by gender, the project, plan, or policy will be based on an overall vision that disregards the differences between genders. It will thus suggest common answers to problems that, in practice, are different, and as a result it will risk deepening those differences and inequalities.

The urban agriculture research or development project cycle,requires consistent application of gender tools at every step. At the outset, gender analysis allows for planning and implementation of 'people-centred' development interventions, based on exact information of who these 'people' are. 'The people' never comprise one homogeneous group, but rather consist of diverse sub-groups as a consequence of differences in age, culture, geography, socio-economic positions, and ethnicity. Without such data on the differentiations within the population with whom and for whom they are working, research and intervention projects are not as relevant as they could and should be.

The aim of gender analysis is to understand and document the differences in gender roles and relations in urban agriculture in a particular location as a basis for the design of gender-responsive research, policies, and projects that increase men's and women's participation in and benefits from development. Gender-sensitive research should be conducted during all phases of the project cycle, including diagnosis, design, planning, implementation, monitoring, and evaluation and policy development (GWA, 2003). It is a cross-cutting issue that is only truly mainstreamed if applied at every stage. If it is not included in the earliest stages of a research initiative or a project, the result may be inadequate scientific knowledge or an entirely different project with different goals, strategies, and activities.

When making policy recommendations with regard to urban agriculture, it is important to conceptualize gender as a social, political, economic, and cultural issue, and as a human-rights issue, to which different actors (local governments, institutions, and organizations) must respond. Urban agriculture policies should be based on an acknowledgement of the real value of women's contribution to production, and on a recognition of women's economic rights. Attention should be paid to trying to establish the impact of urban agriculture on the well-being, dignity, and feelings of self-respect of both women and men. Promoting increased self-respect for women should be seen in the wider context of working towards gender equality as a universal human right. Thus development programmes and policies should strive for full and free participation of women in all decision making that affects their lives; they should target both women and men as beneficiaries.

Attempts have been made to develop appropriate methods for researching and promoting gender issues in urban agriculture (for example, Hovorka, 1998; UH–RUAF methodology workshop in Nairobi 2001; and the Women Feeding Cities meeting in Accra in 2004). Yet the development and implementation of gendered guidelines and instruments specific to urban agriculture remain a challenge. The cases presented in this book reflect an attempt to further mainstream gender in the urban agriculture realm by encouraging insightful research on gender dynamics (Hovorka and Lee-Smith, 2006). Creating a foundation for gender mainstreaming in urban agriculture requires a solid research base which explores conceptual issues and provides empirical evidence of men's and women's differential and often inequitable experiences of food cultivation and livestock rearing in different cities around the world. These case studies reveal differences between men and women, identify the mechanisms that often keep women in a disadvantaged position, and establish the significance of urban agriculture in people's everyday lives.

In turn, research on gender and urban agriculture provides a springboard for programming, planning, and policy initiatives whereby researchers can identify the practical and strategic needs of men and women in order to formulate action plans for urban agriculture activities (Hovorka and Lee-Smith 2006). The cases featured here go a long way towards making recommendations for research and development to enhance and achieve gender equity. For example, the Accra case study (Chapter 4) calls for the development of women-friendly technologies in urban vegetable farming to create a better gender balance, while in Manila (Chapter 3) action taken on health campaigns is seen as a way forward in addressing women's exposure to harmful pesticides. In Kampala (Chapter 5) land access and protected rights to ownership for women farmers are encouraged by creating gender quotas in land allocation, while in Kisumu (Chapter 7) the focus is on legislation and policies pertaining to customs, practices, inheritance, and succession to ensure women's equal rights in and out of marriage.

Additionally, calls for further research by authors of cases featured in this book draw attention to, for example, in-depth gendered studies of the marketing of vegetables in Accra, in order to understand marketing dynamics and the lack of men engaging in marketing activities. To ensure equitable division of labour and responsibilities in urban agriculture, the Kisumu case demands gender-disaggregated calculations of the contribution of urban agriculture to the overall Kenyan economy. Numerous case studies recognize the importance of acknowledging cultural norms and power relations that are fully embedded in society and have implications for men's and women's circumstances and experiences of urban agriculture. As stated in the Kampala case study, 'The methodologies used to integrate gender into urban agriculture projects need to take into consideration the cultural values that favour men and often render women inferior within the household and the larger community'.

In addition to the experiences and insights garnered through the case studies in this volume, the urban agriculture development community must look towards two key elements to further the gender- mainstreaming agenda. First, political will and commitment among key stakeholders at all scales is essential, given that gender issues become meaningful and applicable only if and when the organizations and institutions promoting them actually support them (Hovorka and Lee-Smith, 2006). Wherever possible, gender mainstreaming must be a stated developmental or organizational goal, supported by those in leadership positions. Gender mainstreaming requires a concerted effort by researchers, practitioners, and decision makers in order to strengthen links between research, programming, and policy/planning initiatives in the field of urban agriculture. This includes involvement and promotion of women's groups engaged in urban agriculture and their collective practices so that they will be recognized as social and political actors, thus converting urban agriculture into a citizen's concern. Second, capacity building and resources must be committed and allocated to achieve gender mainstreaming; logistical support and material requirements are essential at municipal, regional, national, and international levels. Gender mainstreaming demands expertise, which in turn requires resources, and until practitioners and organizations back up their promises with money, inaction will persist.

References

Argenti, O. (2000) 'Food for the Cities: Food Supply and Distribution Policies to Reduce Urban Food Insecurity', FAO, Rome

Bopda, A., Nolte, C., Tchouendjou, Z., Dury, S., Temple, L., Brummett, R., Gockowski, J., Soua, N., Elong, P., Kana, C., Ngonthe, R., Kengue, J. and Foto-Menbohan, S. (forthcoming) *Urban Farming Systems in Yaoundé – An Overview* (provisional title), Urban Harvest, Lima.

Brazil Government (2008) *The Fight Against Hunger*, Ministry of Social development, Brazil. Available from www.mds.gov.br

Dahlberg, K.A. (1998) 'The global threat to food security', *The Urban Age* 5 (3): 24–26.

ECOSOC (1997) *Gender Mainstreaming In the United Nations System.*

FAO Regional Office for Latin America and the Caribbean (2008) 'Urban and Peri-urban Agriculture is an Alternative Choice for Improving Livelihood of Poor Neighbourhoods: *Response to Rising Food Prices and Climate Change.* Santiago, Chile'.

Feldstein, H.S. and Jiggins, J. (eds.) (1994) *Tools for the Field: Methodologies Handbook for Gender Analysis in Agriculture*, Kumarian Press West Hartfort, Connecticut.

Gender and Water Alliance (GWA) (2003) 'Mainstreaming Gender in the Project Cycle', Training of Trainers Package on Gender Mainstreaming in Integrated Water Resources Management, GWA, Delft, The Netherlands.

Gonzalez Novo, M. and Murphy, C. (2000) 'Urban agriculture in the city of Havana: a popular response to crisis', in N. Bakker et al. (eds.), *Growing Cities, Growing Food, Urban Agriculture on the Policy Agenda*, pp. 329–347, DSE, Feldafing.

Haddad, L., Ruel, M., and Garrett, J. (1999) 'Are urban poverty and undernutrition growing? Some newly assembled evidence', *World Development* 27 (11).

Hetterschijt, T. (2001) 'Our Daily Realities: A Feminist Perspective on Agro Biodiversity in Urban Organic Home Gardens in Lima, Peru', Master's thesis, Wageningen University, Wageningen.

Hovorka, A. (1998) 'Gender resources for urban agriculture research: methodology, directory and annotated bibliography', in *Cities Feeding People Series*, Report 26, IDRC, Ottawa, Canada.

Hovorka, A. (2001) 'Gender and Urban Agriculture: Emerging Trends and Areas for Future Research', Graduate School of Geography, Clark University, Worcester MA, USA.

Hovorka, A. (2005) 'The (re)production of gendered positionality in Botswana's commercial urban agriculture sector', *Annals of the Association of American Geographers* 95 (2): 294–313.

Hovorka, A., and Keboneilwe, D. (2004) 'Launching a policy initiative in Botswana', *Urban Agriculture Magazine 13, Trees and Cities – Growing Together:* p.46.

Hovorka, A. and Lee-Smith, D. (2006) 'Gendering the urban agriculture agenda' in R. van Veenhuizen (ed.) *Cities Farming for the Future*, pp.125–136, International Development Research Centre & Resource Centre for Urban Agriculture and Forestry, Ottawa & Leusden..

IBRD/World Bank (2008) *World Development Report 2008: Agriculture for Development*, World Bank, Washington.

Jiggins, J. (1994) *Changing the Boundaries: Women-Centered Perspectives on Population and the Environment*, Island Press, Washington.

Lee-Smith, D. (1994) *Gender, Urbanisation and Environment: A Research and Policy Agenda*, Mazingira Institute, Nairobi, Kenya.

Maxwell, D. (1994) 'Internal Struggles over Resources, External Struggles for Survival: Urban Women and Subsistence Household Production', paper presented at the 37th Annual Meeting of the African Studies Association, Panel on Urban Provisioning and Food, The Royal York Hotel, Toronto, Canada, November 3–6 1994. Legon, Ghana: Noguchi Memorial Institute, University of Ghana.

Maxwell, D. and Armar-Klemesu, M. (1998) 'The Impact of Urban Agriculture on Livelihoods, Food and Nutrition Security in Greater Accra', paper presented at the IDRC Cities Feeding People Workshop on Lessons Learned from Urban Agriculture Projects in African Cities. Noguchi Memorial Institute, University of Ghana, Legon, Ghana.

Moser, C.O.N. (1993) *Gender Planning and Development: Theory, Practice and Training*, Routledge, London, UK.

Mougeot, L. (2006) *Growing Better Cities: Urban Agriculture for Sustainable Development*, International Development Research Centre, Ottawa, Canada.

Moustier, P. and Danso, G. (2006) 'Local economic development and marketing of urban produced food', in R. van Veenhuizen, *Cities Farming for the Future: Urban Agriculture for Green and Productive Cities*, RUAF /IDRC/IIRR, Leusden, The Netherlands.

Nabulo, G., Nasinyama, G., Lee-Smith, D., Cole, D. (2004) 'Gender analysis of urban agriculture in Kampala, Uganda', *Urban Agriculture Magazine 12: Gender and Urban Agriculture:* 32–33.

Obuobie, E., Drechsel, P., Danso, G. and Raschid-Sally, L. (2004) 'Gender in open-space irrigated urban vegetable farming in Ghana', *Urban Agricultural Magazine 12, Gender and Urban Agriculture*: 13–15

Ofei-Aboagye, E. (1997) 'Memo on gender analysis of agriculture in Ghana',. sent to Kathleen Clancy, Gender and Sustainable Development Unit, IDRC, 14 February 1997. Report for IDRC Project No. 96-0013 003149, IDRC, Ottawa, Canada.

Olarte, M. de (2004) 'When the women decided to work the gardens', *Urban Agriculture Magazine 12: Gender and Urban Agriculture:* p.12

Palacios, P. (2002) 'Why and how should a gender perspective be included in participatory processes in urban agriculture', in *Latin American Training*

Course on Urban Agriculture, Session 2, Proceedings, PGU–LAC, Quito, Ecuador.

Peters, K. (1998) *Community-based Waste Management for Environmental Management and Income Generation in Low-income Areas: A Case Study of Nairobi, Kenya,* City Farmer, Canada, in association with Mazingira Institute, Nairobi, Kenya.

Rakodi, C. (1988) 'Urban agriculture: research questions and Zambian evidence', *The Journal of Modern African Studies*, 26 (3): 495–515.

Sapkota, K. (2004) 'Gender perspectives on peri-urban agriculture in Nepal', *Urban Agriculture Magazine 12, Gender and Urban Agriculture:* 38–39.

Smit, J., Rattu, A., and Nasr, J. (1996) *Urban Agriculture: Food, Jobs and Sustainable Cities,* Publication Series for Habitat II, Volume One, UNDP, New York.

Talukder et al. (2001) *Improving Food and Nutrition Security through Homestead Gardening in Rural, Urban and Peri-urban Areas in Bangladesh,* Helen Keller International, Asia–Pacific Regional Office, Indonesia.

UNDP (1996) *Urban Agriculture: Food, Jobs and Sustainable Cities*, United Nations Development Programme, New York.

United Nations (1995) *World Summit for Social Development Copenhagen Declaration on Social Development* [online]. Available at http://www.un.org/esa/socdev/wssd/text-version/index.html

Van Veenhuizen, R. and Danso, G. (2007) *Profitability and Sustainability of Urban and Peri-urban Agriculture,* Agricultural Management, Marketing and Finance Occasional Paper, FAO, Rome, Italy.

World Food Programme (2008) *WFP Crisis Page: High Food Prices* [online]. Available at http://www.wfp.org/english/?ModuleID=137andKey=2853 [accessed July 28, 2008].

WHO (2001) 'Urban and Peri-urban Food and Nutrition Action Plan, Elements for Community Action to Promote Social Cohesion and Reduce Inequalities through Local Production for Local Consumption', WHO, ETC, Denmark.

WHO / UNEP (2006) *WHO Guidelines for the Safe Reuse of Wastewater, Excreta and Grey Material,* Geneva, Switzerland.

PART I
Case Studies

CHAPTER 2

Gender dimensions of urban and peri-urban agriculture in Hyderabad, India

Gayathri Devi and Stephanie Buechler

Abstract

This chapter attempts to understand and analyse the gender dimensions of urban and peri-urban agriculture in the case-study areas of Hyderabad, India. A number of households, especially migrants from drought-prone rural areas, depend on wastewater-irrigated urban or peri-urban agriculture for their livelihood and food security. Data have been collected by participatory rapid appraisal tools for this study; the tools included needs assessment; activity profile; access and control profile; and decision-making matrix. The important lessons learned from the study are (1) that the culturally prescribed gender division of labour in urban agriculture does not accord completely with what men and women do in practice; (2) that women contribute a significant part of the household income; (3) that crops and activities that have higher rates of return tend to be controlled by men, whereas women farmers view urban agriculture more as a way of securing food security for the households; (4) that the location of an activity influences the degree and type of involvement of women in agriculture and dairy activities; (5) that men have better access to resources such as credit and land than women, owing to their cultural advantage in a patriarchal society; and (6) that affiliations such as caste, class, and ethnicity affect gender relations and gender roles in urban agriculture. Recommendations proposed include gender-positive / gender-sensitive allocation of land, training, education and capacity-building, projects, policies, and research which contribute to economic growth as well as to social equity.

Introduction

Agriculture has always been associated with rural areas, even though in many parts of the world it has been practised in urban and peri-urban sites for centuries, contributing to the employment and food security of those involved in its production and sale. The purpose of this case study is to understand the gender issues in urban food production and household food security in the study area and to identify the strategies that the South Asia regional office of International Water Management Institute (IWMI) should apply in order to mainstream these issues in its future research projects and policy recommendations.

Study location

The study was conducted in Hyderabad, India, one of the fastest-growing cities in the world. The city is spread out over an area of 625 km², with a population of 6.7 million (*The Hindu* newspaper, 16 April 2007). Various crops irrigated with wastewater are cultivated in the urban and peri-urban areas of the city along the Musi River, which flows right through the centre of the city. Within Hyderabad, an urban location called Kachiguda and two peri-urban locations called Pirzadiguda and Parvathapuram (see Figure 2.1) were chosen for the study in order to understand the contributions of urban and peri-urban agriculture (UPA) to the livelihoods of the people, and especially the differences in gender roles and benefits.

Kachiguda is an urban location almost in the centre of the city. Most of the urban farmers who farm along the Musi River live in the Kachiguda neighbourhood. Many varieties of crops are grown on the river bed (which usually floods in the monsoon season, but remains dry for the rest of the year), on a 5 km stretch along the Musi River. The average landholding size here is one acre (0.4 ha) of irrigated land (Buechler and Devi, 2002). The main crops grown in urban areas are Para grass, green leafy vegetables, banana plants, and coconut palms. Para grass is cultivated as fodder for cows and buffaloes that are raised in urban areas for milk. Much of the land dedicated to fodder production is rented to dairy producers from the *yadava* caste, whereas most of the land in the urban areas on the Musi flood-plains is owned by people from the *kachi* caste, whose members came originally from Uttar Pradesh state and are Hindi-speaking. At present there is no provision to transfer the land titles, because the government stipulates that this land is a riverbed and cannot be bought or sold. Therefore, the names on the land titles have not changed even with the deaths of two or three generations of titleholders (Buechler et al., 2002).

Pirzadiguda and Parvathapuram, two peri-urban study areas in Hyderabad, are located about half a kilometre away from each other. Many of the Muslim and Hindu families in these areas were dependent on urban agriculture in 2004. About 150 households have migrated in the last 15–20 years from the district of Kurnool, a drought-prone area. Although they originally came as field labourers who cut wastewater-irrigated fodder grass, many have been able to buy livestock and rent land for fodder grass and vegetable production. As the city increases in size, most of the land under cultivation in this location is now being converted into plots for construction (Buechler and Devi, 2003). The key farming activities in the urban and peri-urban areas are still dairy, Para grass (fodder grass), and leafy vegetable production. Most farmers in this area lease between half an acre and one acre of land to cultivate leafy vegetables or Para grass. Since 2004, however, this scenario has changed considerably. There is a significant decrease in wastewater-irrigated areas in the urban and peri-

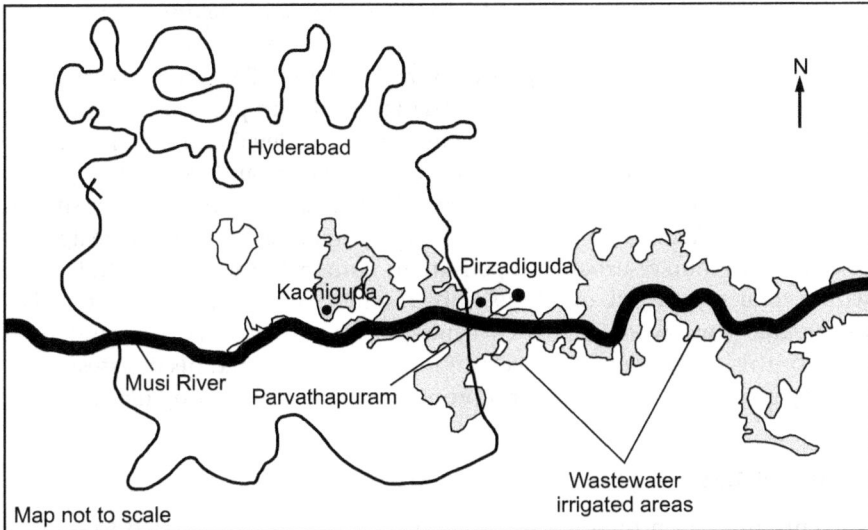

Figure 2.1 Urban and peri-urban study sites and area under wastewater irrigation in 2007

urban study locations, for various reasons: the increasing value of land for real estate; the booming economy of the city, which offers numerous employment opportunities; and the high opportunity cost of working on a farm versus other types of employment. Exact estimates of the land lost under cultivation to real estate are not available, but the current research approximates that more than 50 per cent of the land previously under vegetable and Para grass production in the study locations is now lost to real estate and other commercial purposes and displaced to areas farther away from the city.

Methodology

Data collected over different years have been used to present the results in this study. From August to October 2002, for a larger study sponsored by the UK government's Department for International Development (DFID),on livelihoods along the Musi River, 105 questionnaires were used to gain information from wastewater users (one male and one female member of each household) as well as to vendors in urban, peri-urban and rural sites. From March to April 2003, data were collected about household food security, through the application of a new questionnaire on income and food expenditure patterns in the urban and peri-urban sites of the city. In August 2004, new questionnaires were applied to a sub-sample of the respondents who had been interviewed previously, with the breakdown as follows: 22 respondents (12 women and 10 men) for the urban sites and 30 respondents (18 women and 12 men) for the peri-urban sites. The activity profile, access and control profile, and the decision-making matrix were applied to all the respondents interviewed for the urban sites in August 2004, and the reasons for their responses were discussed. The results of the two profiles and the decision matrix were triangulated with previous interviews in 2002 and 2003, and all the different views are considered and incorporated in the analysis. This chapter is based on the knowledge and experience gained from the previous studies and the new data collected from the study sites in January 2008. Additional participatory rapid appraisal (PRA) tools are applied to a smaller group of urban and peri-urban farmers in the current study sites, with a total sample size of 52 (29 women and 19 men) farmers and four key informants (two women and two men). The participatory appraisal tools used for this study are preference ranking; problem matrix; needs assessment; activity profile; access and control profile; and decision-making matrix.

Gender analysis in peri-urban study areas

The current section presents the outcomes of the analysis related to each of the tools used in this study.

Gender differences in farming preferences

By direct matrix ranking, the participants identified and prioritized their farming activities and discussed their related preferences, opportunities, and constraints associated with them. Understanding preferences is critical for choosing appropriate and effective interventions (World Bank, 2001). Results show that men gave their first preference to dairy, their second to vegetable production, and their third to both Para grass and paddy. Higher profit margins and marketability of the product were important criteria for men. Women gave their first preference to leafy vegetable production, their second to Para grass production, and their third to dairy. Important criteria for the women's ranking were the degree of risk and the ability to manage the activity independently while still being able to manage their household chores and child-care responsibilities.

Gender division of labour

The livelihoods of urban farmers are dynamic, and the activities of the farmers change rapidly in response to changes in the local economy, availability of new opportunities, and rising education levels of the family members who want to move out of farming to desk or office-based jobs. However, little change has been seen in the division of labour between men and women or the kind of activity that they perform in agricultural fields, in the household compound, or in the community in general. Key-informant interviews and data from the questionnaires have been analysed to produce a profile of the division of labour between female and male urban farmers (see Table 2.1).

Table 2.1 Gender division of labour in urban agriculture in the study areas

Socio-economic activity	Females (♀)			Male (♂)			Locus
	Child	Adult	Elder	Child	Adult	Elder	
1. Production of goods and services							
a) Para grass							
Irrigation		+			+++	+	Within the field
Harvesting		++			+++	+	Within the field
Transport as head loads		++			+++		Field to transport vehicle
Driver of fodder transport vehicle					+++		Field to market yard
Sale in the market					+++		In market yard
b) Dairy							
Bringing fodder from market to cattle shed					+++		Market to home
Feeding the cattle		++			+++		Within home/cattle shed
Cleaning the shed		++			++		Within home/cattle shed
Milking the cows		+			+++		Within home
Cleaning the milk cans		+++			+		Within home
Taking cattle to river for bathing		+		+	+++		Outside the house
Selling the milk		+++	+		+		At home
Selling the milk		+		++	+++	+	Outside the house
c) Leafy vegetables							
Land preparation and planting		++			+++		Within the field
Weeding	++	+++		+	+		Within the field
Irrigation		++			+++		Within the field
Harvesting	++	+++		+	+		Within the field
Transport		++			+++		Field to market
Sale in the market	++	+++		+	+		In market
2. Social reproduction							
Child care	++	+++	++		+		Within home
Care of sick children	+	+++					Taking children to hospital
Care of the elderly		+++			+		Within home
Care of the elderly		+			+++		Taking elderly to hospital
Household management	++	+++	++				Within home
Collect water	+++	+++	+	+	++		In house, nearby water tap
3. Community management							
Participation (Part.) in Hyderabad Farmers' Association		+			+++	+	In the local community
Part. In Kachi Association					+++	+	In the local community
Part. In Kachi Women's Association		+++					
Part. In Yadava Caste Association					+++	+	In the local community
Part. In Fodder Grass Farmers' Committee					+++		In the local community
Part. In Uppal Farmers' Association					+++		In the local community
Part. In Self Help Saving Groups		+++					In the village

A child is a girl or boy below the age of 16 years. An elderly person is a person 60 years old and above

+++ indicates that frequency is high

++ indicates that frequency is medium

+ indicates that frequency is low

No major differences were noticed in the activity profile of men and women in the urban and peri-urban areas, except in dairy production, and they are therefore not listed separately.

Production of goods and services

- **Para grass**: In Para grass production, men are more involved than women. Women harvest the crop and also carry it to the truck that transports the grass to the market. A man carries a head-load of 70 kg per trip, and a woman carries 30 kg of head-load per trip from the field to the truck. Women's participation in the grass market is limited by a number of factors. The profit margins are higher for fodder grass than for other crops such as leafy vegetables. Therefore men want to control the income from fodder grass. It would not seem to be a limiting factor that heavy loads of grass have to be loaded and un-loaded from transport vehicles, since women carry loads of grass from the field to the trucks. However, the kind of garment that is culturally acceptable for women to wear (the sari) limits movement of the legs, restricting women's ability to climb into and out of the trucks to unload the fodder grass (Buechler et al., 2003).
- **Milk**: In milk production, men and women play equally important roles. In intra-urban areas most dairy producers belong to the *yadava* caste community. The economic status of the household influences the extent to which women are involved in production activities. In intra-urban areas, since the *yadavas* are reasonably well-off, women are less involved, whereas in the peri-urban areas, where households are

Man carrying Para grass in Hyderabad
By Henk de Zeeuw

classified as poor, the involvement of women in dairy activities is greater than in intra-urban areas.

- **Vegetables**: Most of the land under vegetable cultivation is rented, and the vegetable farmers are usually women, assisted by young boys and girls (Buechler et al., 2003). Most of the urban and peri-urban vegetable farmers are women, because production of leafy vegetables is economically viable when undertaken at a small scale, the plants can be grown in small, manageable plots, and all operations from ploughing to harvesting can be done by women without any help from the male family members (Buechler and Devi, 2008a, 2008b). In addition, agriculture is the only skill that these women have, due to the strong gender bias in favour of boys in the provision of formal education and training (for occupations such as driving, tailoring, technical/industrial training) which would otherwise help them to procure more profitable jobs.

Social reproduction

All activities associated with social reproduction (see Table 2.1) such as cooking, house cleaning, washing dishes and clothes, child care, collection of water, and care of the elderly are done by women. Young girls between the ages of 10 and 17 years (or sometimes even younger) help their mothers in all household chores. Young boys also help their mothers in buying groceries and sometimes fetching water by bicycle if the distance to the public tap is great. Men very rarely participate in these activities. Culturally, women are supposed to take care of all activities within the boundaries of the house, and men are responsible for all activities outside the house. However, these prescribed roles and responsibilities are changing with time.

Community participation

Most community activities are dominated by men. In the urban study location there are four associations – the Hyderabad Farmers' Association, Kachi Association, Kachi Women's Association, and Yadava Sangham. The Hyderabad Farmers' Association is an informal organization formed by the farmers who own land along the Musi in Hyderabad city. It was mainly formed to fight against the government when the latter wanted to take away their land and convert it into a public area for parks. Since the criterion for membership was land ownership, there are very few women members in this group; the few women who are members do not participate in any of the meetings, since women perceive the association as a men's organization, and because they face social barriers which discourage them from speaking in public.

The Kachi Association and Yadava Sangham are caste-based associations, and the members are all male. The Kachi Women's Association is exclusively a women's association, formed in 2004 to help the women belonging to the

Kachi caste to solve their domestic problems. Asked why the Kachi women had to form a separate association, when the *kachis* already had an association, the Secretary of this association, Ms Madhumathi Bai, said, 'Kachi Association is a men's association and women cannot talk freely about their problems in front of the men. So the chairman of the Kachi association encouraged us to form a separate association where women could freely discuss and solve their problems.' This clearly indicates the cultural barriers that limit women's mobility and freedom of speech in public gatherings.

In the peri-urban area, there are three associations: the Fodder Grass Farmers Committee (FGFC), the Uppal Farmers Association (UFA) and the DWACRA women's Self Help Groups (SHGs) in Parvathapuram. All of the members of FGFC and UFA are male. Formation of the DWACRA women's SHG has been a successful government initiative, aimed at encouraging savings and credit for the productive and reproductive needs of women and their families in Parvathapuram. Most women in the study area are members of these SHGs and are benefiting from their membership. It was also found that the loans taken by women from households dependent on dairy were spent on buying more cattle, since most of the dairy activities were done and controlled by men. Men have a strong influence on decisions taken about their households' finances and credit utilization.

Access to and control of resources and benefits

An access and control profile of the male and female farmers in relation to various resources was constructed after the analysis of data from questionnaires and key-informant interviews (see Table 2.2). The access profile helps researchers to understand the extent to which a resource is available for use by a household member, and a control profile tells us who has the power to make decisions about how a resource can be used.

Table 2.2 Male and female access to and control of household resources in the study areas

	Access		Control	
Resources	Female	Male	Female	Male
Productive resources				
Land	xx	xxx	x	xxx
Credit (mainly through friends or relatives or moneylenders)	xx	xxx	xx	xxx
Labour	xx	xxx	x	xxx
Information	x	xx	x	xx
Benefits of resources				
Income from sale of milk	x	xxx	x	xxx
Income from sale of leafy vegetables	xxx	xx	xx	xx
Income from casual labour	xxx	xxx	xx	xx

xxx Indicates complete access / control
xx Indicates partial access / control
x   Indicates limited or no access / control

Access to and control of productive resources

The various resources essential for production are land, capital, labour, and information (related to market prices, new technologies of production, etc.).

- **Land**: Land is considered to be a resource for men. Legally, the land inherited by a person should be equally distributed between the son(s) and the daughter(s). But the land title is usually in the name of the man, and after him it is inherited by the male members of the family (sons). Indians still follow the dowry system, whereby a bride's father has to pay the family of the bridegroom before/during the wedding. Parents of the bride present cash and jewellery to the bridegroom and retain land for their son, because he is the one who will support them in their old age. Women usually do not file a case against their father or brothers, even if they do not get their share of the land. The main reason behind it is that a father pays a dowry to the bridegroom for the wedding of his daughter, and that is supposed to compensate for the land that goes to the son. Culturally, women are taught that land is a man's property. Women get land titles only if the husband dies and the son is too young (less than 18 years old) to work the land. Divorce is not a common phenomenon in the study area, and even in the rare event of a divorce the land remains with the husband. Renting land for agriculture (mainly for leafy vegetables) is done by a male or a female farmer irrespective of sex. Women renters are considered more reliable.

- **Credit**: In both the urban and peri-urban locations, it was found that there are no formal sources of credit. People lend and borrow among themselves. The amount of money that is usually borrowed ranges from Indian Rupee (INR) 500 to INR 5,000 (approx Euro 10 to 100). There are a few moneylenders, who charge very high interest rates of five to ten per cent per month. The high rates are due to the fact that the poor farmers do not have collateral to pledge and have no other source of credit. In extreme cases, women pledge their gold ornaments as collateral for small debts owed to the local moneylenders, who are the traditional informal moneylenders. Female farmers are considered more credit-worthy and reliable than the men. But male farmers have better access because of the higher social capital that they control. Men know greater numbers of people and have a wider network of friends; therefore their reach is greater than that of female farmers. In terms of control, the female farmers have only partial control over the money that they borrow if it is a male-headed household. This is because the male head usually controls all household resources and can divert them for things that he considers to be a priority, rather than for the purpose for which the woman borrowed it. The implication of this is that women farmers have to compromise on the extent of their investments in urban agriculture activities and therefore have to forego the benefits which they otherwise would have reaped with increased investments. However, the male farmers would not face this problem; they have higher chances of realizing the full potential of their entrepreneurial skills and will have higher risk-taking ability.

- **Information**: Information gives people the capacity to negotiate and make informed choices. The education level of the farmers, the extent to which they are exposed to information, and their ability to use information influence the extent to which they can access and use information appropriately. Sixty to 70 per cent of the farmers are uneducated, and 30 per cent are educated only to secondary level. Men have greater access to and control of information than women, due to their higher education levels and greater social capital. this suggests that it is more probable that, compared with female farmers, male farmers will make informed choices and benefit from them. The information needs of the men and women farmers are further elaborated in Table 2.3.

- **Labour**: Differential wages for men and women in the agricultural sector are common in rural India (Labour Bureau of India, 2003), and the same is true for wage labourers in urban agriculture. In the study locations, male casual labourers are paid more for their labour (INR 100–120 for eight hours of work) than female labourers (Rs 60 for eight hours of work). However, in urban areas women labourers find more days of work (30–40 work days per year) than male labourers (10–15 work days), due to the kind of work for which the women are employed (weeding and harvesting of vegetables) and the nature of the crops (mainly leafy vegetables). In peri-urban areas, where Para grass production is great, both male and female casual labourers find employment year-round and are paid on a piece-rate basis (INR 4 per 15 kg bundle of grass harvested). A married couple earns INR 200–250 per day in a Para grass field.

Access to and control of the benefits of the resources

From the interviews, it was seen that women had complete access to the income that they receive directly, but had only partial control over decisions on how to spend it. The income that men collected for the women was controlled more fully by men. For example, women who work as casual labourers and women who sell vegetables in the market receive money directly from their employers or customers. In the case of milk, it is men who collect the money from their customers on a monthly basis, and women have less access to and control over this money. The interviews showed that women can make decisions about spending money if the amount spent is a small sum. For all larger expenses, they have to get the permission of the household head, who is usually a man (Buechler and Devi, 2008a). Control over resources can be better understood by analysing the decision making in a household.

Table 2.3 Decision making in the farm-households in the study areas

Decisions	Male	Male/female member jointly				Comments/explanation
		Male dominates decision	Equal influence	Female dominates decision	Female	
Inputs						Combined decision. If there are young children at home, the man spends more time in the field.
Use of family labour			x			Both men and women. Also, the crop determines what inputs to buy.
What inputs to buy?			x			Men have more say, but also depend on the availability of crop and household labour and its affordability for the family.
Hire additional labour		x				
Production						Combined decision; also depends on soil conditions, water availability, demand for the crop, and economic capacity of household.
Which crop to grow?			x			Decision taken by men and women together, depending on soil conditions, water availability, and demand for the crop.
Where to plant what?			x			This decision is not in people's hands. It depends on when the crop is ready for harvest.
When to harvest?			x			Household head decides, depending upon the availability of household labour, skills of household members, and prior knowledge of the crop/animal production; but the women in household also have a say and can influence this decision. Cash needs and market demand for the products also influence this decision.
Choice between crop production or dairy		x				Head of the household, who is generally the man, decides the number of cattle to be kept, depending upon the household labour and capital available.
Number of milk cattle to be bought		x				
Marketing						
What part of the harvest is sold and how?				x		Women decide how much of the produce (vegetables and milk) will be kept for household consumption.

	Male	Male/female member jointly				Comments/explanation
	Male	*Male dominates decision*	*Equal influence*	*Female dominates decision*	*Female*	
When should a cow/ buffalo be sold?		x				Men generally take any decision that involves a lot of money e.g. selling a cow, but women can influence this decision.
Investments						
Equipment and tools	x					Men are more aware of tools (what tools required, where to buy, and prices) and equipment and therefore they buy them.
Loan		x				It depends on the purpose of the loan, but in general the man dominates the decision making
Buy or rent additional land		x				All land-related decisions are taken by men as they involve major investment, but these decisions can be influenced by women.
Buy more animals		x				Depending on labour availability in the house, space to keep animals, and availability of capital, men decide to buy more cattle; but their decision is influenced by women.
Reproduction						
Whether a child goes to school or not		x				Both mother and father decide, but the father has the upper hand in the decision. If it comes to prioritizing between the education of a girl and a boy, a girl's education is given a lower priority.
To go to a doctor or not?		x				Men dominate this decision, because consulting a doctor is expensive.
Whether or not to apply birth control		x				It used to be a man's decision. With changing attitudes and awareness among women and men of the importance of family planning, this is changing and women have more say.

Decision-making powers and their distribution in a household

A decision-making matrix illustrates power relations between the men and the women in a household. The matrix in Table 2.3 has been constructed on the basis of field interviews and field observations.

Men make most decisions related to production and all decisions related to investment and reproduction. There are very few decisions in which women have an equal say, and most of these decisions are by default, not by design (as noted in Table 2.3). The women respondents said that most decision making is male-dominated because men are the household heads, and culturally the household head makes all decisions.

Gender-differentiated problems/constraints/needs and opportunities

Men and women perceive situations and associated problems differently. Social roles and responsibilities, awareness levels, mobility, and cultural taboos strongly influence their perceptions and preferences.

Scarcity of land for cultivation, lack of credit facilities, and poor quality of irrigation water were mentioned by both men and women as the main constraints limiting agricultural production. Real-estate developments have pushed agricultural land farther and farther away from the city. This increasing distance between the fields and the residential areas was mentioned as one of the important problems for the women farmers, because they need to go back and forth frequently between the fields and their homes to perform their multiple tasks of household work, child care, care for the elderly, and working in the fields (tasks that together constitute a triple day for women). Further, cultural norms (for example: women should be confined to the private domain, or if they go outside of the home they must go with an escort; education is not important for women; women should not talk with strangers, especially men; women's main role is to serve the husband and take care of children; women should not speak out in public) restrict women's mobility and affect the business and entrepreneurial capabilities of women farmers. While male farmers expressed the need for knowledge about more profitable crops, new skills, and credit to enable them to move out of agricultural activities to more profitable occupations, women farmers wanted information on non-land-based occupations, how to avoid problems of poor-quality irrigation water, and education and health facilities for their children and family.

As the city population grows, demand for fresh vegetables and milk is on the rise. This provides a good opportunity for men and women farmers, especially women vegetable farmers, to tap new markets and increase their incomes. Various NGOs and financial institutions are promoting micro-credit, working especially with women's Self Help Groups. Women farmers and vegetable vendors could tap this resource for investments in micro-enterprises and purchase of inputs for their crops. With deteriorating conditions in the rural areas, irrigated urban agriculture will seem even more attractive; a further advantage is that it offers employment throughout the year.

Lessons learned and recommendations

The findings and lessons learned in this study were disseminated at the Multi-stakeholder Policy Design and Action Planning workshops undertaken in the context of the RUAF–Cities Farming for the Future Programme in Hyderabad in order to stimulate gender-sensitive action planning for urban and peri-urban agriculture.

The key lessons learned from the study are as follows.

- The culturally prescribed gender division of labour in urban agriculture does not match completely with what men and women do in practice. For example, women do perform activities that are culturally defined as men's work, such as irrigation, whenever the need arises. However, women will often not admit to performing work that is culturally defined as a male preserve. This shows that women are capable of doing 'male' work but often refrain from performing such tasks simply because it is socially unacceptable.
- Having only one income in the household is not sufficient for a minimum livelihood; therefore even in households where the husband or father has a job in another sector, women's income from urban agriculture is important for the maintenance of the household and the class position of the household.
- Crops that have higher rates of return (such as fodder grass) tend to be controlled by men, whereas women view urban agriculture more as a way of ensuring food security for their households.
- Location influences the degree and type of involvement of women in agriculture and dairy activities. For example, in the urban area, the sale of dairy products takes place mainly outside of the home and is therefore a male-dominated activity, whereas in the peri-urban areas most of the sales occur from the home and most often are managed by women. This has implications for who (the man or the woman) has access to and control over the income from dairy products.
- Men involved in urban agriculture are better connected with influential persons in the local government and can thus more easily obtain the goods and services that they need, whereas women cannot negotiate with influential persons so easily, because of the limitations placed on their freedom of movement and speech in the public arena.
- Affiliations such as caste, class, and ethnicity affect gender relations and gender roles in urban agriculture (as in other sectors), although these effects are partly determined by educational attainment, exposure to media, changing environmental and policy conditions, availability of opportunities, and other urban influences (Buechler and Devi, 2008b).

The following recommendations were made to the Multi-stakeholder Platform on Urban Agriculture for the development of a gender-positive

Strategic Agenda for the sustainable development of urban and peri-urban agriculture and alleviation of urban poverty.

- **Allocation of land:** allow poor farmers (especially tenant farmers, a large proportion of whom are women) to cultivate land in the green-belt areas around the city and any unused land areas within the city. Authorities could lease out vacant land especially to (groups of) female-headed households on a priority basis to encourage and support them.
- **Credit facilities:** stimulate the organization of female producers without collateral in Self Help Groups and enhance provision of credit to these groups by NGOs (now mainly operating in rural areas) and government organizations.
- **Training:** tailor the training programmes of the Ministry of Agriculture to the needs of women as intra-urban and peri-urban farmers, and encourage agricultural training activities as part of the Support to Training and Employment Projects (STEP) of the Department of Women and Child Development (DWCD). Also as part of the Socio-Economic Programme (SEP), poor female urban producers could be trained in value-added processing and packaging of products in food-processing units (Surinder, 1998).
- **Education and capacity building:** most urban and peri-urban female farmers are illiterate, a fact which impedes their capacity to influence the decisions taken in their households and the capacity to use information effectively. The DWCD scheme of condensed courses of education and vocational training for adult women could reach out more effectively to illiterate women farmers in order to build their capacities through intensive courses.
- **Projects:** all urban agriculture projects should be developed with a gender focus in order to ensure that they contribute to women's welfare across the four dimensions of gender-sensitive poverty alleviation. That is to say that such projects should (World Bank 2001) increase women's *opportunities* to access resources and gain employment; increase women's *capabilities* and skills to perform more efficiently and gain from them; strengthen women's *security* (their risk-bearing capacities); and *empower* women at the household and community levels.
- **New ways of coping with land and water scarcity:** methods that increase the productivity of crops (more yield per hectare), as well as the efficiency and safety of wastewater irrigation, should be developed together with the female farmers, who generally operate on a small scale and have less access to land and water than male farmers. Special crops such as mushrooms and herbs could be introduced, because they can be grown in closed rooms or in small spaces and at the same time generate high income.

Men and women experience poverty differently. When policymakers and planners realize this, they can make significant contributions to the lives of the poor men and women. Evidence is growing that gender-sensitive research, development initiatives/projects, and policies contribute to economic growth as well as to equity objectives.

References

Buechler, S. and Devi, G. (2002) 'Highlighting the User in Wastewater Use Research: A Livelihoods Approach to the Study of Wastewater Users in Hyderabad'. Presentation at an IWMI–IDRC Wastewater Experts Meeting: 'Wastewater Use in Irrigated Agriculture: Confronting the Livelihood and Environmental Realities', Hyderabad, India, 11–14 November.

Buechler, S. and Devi, G. (2003) 'Household Food Security and Wastewater-dependent Livelihood Activities Along the Musi River in Andhra Pradesh, India'. Report used as an input for review and publication of the second volume on guidelines for wastewater use for agriculture, Geneva, Switzerland: World Health Organisation. Available from http://www.who.int/water_sanitation_health/wastewater/gwwufoodsecurity.pdf

Buechler, S. and Devi, G. (2008a) 'Gender, Water and Agriculture in the Context of Urban Growth in Central Mexico and South India', special issue on gender and water, *Gender, Place and Culture*.

Buechler, S. and Devi, G. (2008b) 'Highlighting the user in wastewater research: gender, caste and class in the study of wastewater-dependent livelihoods in Hyderabad, India', in S. Ahmed, S. Rimal Gautam and M. Zwarteveen (eds.), *Engendering Integrated Water Management in South Asia: Policy, Practice and Institutions*, Sage Press, New Delhi.

Buechler, S., Devi, G., and Devi, R. (2003) *Making a Living Along the Musi River near Hyderabad, India*, Video co-produced by IWMI, with funding support of DFID and RUAF Foundation of ETC, the Netherlands.

Buechler, S., Devi, G., and Raschid, L. S. (2002) 'Livelihoods and wastewater irrigated agriculture along the Musi River in Hyderabad City, Andhra Pradesh, India', *Urban Agriculture Magazine 8, Wastewater Use for Urban Agriculture*: 14–17.

Labour Bureau Government of India (2003) *Wage rates in rural India for year 2002–2003* [online]. Available from http://www.labourbureau.nic.in/wrr2t5a.htm

Surinder, S. (1998) 'India's revolutionary progress in food production', *India Perspectives*.

World Bank (2001) *Engendering Development: Through Gender Equality in Rights, Resources and Voice*, Oxford University Press, New York. Consultation version available from http://www.worldbank.org/gender

About the authors

Gayathri Devi is a Researcher at the International Water Management Institute, Hyderabad, India.

Stephanie Buechler is a Research Associate at the Bureau of Applied Research in Anthropology, University of Arizona, Tucson, Arizona.

CHAPTER 3

Gender in jasmine flower-garland livelihoods in peri-urban Metro Manila, Philippines

Raul M. Boncodin, Arma R. Bertuso, Jaime A. Gallentes, Dindo M. Campilan, Rehan Abeyratne and Helen F. Dayo

Abstract

This chapter analyses the role of women in jasmine flower-garland livelihoods, with emphasis on division of labour, household decision making, access to and control over resources, health, and other interlocking issues relating to gender and urban agriculture. It draws from results of collaborative research in the Philippines by Urban Harvest and UPWARD, along with the University of the Philippines Los Baños–National Crop Protection Centre and the local and city governments of Laguna and Quezon provinces. In addition to gender issues, the chapter briefly tackles salient problems that affect the jasmine flower-garland livelihoods in peri-urban and urban Metro Manila, Philippines.

Introduction

Background of the case study

This case study is part of a project on *sampaguita* (Jasmine) production in Metro Manila, implemented by the International Potato Center–Users' Perspectives With Agricultural Research and Development (CIP–UPWARD), the Consultative Group on International Agriculture Research (CGIAR)–Urban Harvest programme, and the National Crop Protection Center and Department of Horticulture of the University of the Philippines Los Baños (UPLB), in partnership with the local and city governments of Laguna and Lucena City, as well as the sampaguita farmers.

The project's overall objective is to improve sampaguita-based livelihoods through technological, socio-economic, and institutional innovations that benefit poor urban and peri-urban households in Metro Manila. Since 2001, the project has undergone five key phases, covering socio-economic and technical assessment, technology development, market assessment, and farmer training and capacity building (Table 3.1).

Gender considerations were incorporated in project design and implementation only after the livelihood assessment, which revealed women's distinct roles. Gender-disaggregated data were targeted in subsequent field assessments of flower/garland production, management, and marketing. In the conduct of farmer field schools (FFSs), gender balance among the participants

Table 3.1 Key results of the five project phases

Project phase	Results/findings
Sampaguita flower production and garland-making enterprise in peri-urban Metro Manila: assessing livelihoods and pest-management issues (2001–2002)	• Eight major actors are involved in the sampaguita livelihood system: farmer, flower picker, supplier, vendor, abaca fibre cleaner, garland-making contractor, garland maker, and garland vendor. • The garland-making activity in peri-urban San Pedro serves as the focal point linking rural economies to urban centres. • Sampaguita livelihood is a viable peri-urban enterprise that provides employment, income, and non-monetary benefits to the poor. • Key problems are associated with crop production, post-production, production, and economics, requiring the help of research and academic institutions. • Farmers heavily rely on extensive use of chemical pesticides for pest management.
Assessment of pesticide-residue level in the production of sampaguita flowers (2002–2003)	• Pesticide residues were detected in sampaguita flowers from traders, vendors, garland makers, and garland sellers. • There was higher frequency of detection and higher residue concentrations during lean months, compared with the peak season of sampaguita flower production. • Organophosphate, carbamate, and pyrethoid were the primary residues detected.
Development of pest-management schemes for peri-urban sampaguita production (2003–2004)	• Seasonal occurrence and abundance patterns of various insect pests associated with sampaguita were determined and compared in relation to flower production and cultural management practices of sampaguita growers. • Participatory on-farm trials identified appropriate practices relating to judicious application of chemicals, while promoting use of biological control agents.
Sampaguita garland-selling business in Metro Manila: rapid market assessment (2005)	• Garland selling makes a significant contribution to livelihoods of poor urban and peri-urban households, particularly women and children. • Garland marketing is beset by problems of: health and safety for street children selling garlands, competition among vendors, lack of institutional support, and poor quality of garlands.
Training and information support to improve crop-management practices (2006–2008)	• A field manual, synthesizing results of earlier pest-management experiments, was developed to serve as key reference for farmers and extensionists. • Farmer field schools served as an effective training strategy to improve knowledge and practices of jasmine producers.

was a key consideration, so that, for example, both husbands and wives were invited to participate. Gender analysis was also included in the FFS baseline and preparatory activities.

This chapter is based on three sets of data:

1. The case study of gender roles in sampaguita production implemented in 2004 by interviewing eight key informants.
2. Review of the reports on studies undertaken during 2003–2005, including:
   - the rapid assessment of the livelihoods and marketing system of sampaguita growers (survey among 65 growers plus case studies);
   - a study among 20 farmers of the way they produce knowledge (participant observation, case studies, network analysis);
   - focus-group discussions (applying participatory tools such as activity analysis, seasonal calendar, resource and benefits analysis chart, problem ranking) during farmer field schools in the years 2005–2007.
3. A gender-analysis workshop with 18 participants implemented in 2008, applying similar tools.

Women producers analysing gender aspects of jasmine production during a workshop in Manila
By CIP–UPWARD

Jasmine production in Metro Manila

Sampaguita, a local name for the Jasmine plant (*Jasminum sambac* (L.) Ait), bears white, dainty, and fragrant flower buds. Strung together into garlands, they are widely used by Filipinos to venerate religious icons in churches and homes, adorn wedding and funeral ceremonies, welcome visitors, and congratulate new graduates. The small size and simplicity of sampaguita, considered the Philippine national flower, belie its role as an important source of livelihood for poor peri-urban and urban households in Metro Manila.

A livelihood-system analysis conducted by De Guzman (2003) described the extent of this agricultural activity. The analysis identified eight key actors in the system, from the peri-urban farmers producing floral buds to garland sellers in the streets of Metro Manila. Sampaguita flowers are produced in nearby provinces surrounding Metro Manila. The nexus of the entire livelihood system, however, is in San Pedro, Laguna, a suburban town 29 km south of Manila, where most sampaguita flowers and other raw materials for garland making are traded. It is also here where the big garland-making contractors and garland makers are located.

Nearly 3 million flower buds are traded daily in San Pedro, resulting in multifarious livelihood activities that provide financial and socio-economic benefits to various livelihood actors. In terms of financial benefits, farmers and garland-making contractors are the highest earners, while flower pickers and fibre cleaners earn the least money. Livelihood actors face various problems: pests and diseases; poor crop-management practices; low prices of flowers during the peak flowering season; sick and old sampaguita plants; skin allergies for flower pickers, garland makers, and garland vendors; hand injuries for fibre cleaners and garland makers; irregular price fluctuations and competition among suppliers; the need for better methods for storing flowers for suppliers and dealers; and lack of capital to increase business and competition in the sourcing of raw materials for garland makers.

Women play vital roles in this livelihood system, but their impact in the system has not been particularly studied. Previous studies analysed neither the gender differentiation among the actors nor the decision-making processes within households. Thus, this chapter aims to examine gender differentiation in labour, access to and control over resources and benefits, and decision making.

Laguna and Quezon provinces, which are part of peri-urban Metro Manila, are major sites for flower-buds production. Project sites in Laguna include the municipalities/cities of San Pedro, Cabuyao, Calamba City, and Santa Cruz. San Pedro, the nearest town to Metro Manila, is the regional market centre for buds. In Quezon, on the other hand, project sites are located in the outskirts of Lucena City, the province's business capital. Metro Manila, the capital city of the Philippines with about 11.5 million inhabitants spread out over an area of 636 km² (NSO, 2008), serves as the 'receiving hub' of the garlands, the final product of the jasmine flower-garland livelihood system, where the garland-

Flower-production areas *Laguna (Calamba City,* *Cabuyao, Sta. Cruz)* *Lucena City (outskirts,* *villages)*	➡	Flower-trading and garland-making centre *San Pedro, Laguna*	➡	Garland-seling areas *Metro Manila (major* *thoroughfares, churches,* *residential areas, local* *markets)*

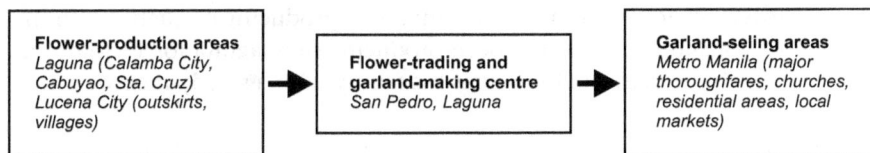

Figure 3.1 Sampaguita flower production and marketing sites

selling enterprise is widely prevalent, particularly in major thoroughfares, churches, and residential establishments.

Gender analysis of the sampaguita livelihood system

In most cases, the members of a single household pursue multiple activities: sampaguita production, picking/harvesting, packaging, transport, and garland making. This study follows the definition of 'household' proposed by Rudie (1995 as cited by Niehof, 1998): 'a co-residential unit, usually family-based in some way, which takes care of resource management and primary needs of its members'. Niehof (1998) further stipulates that household members may share household budgets and resource management without living under one roof.

Four typical households

Below we feature four households, to show the diversity of activities and roles played by the members of a household involved in the sampaguita livelihood system, and to highlight the gender role differences, household resource-allocation, and decision-making processes.

Household 1

Renato and Gina Ernas are a married couple in their mid-30s, both of whom are high-school graduates. Aside from a 2 ha rice farm, the family cultivates a 3,000 m² sampaguita farm in Santa Cruz, Laguna (80 km south of Manila). They are tenants of these lands, and Renato annually negotiates land use with the owners. Farming is their main source of livelihood, providing enough income to raise three school-age children.

Renato and Gina consider sampaguita farming as their primary source of income, while rice farming is secondary. They harvest 60–70 *tabos* (a one-litre motor-oil can, which can contain approximately 1,000 flowers) per day (during the peak season from February to May), which earns a net profit of US$ 125 per month during the peak season, and $36–54 per month in the lean season.[1] This constitutes about 70 per cent of the family's income. Renato also earns $57 per month by serving as a village councillor. Renato says that he barely earns any money from rice farming, only producing enough rice for his household consumption. He has been producing sampaguita for 12 years and plans to continue, because it provides his family with a steady income.

Household 2

Mario and Elizabeth Alemania are a married couple in their mid-60s who are both high-school graduates. They own and cultivate a 1,500 m² plot of land in Mamatid, Cabuyao, Laguna (40 km south of Manila). Mario has used the land for sampaguita farming for the last 15 years, but on account of his advancing age he has gradually shifted to growing *camia* (a fragrant white flower, often used as pendants in sampaguita garlands). Sampaguita is more labour-intensive to maintain than camia. At present, half of the land is planted with sampaguita and the other half with camia. Mario plans to switch entirely to camia production in the next few years, finding it more profitable than sampaguita production. Camia, however, is a seasonal crop, unlike sampaguita, which flowers year-round. Being the main camia producers in Cabuyao, where only a few farmers have planted the crop, Mario and Elizabeth can sell the flowers for higher prices in San Pedro.

Mario and Elizabeth earn daily an average of $9 from the combined production of sampaguita and camia. They have also started raising 200 ducks for egg production.[2] The family is financially secure. Two of their three children have finished college and earned their degrees; they are married and are no longer living with their parents. The youngest is still studying in the university and is staying with Mario and Elizabeth.

Household 3

Norlito and Erlinda Ramos are a married couple in their 50s, both of whom completed elementary education. They cultivate a strip of land along an irrigation canal owned by the government in Victoria, Laguna (70 km south of Manila). Norlito does not pay rent, but he does have permission from the National Irrigation Authority to use the land, provided that he keeps local irrigation canals clean.

Within this land, he cultivates a 2,000 m² plot of sampaguita. He produces on average 40–50 *tabos* per day, which earns the family between $89 and $268 monthly, depending on the season. This represents his net income, factoring in the costs of farm maintenance and workers' wages that are deducted from the total. This is the family's main source of income. Erlinda earns additional money from acting as supplier for nearby sampaguita farmers: she transports their daily produce, getting a 10 per cent share of their total sale.

The eldest daughter, Babylin Puno, is married and lives in a nearby house. She is a garland maker, and Norlito provides her with free flowers (1–3 *tabos* per day) so that she can earn income to support her own family. Her husband is unemployed. She earns about $5–$7 per day.

Norlito used to be a carpenter, but he switched to sampaguita production 10 years ago because it was a more profitable venture. Carpentry is a very seasonal job, and often he found himself unemployed. He observed sampaguita farming while doing a carpentry job in Santa Cruz and asked a local farmer how to plant it. He then experimented and was able to develop his own way of managing the crop. He plans to continue sampaguita farming, because he has enjoyed consistently high profits over the last three years.

Household 4

Nestor and Maria Avenue are a married couple in their 40s, both of whom are high-school graduates. They live on government-owned land next to a railroad track in Banlic, Calamba, Laguna (40 km south of Manila). The Avenue family is what the Philippine government refers to as 'informal settlers' (squatters), as they neither own the land nor pay rent to occupy it.

Maria runs a garland-contracting business. She buys sampaguita flowers daily from a neighbouring farmer and then contracts five of her neighbours, all unemployed housewives, to produce garlands. Once the garlands are ready, Maria transports them to downtown Calamba, where she sells them to local sampaguita sellers.

She purchases two *tabos* per day (about 2,000 flowers), which she supplies to garland makers who make about 1,000 garlands daily. Her neighbours earn as much as $1 each daily from this garland-making activity.[3] On the other hand, Maria earns a net income of $7 daily from the garland-contracting venture, which constitutes 60 per cent of her family's income. The remaining income comes from a small food canteen that Maria runs from her house. Her husband has been ill for the past two years and unable to work, but he helps to run the canteen.

The Avenue family is economically secure, as two of their children have earned their degrees and are already married. The youngest daughter, who still lives with them, is in her final year in college. Maria has been involved in the garland-contracting business for 10 years, and she plans to continue. The family used to grow sampaguita on the land near the railroads, but that land became non-tillable in 1994, due to an increase in gravel deposits caused by railroad repairs. This change prompted Maria to shift to garland-making contracting, because her family could no longer produce sampaguita.

Division of labour in the sampaguita-producing households

In all on-farm activities, there was a consistent division of labour among household members. Male members are responsible for farm maintenance, while women usually transport the daily harvest of flowers to the San Pedro market and pick up the payment from traders. Elizabeth Alemania and Erlinda Ramos act as local suppliers in their vicinities. They transport the sampaguita harvests of several farmers and receive 10 per cent commission from the total sales.

In the case-study households, flower pickers tend to be women and young children, as the task requires small and agile hands. Renato Ernas hires 15 women and children to pick sampaguita flowers on his farm on a daily basis. Norlito Ramos employs two of his nephews to help on the farm, although his wife and daughter assist in flower picking. On the other hand, Mario Alemania hires male pickers, because camia is a tall shrub and picking camia flowers is considered too dangerous for women. Nevertheless, he hires five women to sort and bundle these flowers. Women working in farming-related activities earn modest incomes, but their families are often very poor, and these incomes make a significant contribution to their households. For reference, the latest official Annual Per Capita Poverty Thresholds figure for Metro Manila is set at $285, while for the province of Laguna it is $253 (NSCB, 2004).

In the garland-contracting business, the gender roles are not so well defined. Garland-making contractors and traders may be men or women, and there is no established gender differentiation among these roles. But in the case of her household, Maria Avenue is solely responsible for the activities related to garland contracting. Garland makers, however, tend to be women, usually unemployed housewives. As mentioned, Maria contracts five women to make garlands; they work for approximately four hours per day.[4]

In all cases, the wives receive and handle the money generated from sampaguita livelihood activity. It follows that women are responsible for most of the household reproductive tasks. In general, the women buy and cook the food, clean the house, and take care of the young children. They are also responsible for monitoring the children's education. Children, even adult ones, are dependent on their parents, especially their mothers, for as long as they live in the same house.

Data gathered from the gender analysis workshop support the above findings. Men and women farmers, including their children, contribute to sampaguita production, making the activity a household enterprise. Men are responsible for tasks requiring heavier physical exertion, as well as for seasonal activities such as clearing the land and land cultivation. Men spend more time in preparing planting materials, planting, weeding, watering, fertilizer application, pesticide spraying, and hilling up. Meanwhile women make significant contributions in terms of daily tasks such as flower picking and preparing the flower buds for market. In addition, tasks such as leaf stripping are primarily women's work.

Both men and women agreed that clearing and cultivating land are men's domain. Men's labour contribution to the preparation of planting materials was estimated at 70 per cent, compared with 30 per cent for women. However, women tended to underestimate their contribution to tasks that are mainly their responsibility; for example, for flower picking, women claimed their contribution as only 60 per cent, but men estimated it as 80 per cent. Meanwhile for men-dominated tasks such as planting and weeding, women's

Table 3.2 Comparison of male and female responses on the gender division of labour

Activity	Male (% of contribution)				Female (% of contribution)			
	Men	Women	Male children	Female children	Men	Women	Male children	Female children
Preparation of planting materials	70	30			70	30		
Planting	70	30			70	30		
Weeding	50	50	*	*	25	25	25	25
Watering	30	30	20	20	90	10	**	**
Fertilizer application	60	40			90	10		
Hilling up	90	10			70	30	***	
Flower picking	10	40	10	40	20	40	20	20
Leaf stripping	10	90			10	90		
Pruning	90	10			100			
Selling	10	90			30	70		

* Occasionally, 25% for male and female children
** Occasionally, 20% for male and female children
*** Occasionally, 10% for male children

estimations of their labour contribution were higher than the estimates made by men. Furthermore, only the men mentioned hired and contract workers as an additional labour source.

Gender division of access to and control over resources

In all the case-study households, women have a high level of control over productive resources. While the men in sampaguita-farming households control most of the assets (such as land, livestock, water, equipment), the women control much of the market information and the cash income.

Elizabeth Alemania and Erlinda Ramos deliver the daily harvest of flowers to San Pedro, and they therefore decide what price to pay and they negotiate market prices with the traders. Moreover, they receive the payment from the traders and are responsible for allocating this income as they see fit. Although women are mainly responsible for the finances, major expenditures are a matter for consultation with their husband. Gina Ernas also receives daily payment from San Pedro[5] and, much like the above-mentioned women, she spends the money on supplies and food, allocates some of it to farm maintenance, and pays the children's tuition fees.

Loans for as much as $179 can be obtained from the traders in San Pedro. Both husband and wife decide when to access loans, which are often used for the educational needs of the children.[6] It is the women, however, who negotiate loan terms with the traders, usually repayment within six months, with traders entitled to a certain percentage of the farmer's daily sale. These women may not control the land, but they do control the income and how it is consumed. Thus, the sampaguita livelihood system empowers them with important decision-making capabilities.

Maria Avenue has complete control over all productive resources. She personally buys raw materials for the garlands, hires garland makers, and sells the finished products in Calamba. She alone decides how much to spend on materials and how to allocate the income that she earns. As a garland-making contractor, Maria has critical decision-making capabilities that affect her entire community. The garland makers and their families rely on her for their income and well-being.

The garland makers too have control over their income and situation, albeit on a much smaller scale. They can only produce what Maria (the contractor) provides, and they are paid on a piecemeal basis. However, in times of urgent need, the garland makers can take loans from Maria or be paid in advance. The fact that this money, plus the amount that they earn from flower picking, constitutes the total household income gives these garland-making housewives a high degree of discretion in allocating their household's meagre resources.

During the gender analysis workshops, both male and female farmers mentioned that resources such as land, inputs, and information services (including training) are accessed less by women and controlled more usually by men. Men, however, have less access to the benefits, particularly income, which are more typically controlled by women (Table 3.3). The wife is mainly responsible for budgeting the sales and household expenses. Most of the income is spent on production inputs for the farm, food, and the children's school allowances and expenses. However, both male and female respondents said that 10 per cent of the income is automatically given to the husband, and the use of this money is solely decided by him: the wife has no say about it. They said that this money is mostly spent on cigarettes and drinking with peers. For their part, women occasionally use the income for social gatherings

Table 3.3 Resource analysis according to male and female farmers

	% Access				% Control			
	Accord. to men		Accord. to women		Accord. to men		Accord. to women	
	Men	Women	Men	Women	Men	Women	Men	Women
Resources								
Land	70	30	70	30	70	30	70	30
Inputs								
– planting materials	80	20	70	30	100	–	100	–
– fertilizer			70	30			100	–
– pesticide	80	20	70	30	80	20	70	30
– water (irrigation)	–	–	70	30	–	–	100	–
Information sources	100	–	50	50	100	–	70	30
Benefits								
Income from sales	10	90	10	90	10	90	10	90

such as birthdays. Both husbands and wives agree that these are essential aspects of socialization in the community.

There are also slight differences in men's and women's perceptions of the access to and control over resources. Male farmers estimated husbands' access to and control over resources more highly than the female farmers did. A similar difference is very clearly highlighted in replies to questions about access to and control of information services: according to men, they have full control, but women said otherwise.

Gender differences in estimations of access to and control of information depend on the role that the husband and wife play in their sampaguita production. In some households, the husband played a greater role, while in some cases women are more actively involved; thus their access to and control over resources and benefits varies.

Problems encountered by male and female farmers

The gender-analysis workshop identified three key problems in sampaguita production: (1) lack of financial capital, (2) pests and diseases causing rotting and stunting, and (3) high input costs (see Table 3.4).

Table 3.4 Gender-disaggregated problem ranking

Problems	Score			Rank		
	Male	Female	Total	Male	Female	Total
1. Occurrence of pest and diseases	5	11	16	3	2	2
2. Weather disturbances cause poor quality of flowers	1	3	4	7	5	7
3. Lack of or limited finance	10	18	28	1	1	1
4. Seasonal flowering	0	6	6	0	4	6
5. Price and market	4	7	11	4	3	4
6. Yellowing, too much water	1	7	8	7	3	5
7. Expensive inputs (pesticide and fertilizers)	8	7	15	2	3	3
8. Land security (most are tenants)	1	0	1	7	0	9
9. Pest identification	2	0	2	6	0	8
10. Judicious use of pesticides	3	1	4	5	6	7
Total	35	60	95			

Gender-related health risks in sampaguita farming

One of the main concerns of the sampaguita livelihood system is the overuse or misuse of pesticides, because farmers spray as often as four times a week. Farmers are often oblivious to the negative health effects that can result from using very toxic or unsafe pesticides, or from simply using too much pesticide (Navasero et. al., 2004). The women and children, who pick the sampaguita buds, re-enter the farm less than 12 hours after spraying and are therefore the most exposed to health risks associated with pesticide use, including skin rashes and difficulties in breathing.

Conclusions and recommendations

Lessons learned

- The initial case studies of households provide a deeper understanding of gender differences in sampaguita livelihoods, although the data were not completely gender-disaggregated. Thus, additional gender-analysis tools were used to elicit information from men and women. The use of gender-analysis tools in a participatory manner facilitated interactions in same-sex groups of farmers.
- Gender analysis focused on sampaguita farmers and flower pickers. A similar analysis for traders, garland makers, and vendors would help to complete the picture for the entire livelihood system.
- The mainstreaming of gender concerns in project activities related to sampaguita production is still in its initial stage. Preliminary results indicate its potential value for group learning activities – from design and implementation to documentation and dissemination.

Recommendations

The full extent of the sampaguita garland industry and its total contribution to the economy has yet to be fully recognized by policy makers and the general public. Towards achieving this goal, further research and development efforts on the following are recommended:

- Additional technology components for crop management are needed to improve sampaguita flower production, and for better environment and health protection. There is an on-going project to develop integrated pest-management schemes for sampaguita to rationalize the use of pesticides near urban households. The immediate aims are to reduce the frequency of pesticide applications and to encourage selection of less toxic pesticides. Non-pesticide-based management techniques are also being introduced, such as the use of biological pest controls, better farm-management practices that minimize pest infestations, and replacement of sick and old sampaguita plants. Furthermore, to address gender issues there should be gender-balanced participation so as to inform those who are most significantly affected by improper pesticide use (for example, men when spraying, women and children when picking).
- Follow-up initiatives to influence policy would provide an enabling environment to reduce exposure of sampaguita actors, especially women and children, to harmful pesticides. Less pesticide use will also mean additional income for the household. Health information campaigns are

needed to provide women with better knowledge about the effects of pesticides on their health and environment.

- At the post-production level, there is a need to identify better storage methods. Sampaguita can be stored only for 48 hours with the existing practice of using styrofoam containers with ice. This limits garland makers and contractors to buying only what they can process in one day. Most often, they have to complete the tasks immediately to avoid spoilage, which prevents them from carrying out reproductive tasks. Better storage methods would allow women to manage their time more efficiently.
- The full extent of the sampaguita garland industry is yet to be established and recognized. Thus local government units do not provide any direct support for sampaguita livelihoods. A detailed industry study for sampaguita is needed, which among other benefits will highlight women's roles and contributions.
- Sampaguita market prices are highly unstable. In San Pedro, prices vary from one place to another within an hour of travel, depending on daily supply and demand. There is need to diversify uses/products of sampaguita besides garlands. One possibility is the use of sampaguita in decorating venues for social events, which are presently dominated by roses and chrysanthemums. This would further boost women's roles, since they are traditionally tasked with making the bouquets, corsages, and other decorations required in these social events.

Notes

1. Renato splits the profits from sampaguita farming with his neighbour, who owns the land. Hence, the total profit from sampaguita farming is US$250 per month in the peak season, and US$71–107 in the lean season.
2. Duck eggs, called *balut*, are eaten just before they hatch and are considered a delicacy in the Philippines.
3. Maria pays the garland makers 10–15 centavos per garland. They also pick sampaguita flowers daily in nearby farms, earning an additional $0.89 to $1.07. These women thus earn as much as $2.14 daily, money which is often the only source of family income.
4. Garland makers work from 10:00 a.m. to 2:00 p.m., so that they can cook and take care of children in the afternoon.
5. Gina Ernas does not personally travel to San Pedro, but pays a young man on a motorbike to deliver the flowers for her. When he returns from San Pedro, it is she who receives the payment, rather than her husband.
6. All case-study households said that they use part of the money earned from sampaguita to pay for their children's education. Considering that the sampaguita income accounts for 70 per cent of household income, we can safely say that sampaguita pays for half of children's education.

References

De Guzman, C. (2003) 'Sampaguita livelihoods of peri-urban Metro Manila, Philippines: Key actors, activities, benefits, and constraints', in CIP–UPWARD, *From Cultivators to Consumers: Participatory Research with Various User Groups*, CIP-UPWARD, Los Baños, Laguna, Philippines.

National Statistics Office (NSO) (2008) *2007 census population* [online]. Available at http://www.census.gov.ph/data/sectordata/2007/ncr.pdf

National Statistical Co-ordination Board (NSCB) (2004) *Highlights of the Philippines official poverty estimates for 2000–2002* [online]. Available at http://www.nscb.gov.ph/poverty/2002/2002povTreshold.asp

Navasero, M. V., Navasero, M. M., Daquioag, V., De Guzman, C., Bajet, C., Navarro, M., and Boncodin, R. (2004) 'Sustaining the Sampaguita Flower-garland Livelihood System in Peri-urban Metro Manila'. Paper presented at the Philippine National Academy of Science and Technology 26th Annual Scientific Meeting, 14–15 July 2004, The Manila Hotel, Manila, Philippines.

Niehof, A. (1998) 'Households and the food chain: how do they relate?', in CIP–UPWARD, *Sustainable Livelihood for Rural Households: Contributions from Root Crop Agriculture*, CIP–UPWARD, Los Baños, the Philippines.

About the authors

Raul M. Boncodin is Assistant Manager at the Intellectual Property Management Unit, IRRI, Los Baños, the Philippines.

Arma R. Bertuso is Network Affiliate for the International Potato Center (CIP)–Users' Perspectives With Agricultural Research and Development (UPWARD), Manila, the Philippines.

Jaime A. Gallentes is a Research Fellow at CIP–UPWARD, Manila, the Philippines.

Dindo M. Campilan is Programme Leader for CIP–UPWARD, Manila, the Philippines.

Rehan Abeyratne is a BSc. Economics student at Brown University, Providence, RI, USA.

Helen F. Dayo is a Researcher at the Agricultural Systems Cluster, College of Agriculture, University of the Philippines Los Baños.

CHAPTER 4

Gender and urban agriculture: the case of Accra, Ghana

Lesley Hope, Olufunke Cofie, Bernard Keraita and Pay Drechsel

Abstract

Gender analysis in agricultural production is important for creating a level playing field for both men and women farmers. This is especially important in urban agriculture which is commercialized and characterized by competition for resources. This study was conducted in Accra, Ghana, with a focus on open-space urban vegetable production. Gender-disaggregated data were collected from farmers at both household and farm levels on key issues such as access to and control of resources, division of tasks, decision-making process, and challenges faced. Data were collected by using qualitative methods such as semi-structured interviews and focus-group discussions, in which participatory tools were used. Two clear facts emerged from the study of open-space vegetable farming: male dominance in farming and female dominance in marketing. (There are exceptions to both these general findings.) While the general differentiation is attributed to societal norms that prevail in marketing, women farmers feel mostly constrained by existing irrigation practices, which are not women-friendly and consume time that is required equally at household level. Men feel significantly oppressed by their dependency on credit and prices dictated by market women, but increasingly they are making ground in certain commodities and those areas of wholesale trading where overland transport is required. Improved irrigation technology and other practices appear to facilitate a better gender balance on farm. These initiatives should be supported in gender-sensitive policies.

Introduction

The recognition and integration of gender concerns into national and international policies and programmes have increased over the past decade. In the domain of urban agriculture considerable progress has been made in terms of gender mainstreaming. However, it is still necessary to understand and assess the contributions of women and men in urban agriculture development and the impact of this development on both. Gender analysis in urban agriculture

Table 4.1 Production systems in urban agriculture in Accra

Production system	Description	Who is involved	Value chain
Open-space vegetable production	Farming is intensive on small plots of land, usually 0.01–0.02 ha per farmer. About 60% of vegetables grown are exotic, such as lettuce, cabbage, and spring onions, for which there is high market demand.	About 800–1000 farmers involved, 80–90% of them men. However, almost all vegetable marketers are women.	Vegetables are sold in the farms to marketers They are then taken to a central point for sale and distribution. Very lucrative, with annual incomes US$400–800 per farm. Vegetables produced are eaten by about 200,000 Accra residents daily as a supplement to certain street-food dishes.
Animal rearing	A wide range of animals is reared in Accra. They include livestock such as cattle, sheep, goats, and pigs, poultry, and small ruminants such as grass cutters. Poultry farming and pig farming are most common, mainly for commercial purposes.	Mainly men and migrants from rural areas. Practised mainly in low- to middle-income areas.	Animal products such as meat, milk, and eggs mainly sold in local markets. Farmers producing larger quantities of animal products supply directly to large establishments such as hotels and schools. Manure, especially from poultry, is in great demand for urban vegetable production.
Backyard farming	Involves both animal and crop farming around households and mainly for subsistence purposes. Land sizes vary but on average are smaller than those used in other production systems.	Practised in two of every three households in middle- to high-income areas and also compound houses. Involves mostly women and children.	Improves household nutrition and saves money which can be used to buy food from markets. Generally perceived 'safer and healthier', due to controlled farming conditions. However, seen as a nuisance by non-farmers.
Ornamental production	Wide range of ornamental plants grown for decoration and ceremonies such as weddings and funerals. Done in small plots along major roads.	Men and women are involved. Not very extensively practised.	Flowers are sold directly to households and commercial establishments. Manure from animal rearing extensively used in production.
Mushroom production	One of the emerging production systems in Accra. Done in closed environments, usually near households and for commercial purposes.	Involves 250 farmers, of whom about 100 are very active. About 40% of all farmers are women.	Sold directly to households and commercial establishments. Used also for medicinal purposes.

Source: Cofie et al., 2005

is essential for policy formulation and programme planning, to ensure equity in resource allocation and a balanced development which benefits both male and female urban dwellers.

The objective of this study in Accra was to gather gender-disaggregated data to complement existing studies within urban and peri-urban households, in order to make recommendations for the formulation of appropriate policies on urban and peri-urban agriculture, such as the municipal by-law revision supported by the Resource Centres on Urban Agriculture and Food Security (RUAF).

Accra, the capital of Ghana, covers an area of about 240 km². Its current population is about 2 million (51 per cent women and 49 per cent men), with an annual growth rate of about 3.4 per cent. Accra has a hot humid climate. Mean temperatures vary from 24°C in August to 28°C in March. The rainfall pattern is bimodal, with the major season occurring between the months of March and June, and a minor rainy season around October. Natural drainage systems in Accra include streams, ponds, and lagoons. Floodwater drains and gutters are used as open grey-water channels, draining into natural streams and the ocean.

A number of studies have identified several urban production systems in Accra, although the clustering of production systems has been based on the objectives of each individual study. These systems include open-space vegetable farming, seasonal crop farming, small-ruminant rearing, livestock farming, poultry production, mushroom farming, the customary land-rights system, floriculture, and backyard farming (Zakaria et al., 1998; Armar-Klemesu and Maxwell, 1998; Danso et al., 2002; Cofie et al., 2005; Table 4.1).

For this study, we focused on the production and marketing of vegetables in inner-urban open spaces, as described in detail by Obuobie et al. (2006).

With seasonal variation, about 50–100 ha are under vegetable production in Accra, distributed over many open spaces, including seven larger sites, some of them in use for more than 50 years (Obuobie et al., 2006). Vegetables commonly grown include lettuce, cabbage, cauliflower, green pepper, spring onions, onions, Ayoyo, Alefi, and Gboma mainly during the dry season, while in the wet season maize and okra are cultivated in addition. Besides these open spaces, there are also about 80,000 tiny backyards in Accra, some of them supporting the cultivation of vegetables.

Study methodology

Farm and market surveys reported in this chapter began in 2004 and have been updated since then via various student surveys and focus-group discussions. Three key qualitative data-collection methods i.e. interviews, focus-group discussions, and observations, were used, with special consideration of methodological issues highlighted by Hovorka (1998) and RUAF (2001).

Interviews

In 2004, the first gender study began with semi-structured interviews targeting 60 farming households from three major farming sites in Accra (Marine Drive, Dzorwulu, and La). A farming household was defined as a 'family whose main provider and decision maker is engaged in urban farming'. Farmers were sampled by means of the stratified purposive technique. Two of the three sites (La and Marine Drive) have since been significantly reduced in size.

While a number of student surveys followed at farm and market levels, in 2008 a rapid assessment of marketing activities at Agbobloshie, the major vegetable wholesale and retail outlet in Accra, ended the investigations with the interview of key informants, such as one of the founding members of the market as well as women representatives at most meetings.

Focus-group discussions (FGDs)

In 2008, two gender-disaggregated FGDs were held at one central farming site (Dzorwulu) in Accra, to complement the interviews. The small number of farming women participating in the discussions limited the value of the feedback received. The FDGs focused on two types of analysis:

a. **Gender-task analysis:** A trained facilitator made a chart, placing various actors in columns (women, men, joint activities, children, society) and tasks in rows. This was done for both household and farm-based tasks. Farmers identified various tasks. Analysis targeted activities, responsibilities, decision-making process, who defines tasks, etc. First, each farmer had a chance to identify the contribution of each actor to the listed tasks. This was followed by group discussions and refinement. A wrap-up meeting was held in which outcomes from the two FGDs were shared and discussed for further refining.

b. **Problem analysis:** Challenges facing farmers were identified and listed on a board which every participant could see and read. Discussions followed which generated more challenges, which were then ranked using semi-quantitative ranking methods. Ranking by women farmers was done separately. For each problem, coping strategies that farmers had adopted or could adopt to address the challenges were identified.

Results

Corresponding with previous reports from Ghana and most countries in West Africa, three out of four farm households engaged in urban vegetable production were found to be headed by men (Obuobie et al., 2006; Drechsel et al., 2006). Eighty-three per cent of the respondents were aged 40 years and above, while 17 per cent fell between 20 and 29 years of age. None of the farmers was below the age of 30. Farming was the primary occupation of most (90 per cent) of the farmers, although they all have other sources such as trading, teaching, etc., from which they derive supplementary income. Only 23 per cent of farmers interviewed lacked formal education; a greater number had primary (33 per cent) or secondary (37 per cent) education, while 6 per cent had tertiary education. This confirms previous findings that people of all educational backgrounds are involved in urban farming in Accra (Obosu-Mensah, 1999; Danso et al., 2002; Keraita, 2002). However, the gender ratio changes when wholesale and retail activities are under consideration. While in the wholesale trade women are facing more competition, the vegetable retail sector remains almost exclusively (over 98 per cent) dominated by women, a fact which is attributed among other things to cultural or societal norms (Obuobie et al., 2006).

Access to and control of productive resources

Access to land

Agricultural land is an important resource in urban vegetable farming. This is due to increasing competition from other sectors such as housing, as more and more space is needed for up-surging urban populations. In this study, 87 per cent of the respondent farmers indicated that men and women had equal access to land for open-space vegetable production. The findings were attributed to the fact that most farming (about 72 per cent) is done on land belonging to government institutions, and access is not gender-based but rather the result of individual lobbying. This is further illustrated in the case of Dzorwulu, where there is a history behind the involvement of three women in vegetable production. They are former employees of the Ministry of Food and Agriculture who worked at the same site. They were laid off during the government's structural adjustment programme. As they had gained some experience in vegetable production, they chose to remain at the site to embark on their own production.

In Dzorwulu, open-space vegetable farming is irrigated and not rain-fed, so access to irrigation water is important for production. While women and men farmers were said to have equal access to irrigation water, women farmers said they preferred having farms closer to water sources. This is because watering is strenuous, as it is in most cases done manually using watering cans which are carried as hand-loads (carrying one watering can in each hand) rather than head-loads, with which women are more comfortable. Farm plots farther from water sources are usually smaller, as the workload increases with water-transportation distance. Control of land was said to depend on each individual arrangement. For example, farmers may or may not be given notice to move off the land when the owner decides to develop it. Compensation arrangements are in such cases an exception.

Access to water

Access to irrigation water largely depends on land access and location. In general, farmers make different kinds of informal arrangement with landowners to access land for irrigated urban vegetable farming in Accra. In most cases, access is achieved through direct contact with the owner or caretaker, or through a third party working with the government institutions in the area. In some peri-urban areas of Accra, where share-cropping is used as payment for cultivating land owned by individuals, landowners or traditional leaders (such as chiefs) prefer that men rather than women cultivate larger plots, hence providing them with greater benefit (Obuobie et al., 2004). The landowners seemed to perceive that men are likely to produce higher yields than women, but there might be crop-related differences and an element of bias in their decisions. Farmers are limited in the kinds of crop that they grow, and many landowners do not permit perennial crops on their land. Also no heavy and long-term investments are allowed, as the installation of infrastructure facilitates land claims. From the farmers' perspective, investments are very risky, owing to the insecure duration of tenure.

Access to credit facilities

Urban vegetable farmers in Accra, as in other parts of the country, do not have access to formal credit schemes. This is mainly due to the fact that farmers cannot meet the collateral demands (usually land) of the financial institutions. Nevertheless, informal credit schemes are common. For example, urban farmers have managed to evolve a win–win situation with the vegetable sellers in terms of access to informal credit. Sellers, who are mostly women, pre-finance farming activities by providing seeds, fertilizer, pesticides, or cash, negotiated through verbal agreement based on trust and confidence. This means that a farmer has to sell to the pre-financing vegetable seller, and prices are negotiated even before cropping is completed. This is a huge problem for most of the farmers (usually male); they are the losers, as vegetable sellers dictate the price. Similar situations have been observed in Lomé, Kumasi, and Benin (Danso and Drechsel, 2003). In Kumasi, male farmers strongly believe that they are 'cheated' by female vegetable sellers. They proposed strengthening of farmers' associations to give them more control over market prices and put them in a better position to obtain state funding and even gain access to formal credit facilities (Boateng et al., 2007). The main means of sourcing credit at the market level is through informal financial institutions normally known as 'susu collectors'. Accessing credit from this body does not depend on gender but is greatly determined by members' savings record and factors such as consistency in saving and payback.

Access to extension services, market information, and markets

Extension services for urban vegetable farming have for long been very limited, because this practice was largely considered an informal activity. However this changed with the decentralization of key ministries and the related creation of district directorates also in the urban districts. Accra's Director for Food and Agriculture is in charge of farming, fishing, markets, slaughterhouses, etc., supported by more than 10 extension officers. Some of their services include advice on best farming practices such as water management, pest control, improving soil fertility, and supporting marketing. Usually an extension-service officer organizes meetings with farmers in each farming site. Thus, access to official information is not gender-biased. Indeed, about 80 per cent of the respondents indicated that men and women have equal access to market information concerning the demand for their produce, as well as access to extension services. Information on market prices of vegetables easily circulates among vegetable farmers, some of whose wives are vegetable sellers. However, women farmers and male farmers whose wives are vegetable sellers seemed to have an upper hand in access to market information, because they had more frequent interactions with sellers. Only a few male farmers tried to market their produce directly (Danso and Drechsel, 2003). Successful examples are cabbage wholesalers in Accra; but other leafy vegetables and the retail market remain controlled and clearly dominated by women.

Division of labour and responsibilities

Household activities

Division of labour and responsibilities among farming households in Accra was found to be highly gendered. Household activities such as cooking, washing of clothes, taking care of children, and general household cleaning are done mostly by women, while men are more involved in the provision of household income. However, both parents undertake joint responsibility for disciplining the children (63 per cent) and in providing clothing for them. So it is important to understand and differentiate the static and dynamic roles and responsibilities that the society plays in determining labour divisions and defining tasks in households. It is expected that there will be a significant difference in roles and responsibilities in non-farming households where both men and women are formally employed, which will tend to benefit from more joint contributions.

Farming activities

Unlike peri-urban and rural areas, where farmers live close to their plots and farming is a family affair, in urban areas farmers might live far from their plots, and farming is not done jointly. Another reason for the tendency for husbands and wives not to farm jointly is that the city offers more alternative

Table 4.2 Division of labour and responsibilities among urban vegetable-farming households in Accra (n=60)

Main activity	Contribution to tasks (% of respondents)			Who defines tasks/responsibility (% of respondents)			
	Women	Men	Joint	Women	Men	Joint	Tradition
Cleaning and washing	75	10	15	10	6	24	60
Cooking	90	5	5	3	3	10	84
Child care	50	30	20	14	16	42	20
Providing household income	10	80	n.a.	10	3	24	63

Source: Authors' survey, 2005

options for income generation. However, there are many cases where husbands do farming and wives do marketing. Both men and women farmers involve their children, especially after school and at weekends. In men's farms, all main activities such as land preparation, irrigation, pesticide spraying, and weeding are done by men. Women (sellers) come in only to harvest and market the produce. Those with larger farms (more than 0.02 ha) hire labour to work the land. Men provide all hired labour. On the other hand, women farmers often hire labour for land preparation and spraying of pesticides. In this study, half of the women interviewees hired labour for land preparation. Women farmers said that land preparation is 'an energy-draining exercise and not good for women', while spraying pesticides requires the sprayer to carry heavy backpacks, which 'men can handle' better. Women with limited financial resources usually cultivate relatively smaller plots which can more easily be managed.

In general, some farming equipment, such as sprayers, watering cans, and hoes, is not women-friendly, and its use is thus largely restricted to men. Nevertheless, women farmers have adapted to this problem in different ways. For example, while many male farmers use watering cans, where possible women prefer using water hoses, which they connect to pressurized water sources to irrigate their plots. They also plant vegetables that are less water-demanding, such as spring onions and local vegetables. Similar observations were made in Kumasi, where watering as a farm activity was rated the 'most arduous and time-consuming' farming activity (Keraita et al., 2003). Obaa Yaa is one of the women vegetable farmers at La in Accra, where more women are farming than on any other site in the city, thanks to the labour-saving possibility of a gravity-supported furrow-irrigation system. She has been farming for the past 11 years; when asked the reason for the relatively small numbers of women farmers, she said:

> *I started with five other women but they have all left because of the difficulty of the tasks involved. Talking about land clearing and preparation, forking of beds, spraying of chemicals etc., it takes much determination to continue cultivating.*

I mostly use men hired labour for land clearing and preparation. When I have not got enough money to hire labour, I do the land preparation myself but then I'm able to cultivate only part of my plot.

Marketing activities

Urban vegetables are sold at farms and harvested by market women, not by the farmers. Farmers hardly engage themselves in direct marketing to consumers, unless the consumers go to farms, which is however very rare. Entry into vegetable trading depends to a great extent on availability and ability to pay for a place in the market. Moreover, society has determined which type of marketing activity each sex can undertake. While females can do any type of marketing, i.e. from retailing to itinerant trading, it is difficult to find men retailing vegetables, because society has labelled this as 'women's work'. As exotic vegetable farming is very time-consuming, farmers are often not able to sell their produce themselves, unless they split the responsibilities. That is another reason why it is mostly women who are seen in the markets (Vorberg, 2004).

Women preparing harvest for the market in Accra
By IWMI Ghana

Rapid assessment of marketing activities at Agbobloshie market, a major centre for trading in vegetables, showed female dominance in both retailing and wholesaling activities. Several reasons have been suggested for this, apart from cultural norms. Vegetable selling was perceived as a quicker way to make money, more profitable and less risky in terms of investments. Women are known to be better at bargaining, hence more likely to obtain better prices than men. Money generated from marketing activities supplements household incomes. Here, women engaged in wholesale trading often have higher profit margins than the male farmers in their family (Drechsel et al., 2006).

However, vegetable cultivation in Accra is limited to certain traditional and exotic leafy species. As a result, other produce (especially cabbage and tomatoes) is imported from other regions, especially the Ashanti and Brong Ahafo regions. This situation has resulted in itinerant trading, which for these commodities is dominated by male wholesalers. The situation is different for lettuce. A large proportion of the lettuce coming from Kumasi and Lomé (Togo) is organized by a small group of between seven and ten female wholesalers who bring their produce on public buses or lorries to Abogloshie market and sell it there to other wholesalers and retailers (Henseler et al., 2005).

Differences in decision making and levels of knowledge

Decision making

Urban vegetable farming is highly individualized, and each farmer makes his/her own decision on farming activities. Despite strong competition among farmers, they often exchange knowledge, share farming equipment, make bulk purchases of farm inputs, or even have joint savings (Danso and Drechsel, 2003). In addition, in times of crisis, such as water scarcity in the dry season and pest invasions, farmers make collective decisions. Therefore in most farming sites in Accra there are farmers' associations (informal and formal), and some have elected leaders and written constitutions. There are also efforts to create a city-level farmers' association (Accra Vegetable Growers Association).

One of the local associations is at Dzorwulu (Accra), where farmers have a formal farmers' association with clear rules and regulations governing the group. All farmers at the site, currently 26 men and three women, support the association. Its main aim is to protect the interests of the farmers. The members make monthly contributions for the purchase of farm equipment such as knapsack pesticide sprayers and water hoses for irrigation. There is no woman in the leadership. The chairman of the association, Mr Fusseini, said that the leadership positions were open to all, but, due to men's numerical advantage and the unwillingness of the women farmers to take leadership roles, all current leaders are men. However, women participate in the meetings, and their suggestions are considered in the decision-making process.

Just as in farming, the selling of vegetables is done on an individual basis, and thus decisions regarding the activity are taken by the individuals involved. However, events that require collective action are first discussed at the association level. One would expect that, due to women's dominance in marketing, most of the executive positions would be held by women, as is the case in urban farming. This is indeed the norm in retail markets, with related 'queen mothers' overseeing particular commodities. However, the situation can also be different. In the case of the Agbobloshie market in this study, only two of nine executive positions in the exotic-vegetable wholesalers' association were held by females.

Knowledge-related differences

Knowledge among farmers is not gender-based, since it largely depends on the individual farmer's academic level, training and experience in farming, including mutual knowledge exchange, and external support, be it from extension services or research projects. Farmers with higher levels of education, especially those with professional training in agriculture, appear more knowledgeable about current farming practices, farming technologies, managing cropping patterns for better markets, and coping with extreme weather conditions such as droughts. On the other hand, farmers with non-agricultural backgrounds can become professional if they have the right entrepreneurial spirit. One such group, at La in Accra, demonstrates how urban vegetable farmers can use knowledge to their benefit.

> *The group has seven members and is taking advantage of their experience and knowledge for higher productivity. The farmers' group is headed by two leaders based on their knowledge: one supervising vegetable production (Production Supervisor) and another one dealing with marketing of vegetables (Marketing Manager). The marketing manager, who has long history of trading non-agricultural products, is responsible for input supplies, marketing of vegetables and gathering information on production technologies and market trends. The group has now developed a cropping pattern that enables the group to sell their produce at peak prices, hence making high profits.* (Adapted from Danso and Drechsel, 2003)

This is however an exception. Most farmers complain strongly about their limited access to information on market prices and channels, and their financial dependence on urban market women. Access to knowledge at the market level is significantly more gender-biased than at the farm level.

Challenges for urban vegetable farmers in Accra

A number of challenges and coping strategies were listed by farmers in Accra (Table 4.3) without any significant gender-specific difference. However, women complained more about the time-consuming task of watering, which conflicted with their household tasks. To reduce the labour burden, relatively

Table 4.3 Challenges faced by vegetable farmers and their coping strategies

Problem	Causes	Coping strategies
Low prices fo output	Abundance of vegetables in the market. Unfair treatment by vegetable sellers	Better cropping patterns. Men involve their wives in the marketing of products. 'We organize ourselves to produce in sequence.'
High cost of inputs	High cost of living	'We buy in bulk as an association to reduce costs of transportation and single purchases.'
Public criticism	Use of polluted irrigation-water sources	Working with researchers to develop risk-reduction strategies. Inviting NGOs and government to provide safe groundwater. Being careful not to use polluted streams at peak pollution times. Inviting public to observe farming activities.
High irrigation-labour demands	Manual irrigation methods used	Planting crops that demand less water. Using pumps and water hoses instead of watering cans.
Unreliable water supplies	For farmers using piped water, the supply from Ghana Water Company is intermittent.	Collecting water at night or whenever there is flow and storing it in drums for irrigation.
Finance	Inability to obtain loans from formal organizations	Investing money earned from other secondary activities.

Source: authors' focus-group discussions, 2008

more women take advantage of the availability of piped water at this site (the only one in Accra with this option). This however led to complaints about the common unreliability of piped water supplies. Male farmers stressed on the other hand the poor market prices and high cost of inputs. For farmers using polluted water sources, there was notable concern about their public image, due to condemnation of their practice by the media and public officials.

Conclusion: gender mainstreaming in urban vegetable production in Accra

This study shows that in urban Ghana both sexes have distinct roles on- and off-farm. Men dominate vegetable farming and women dominate vegetable marketing. In principle, men and women can access either sector, but it does not happen very often. The reasons vary between cultural tradition (which affects marketing) and laborious irrigation practices (which affect farming). However, it appears easier for women to start farming than for men to enter the retail market. Overlap between men and women is observed in wholesale marketing, where men make ground where overland transport is required.

To support women in taking up urban vegetable production requires the provision of easier irrigation methods. On the La site in Accra, where the topography allows furrow irrigation, equal numbers of men and women were observed. Also the increasing trend to lift water by means of small pumps offers opportunities for women – if they have the resources to pay their share when the local farmers' group rents a pump. Listening to complaints, however, one deduces that men farmers feel significantly more oppressed than women, in terms of their dependency on market women who provide credit and dictate prices without sharing essential market information. These complaints can be substantiated by the observation that wholesalers especially (but also many retailers) can make significantly higher profits than farmers (Vorberg, 2004).

Issues like these have to be considered if a full gender mainstreaming is to be achieved, as emphasized in the RUAF–Cities Farming for the Future (CFF) programme. Gender mainstreaming was correspondingly incorporated in the City Strategic Agenda formulated by the Accra Multi-stakeholder Forum for Urban and Peri-Urban Agriculture, with the specific objective 'To promote and harness increased gender representation in all aspects of UA' (IWMI–RUAF Annual Report, 2006). Subsequently, efforts have been made at the city level to make the urban agriculture by-laws (under revision) more gender-sensitive than before. Both male and female farmers and marketers in urban agriculture were involved in the review process (which is nearly completed). For the first time in Accra, and for that matter Ghana, the urban agricultural practitioners were exposed to various by-laws governing their livelihoods and were able to raise their concerns to effect necessary changes, at least on paper.

References

Armar-Klemesu, M. and Maxwell , D. (eds.) (1998) *Urban Agriculture in Greater Accra Metropolitan Area. Project: 003149*, final report to IDRC, Noguchi Memorial Institute for Medical Research, University of Ghana. Available from http://.fao.org/Gender/en/lab-e.htm

Boateng, O. K., Keraita, B. and Akple, S. K. (2007) 'Gyinyase Organic Vegetable Growers' Association in Kumasi, Ghana', *Urban Agricultural Magazine 17, Strengthening Urban Producers' Organisations*: 38–40.

Cofie, O., Danso, G., Larbi, T., Kufogbe, S. K., Obiri-opareh, N., Abraham E., Schuetz, T. and Henseler, M. (2005) *Urban Agriculture in Accra, Ghana: Assessing Livelihood Potentials and Policy Mechanisms*, IWMI–RUAF Working Paper for Urban Agriculture in Accra, IWMI, Ghana.

Danso, G. and Drechsel, P. (2003) 'The marketing manager in Ghana', *Urban Agriculture Magazine 9, Financing Urban Agriculture*: 7.

Danso, G., Drechsel, P., Wiafe-Antwi, T., and Gyiele, L. (2002) 'Income of farming systems around Kumasi', *Urban Agriculture Magazine 7, Economic Aspects of Urban Agriculture*: 5–7.

Drechsel, P., Graefe, S., Sonou, M. and Cofie, O. O. (2006) *Informal irrigation in urban West Africa: an overview*, IWMI, Colombo. Research Report 102.

Available from http://www.iwmi.cgiar.org/Publications/IWMI_Research_
Reports/PDF/pub102/RR102.pdf

Henseler, M., Danso, G., and Annang, L. (2005) 'Lettuce Survey. Project Report. Lettuce Survey Component of CP51, CGIAR CPWF Project 51', unpublished report, IWMI, Ghana.

Hovorka, A. J. (1998) *Gender Resources for Urban Agriculture Research: Methodology, Directory and Annotated Bibliography*, Cities Feeding People Series Report 26, IDRC, Ottawa, Canada.

Keraita, B. (2002) 'Wastewater use in urban and peri-urban vegetable farming in Kumasi, Ghana', MSc. thesis, Wageningen University, The Netherlands.

Keraita, B., Danso, G., and Drechsel, P. (2003) 'Irrigation methods and practices in urban agriculture in Ghana and Togo', *Urban Agriculture Magazine 10, Appropriate (micro) Technologies for Urban Agriculture*: 6–7.

Obosu-Mensah, K. (1999) *Food Production in Urban Areas. A Case Study of Urban Agriculture in Accra, Ghana*, Ashgate Publishing, England.

Obuobie, E., Drechsel, P., Danso, G., and Raschid-Sally, L. (2004) 'Gender in open-space irrigated urban vegetable farming in Ghana', *Urban Agricultural Magazine 12, Gender and Urban Agriculture*: 13–15.

Obuobie, E., Keraita, B., Danso, G., Amoah, P., Cofie, O. O., Raschid-Sally, L., and Drechsel, P. (2006) *Irrigated urban vegetable production in Ghana: characteristics, benefits and risks*, IWMI-RUAF-IDRC-CPWF, Accra, Ghana. Available from www.cityfarmer.org/GhanaIrrigateVegis.html

RUAF (2001) *Urban Agriculture Magazine 5, Methodologies for Urban Agriculture Research, Policy Development, Planning and Implementation*. Available from http://www.ruaf.org/node/187

Vorberg, T. (2004) 'Income Generation with Exotic Vegetable in Kumasi, Ghana', unpublished survey, IWMI, Kumasi.

Zakaria, S., Lamptey, M. G., and Maxwell, D. (1998) 'Urban agriculture in Accra: a descriptive analysis', in M. Armar-Klemesu and D. Maxwell (eds.) 1998.

About the authors

Lesley Hope is Junior researcher at the International Water Management Institute (IWMI), Accra, Ghana.

Olufunke Cofie is Senior researcher at IWMI, Accra, Ghana.

Bernard Keraita is Research officer at IWMI, Accra, Ghana.

Pay Drechsel is Principal researcher at IWMI, Accra, Ghana.

CHAPTER 5

Gender in urban food production in hazardous areas in Kampala, Uganda

Grace Nabulo, Juliet Kiguli and Lilian N. Kiguli

Abstract

Urbanization is an important development process that is linked to land access, food production, and food security. This chapter focuses on a gender-analysis study of urban agriculture, specifically of farmers growing food crops in hazardous areas in Kampala city, investigating the division of labour, relationships, constraints, and initiatives within urban farming households. Such a study is important to ensure that good policy interventions are put in place. A survey of 202 farmers growing food crops on former rubbish tips and wastewater-irrigated wetlands in Kampala city was carried out, using semi-structured questionnaires administered by personal interview. The questionnaires were developed with the help of International Development Research Centre (IDRC) guidelines on gender-analysis methods. Focus-group discussions and key-informant interviews were held with men and women farmers. The study sought to describe the distribution of activities and resources, and the benefits and risks of urban agriculture, based on gender. The main motivating benefit of urban agriculture in Kampala city was food. However, more men were motivated by economic benefits. Women suffered more than men from lack of ownership and control over land. The study also showed that women were more likely to grow food crops on contaminated land, which made them more vulnerable to health risks. Many urban farmers lacked access to land, especially women, while men owned land in small pieces, less than three acres in area.

Introduction

This study was carried out in Kampala city in 2001 as part of a PhD research project entitled 'Assessment of Heavy Metal Contamination of Food Crops and Vegetables Grown In and Around Kampala City, Uganda' and other research on urban farming throughout the city, undertaken by the project 'Urban Agriculture and Access to Land by the Poor in Kampala'. The principal objective of the study was to establish the roles of men and women in urban agriculture and to understand how resources are distributed between them. The study focused on farmers growing food crops in hazardous areas in Kampala city. The project received funding from the International Development Research Centre (IDRC) under the Agropolis Graduate Research Awards. Other stakeholders involved in the project were Makerere University, Municipal Development Programme-Harare, and Urban Harvest, a Consultative Group on International Agricultural Research (CGIAR) Strategic Initiative on Urban and Peri-urban Agriculture.

Urban agriculture in Kampala

In Kampala, urban agriculture is increasingly significant as a source of livelihood for the urban poor, due to the high rate of urbanization, accompanied by the rapid growth of unemployment resulting from immigration. Kampala is divided into five administrative divisions: Nakawa, Makindye, Rubaga, Kawempe, and Central, covering approximately 189 km² of land. The district has a resident population of over 1.2 million inhabitants, with a population density of 7,378 persons per km² (UBOS, 2003). About 35 per cent of the city population is involved in urban agriculture (Maxwell, 1994).

Urban agriculture in Kampala takes place predominantly on private land, in backyards and on undeveloped public land. However, due to lack of access to agricultural land, people are growing crops on hazardous sites that are unsuitable for development. Such places include road reserves, banks of drainage channels, wetlands, and contaminated sites such as scrap yards and dumping sites for solid and liquid waste. Urban agricultural activities often take place on undeveloped land and are therefore not included in the urban planning and development projects. These sites often lack major public services such as clean water, waste-disposal facilities, and transport.

Urban agriculture has emerged as an unplanned activity in most developing cities (Kaneez, 1998); conversely, urban agriculture in Kampala is recognized as an important livelihood activity by the Kampala City Council and is currently a legal activity after a protracted process involving various stakeholders which reviewed the ordinances regulating this activity. Studies have shown clear gender differences in the practice of urban agriculture in the city.

Methodology of the study

The study was developed with the help of IDRC guidelines on Gender Analysis Research Methodology (Hovorka, 1998). It focused on the analysis of the gender division of labour in crop production, processing, transporting, and marketing activities, on the way in which access to and control over resources are distributed between men and women farmers, on the distribution of benefits among men and women farmers, and on the specific constraints, problems, and risks of urban agriculture for men and women.

The multi-disciplinary study team comprised two male and two female research assistants, who received prior training in gender-sensitive survey work. The study team received support from a gender-resource consultant, Diana Lee-Smith, who built their capacity in gender-analysis methodology and social statistics. Two men and two women farmers were involved in the project to mobilize their fellow farmers. Also the local council authorities, including the male chairperson, a male security officer, and a female representative, supported meetings in the area aimed at identifying farming households and mobilizing farmers for the project activities. Both husbands and wives were invited to these meetings, and both men and women farmers were involved in identifying fellow respondents. In households where both husband and wife were farmers, both were invited to participate in the project as respondents. In total, 111 women and 99 men were included in the study.

The study sites were selected from former rubbish tips around the city where farmers are actively involved in agriculture: Kinawataka, Wakaliga, and Lugogo, together with Namuwongo wetland, which is subjected to disposal of wastewater, industrial effluents, and untreated sewage from the city catchment areas. The criteria for selection of these sites were based on the history of waste disposal from industry and municipality, irrigation with wastewater, and application of sewage manure on the land.

Data were collected by use of semi-structured questionnaires through personal interviews, with the help of various gender-sensitive tools like the gender-activity and gender-benefit analysis tools. The questionnaires were administered to farming households in two languages: English and the vernacular Luganda. Data were analysed using a Software Program for Social Sciences (SPSS). Qualitative research was carried out by conducting focus-group discussions with female farmers and key informants, both males and females, from Kampala city. This data were analysed using Atlas-ti software and manually interpreted using thematic and content analyses.

Gender analysis of the local situation

Gender relations in urban food production and access to land for urban agriculture

Access to and control of resources

The study showed that a large proportion of the farmers (59 per cent) had limited access to land for expansion of agricultural activities. These comprised 63 per cent of the women and 55 per cent of the men. As squatters, these producers have only usufruct rights on land for food production and can be evicted at any time. Security of tenure is non-existent; and without it the producers are less concerned with sustainable environmental concerns such as land degradation and development of the land. They live in constant fear of their crops being slashed, and their primary concern is the survival of their families from day to day. Most of these women live in the slums, are generally poor, and have access to polluted land such as rubbish tips and wetlands. They reported: 'Fellow women occupy the wetlands/swampy areas because land is cheap and readily available ...the poor access marginal lands, people with small means resort to the informal areas for mainly agriculture, and then settlements develop in these areas over time' (married women in Kigobe–Rubaga division).

The study showed that landowners prefer to lease land to women rather than to men, because they think that women are less prone to dubious acts than men, who are expected to build structures so that the landowner may have difficulty in reclaiming the land and may eventually lose the ownership. On the other hand, women's produce is consumed mostly by the household, while men sell most of their agricultural produce. This enables men to generate income, as well as to get access to credit, with which they can expand their agricultural production by renting and borrowing more land, and paying for extra labour, which is much more difficult for women. The women seek access to land for urban agriculture through borrowing and searching for free unused pieces of land in the neighbourhood, garbage areas or undeveloped land in valleys. Women tend to concentrate their agricultural activities near to their homes and/or seek areas farther away where they might undertake farming on plots close to each other. Some women travel long distances to claim undeveloped land. In Kololo East, women hold land near a primary school attended by a number of their children; on the way home the children can assist their mothers in the gardens.

Both men and women farmers had limited control over land: 25 per cent of the women having control, compared with 32 per cent of the men. Up to 33 per cent of the land was controlled by landlords and 21 per cent by institutions such as Kampala City Council. The men accessed land through purchase, inheritance, renting, borrowing, or squatting. Most of the women acquired land by squatting or borrowing, or through their husbands, who rented or inherited it. Women's access to and control of land was adversely affected by social cultural practices, as discussed later under the role of external factors on gender in urban agriculture.

The study revealed that farming is a major source of household income in Kampala. Sixty-six per cent of the male farmers and 47 per cent of the women obtained income from other forms of self-employment, but 5 per cent of the men and 15 per cent of the women obtained all their income from farming. Only 11 per cent of the men and 8 per cent of the women acquired their income from a regular salary and supplemented their incomes through urban agriculture. In this study, the majority of farmers had informal employment. This led to the development of a class of unskilled labourers with no formal vocational training, of whom 51 per cent are men and 61 per cent are women.

The study also showed that women would be affected more than men if they were prevented from farming on contaminated sites (Table 5.1).

The major benefit that would be lost if urban agriculture were prohibited in Kampala city was access to food. Statistically, 31 per cent of the men and 37 per cent of the women would have little or no food if they were prevented from

Table 5.1 Impact on men and women of ceasing to grow crops on dump sites

Effect of stopping farming	Frequency	Proportion of men (%)	Proportion of women (%)
No effect	13	14	01
Would suffer financial loss	06	02	04
No school fees	09	05	05
No food	66	31	37
Economic crisis	42	12	30
Would transfer	13	12	03
Would become poor	33	14	20
Not sure	10	10	00
Total	192	100	100

farming on rubbish tips. On the other hand, 30 per cent of the women would suffer an economic crisis, compared with 12 per cent of the men. Generally, most of the farmers would experience a financial crisis, and the women would be more affected than men. The women depended on urban agriculture to feed and maintain their families, while the men had other sources of income.

Sourcing inputs

A much higher proportion of women than men spent income on purchasing seeds for planting. Similarly, 88.2 per cent of the women, compared with 65.5 per cent of the men, obtained seeds from the market, while 12.6 per cent of the men and 2.7 per cent of the women obtained seed from the previous season's crops. This is due to the fact that men grow sugar cane and cocoyam, which can multiply by vegetative propagation, and seeds for planting are available from the previous crop. Seeds for maize and vegetables, grown mainly by women, on the other hand are not readily available, which means that they must be purchased from the market. The rest of the farmers obtained seeds from friends or from the rural areas, depending on the type of crop grown.

Processing, transportation, and marketing of urban agricultural products

Women grow crops that are less demanding in terms of production activities, capital, processing, transporting, and marketing, such as maize, sweet potatoes, and vegetables, which they grow near households and dump sites. These crops are sold in the neighbourhoods either directly to consumers, or to market vendors, or by the roadside. However, crops grown by women fetched less income, because production is on a small scale and products are perishable, hence farmers suffer severe post-harvest losses. Men on the other hand engage in larger-scale farming activities such as growing sugar cane and cocoyam for commercial use. These are bulky and more labour-intensive in terms of harvesting and transport to marketing sites. For example, harvesting cocoyams involves uprooting from wetlands; excess mud and roots are removed from the rhizome before they are packed in sacks and transferred to drier areas, where they are loaded on bicycles. Buyers from various markets around the city collect the produce from the wetlands and transport it to markets, using hired bicycles.

Utilization of farm produce

This study showed that a higher proportion of men than women sold some of the food produced from farming activities (Table 5.2).

A higher proportion of the farmers used the food crops mainly to feed their families, although a small proportion of them grew food purposely for sale. The proportion of farmers who sold all of the food grown on contaminated sites to consumers was higher among women, who consequently used the funds to buy other foodstuffs from the market.

Table 5.2 Utilization of food derived from urban agriculture

Utilization of food	Proportion of farmers (%)		
	Men (n = 80)	Women (n = 101)	All farmers (n = 181)
Consume all	(24) 30%	(34) 34%	(58) 32%
Sell some	(53) 65%	(53) 52%	(106) 59%
Sell all	(3) 5%	(14) 14%	(17) 9%

Exposure to health hazards in urban agriculture

Women as a group, however, face particular problems in most of their traditional roles. For instance, a man's hoe always has a long handle, while a woman's hoe often has a short handle, which necessitates that a woman bends while digging. Moreover, of the 56 per cent of the farmers who did not use any protective clothing while working on contaminated land, 59 per cent were women and 41 per cent men. Additionally, most of the farmers did not have access to clean and adequate water supplies. This has negative implications for food production, the quality of food, and human health. Overall, women were at increased risk of exposure to occupational and health hazards associated with urban agriculture, compared with men.

Gender roles and bargaining power in decision making in the community

With regard to the control of funds, this study showed that both men and women had control over their own funds. Budget decisions were largely controlled by men and women involved in income generation. Eighty-three per cent of respondents made budget decisions by themselves, of whom 43 per cent were men and 57 per cent women, while 13 per cent made budget decisions jointly between husband and wife and only 4 per cent said that budget decisions were made by the head of household (husband, father, or mother).

Women and men contributed income to their households from urban agriculture on a more or less equal basis (Table 5.3).

Assessment of expenditure showed that the majority of the farmers used their income obtained from urban agriculture to meet domestic needs, with a higher proportion of women doing so than men. In fact, the outcome of this survey found that women contributed to all household expenses, including those that were traditionally male responsibilities such as housing. Equal but small proportions of men and women spent income accruing from farming activities on purchasing food. Farmers spent less of their income on food, because what they reaped from the gardens was not valued in monetary terms as it was considered free food.

Table 5.3 Men's and women's utilization of income obtained from urban agriculture

Item	Proportion of men (%)	Proportion of women (%)
Paying school fees	26	12
Buying seeds for planting	5	9
Spending on domestic needs	38	50
Buying food	5	5
Medical treatment	0	3
Paying house rent	3	1
Savings	14	13
Development	9	7
Total	100	100

Gender division of labour

Men are slightly more involved in wastewater-fed cultivating of sugar cane and cocoyam in the wetlands (55 per cent of the men, compared with 45 per cent of the women), crops that are non-perishable and therefore can be transported for longer distances in search of markets. According to the local division of labour, men open up the wetlands by cutting down the papyrus, while women plant the crops.

Women are more involved (55 per cent of the women and 45 per cent of the men) in growing maize, sweet potatoes, and vegetables on former rubbish tips and road sides, crops which require less time in the field, require less capital investment, and are perishable and hence need to be marketed closer by (within the neighbourhood: by the roadside, in nearby markets, or on stalls in front of their homes).

Urban agriculture is being feminized as men move out of agriculture to other informal sectors like petty trading, while women continue as farmers, with a few migrant men working as hired labourers. The study showed that most (67.4 per cent) of the urban farmers are married, 26.3 per cent single (with a large proportion of men), 5.8 per cent widowed, and 0.5 per cent students. The married farming households included 64 per cent of the men and 70 per cent of the women. The male-headed households had larger numbers

Women harvesting vegetables at Luzira, a wastewater-irrigated wetland near Lake Victoria
By Grace Nabulo

of children (0–16), compared with the women-headed households (0–12); on average each household had three children. Most of the women in the study were mothers and engaged in urban agriculture as a source of food and income for their families. The highest proportion of the women (26 per cent) had two children, while the highest proportion of the men (21 per cent) had none. Older children and students take part in household farming activities, providing free and extra labour. Conversely, most of the men engaged in urban agriculture as an income-generating activity.

The men spent longer hours in the garden, because the crops that they grow are more labour-intensive. Since most of the women (70 per cent) are married and have to perform traditional household roles as well, the women tend to grow crops that are less labour-intensive (Table 5.4).

Although women spent relatively less time in the gardens than men, the study revealed that women had much less leisure time than men (women 4 per cent, men 12 per cent leisure time). This is an indication that women engaged in several unpaid, unrecognized, and undocumented labour tasks that men take for granted. Women and men were involved in different activities in addition to agriculture, some of which were specific to either men or women. In Uganda, house keeping, child care, nursing the sick, cooking food, and

Table 5.4 Time spent in the garden daily by men and women

Number of working hours spent in the garden	Proportion of male farmers (%)	Proportion of female farmers (%)
1	6	17
2	16	23
3	21	18
4	19	18
5	9	8
6	11	7
Whole day	18	9
Total	100	100

housework are a woman's responsibilities. Cooking food, for example, is culturally unacceptable for a married man. In addition, some women carried out other income-generating activities such as tailoring, poultry keeping, shop keeping, and serving in saloons and restaurants, as well as selling food products.

Only 9 per cent of the farmers had formal employment and earned a salary. Few women had professional training, and those that did were predominantly nurses and teachers. Men were often carpenters, shopkeepers, builders, drivers, stone breakers, bricklayers, cattle keepers, hawkers, and security guards in addition to being farmers.

Men's and women's interests, constraints, and knowledge of urban farming

The main motivating benefit of urban agriculture for most of the farmers was food. Asked what motivated them to practise urban farming, 41 per cent of the respondents said they benefited from getting food, 21 per cent from easy access to markets, and 9 per cent from economic benefits. As discussed earlier, 35 per cent of the farmers would have no food if prevented from growing food, including 37 per cent of the women and 31 per cent of the men. Moreover, 22 per cent of the farmers (30 per cent of the women and 12 per cent of the men) would suffer an economic crisis if they had to stop urban agriculture.

As indicated before, men and women are generally interested in – and have better knowledge of – different crops: for men it is mainly sugar cane and cocoyam, and for women maize, sweet potatoes, and vegetables.

The role of external factors in urban agriculture

In many countries, religious and cultural traditions restrict women from owning land. In Kampala city, for example, only 7 per cent of the women own land (Kiguli, 2004). According to customary law in Uganda, a woman is not entitled to land ownership. She cannot inherit family land, because she is expected to get married and leave her birth home for her husband's home. Even in her marital home, a woman is only allocated land where she can dig and produce food to feed the family, but the man remains the legal landowner. This puts the woman in an insecure and tenuous position that is inferior to that of her husband.

A minor proportion of the urban farmers in the study (36 per cent of the men and 30 per cent of the women) have access to local organizations that teach them ways to expand their production and improve the quality of their products. Some of these organizations offer credit facilities, but a higher proportion of men than women obtained credit facilities, since most of the women's produce is consumed by the household, while men sell most of their agricultural produce. This enables men to expand their agricultural production by renting and borrowing more land and paying for extra labour, while this is much more difficult for the women.

In addition, the traditional and reproductive roles of women hinder them from obtaining land far away from their homes. Most of the women use land that is available to them near their homes or readily available in sites that are not suitable for farming, such as the rubbish dumps and roadsides. Women's productive capacities were therefore constrained by several negative factors, including limited access to expansion of land available for farming.

Potential risks and disadvantages of urban agriculture

Rubbish dumps are used for cultivation by most farmers because the land is public land, and easily accessible. These areas are considered 'no man's land', which farmers may obtain by squatting or by renting it from other farmers. The waste-dumping sites are perceived by many urban farmers to be fertile because they contain high levels of nutrients; they therefore attract farmers, as they are considered more productive. Also wetlands that receive wastewater from municipal and industrial sources are utilized by urban farmers because of the availability of water for irrigation throughout the year.

Under such conditions and without proper management, urban agriculture may lead to contamination of food crops, a reason why municipalities often judge urban agriculture negatively. Farmers on waste sites are exposed to a variety of health hazards, including chemical and biological contaminants in the soil, physical injury from sharp objects, and psychosocial discrimination. Biological agents may include bacteria, helianthus, viruses, protozoa, and micro-fauna. Contamination of fruits and vegetables by pollutants in the soil and in untreated wastewater poses a major health risk, especially when consumers eat raw vegetables.

The division of labour in urban farming households exposes men and women to health risks in different degrees (Flynn, 1999). Women are more vulnerable to health hazards, although they work less time in the gardens. They perform various roles that expose them to health hazards, including contact with contaminated soil and water in the garden (no protective gloves or boots), contact with contaminated foods during food preparation and marketing, and inhaling motor-vehicle emissions during the long hours when they are selling food at busy roadsides

A high proportion of the farmers in Kampala are aware of the health risks of growing food on these contaminated sites. Asked whether they agree with growing food on waste sites, above 80 per cent disagreed, stating that it was not healthy. Some of the farmers admitted that they sell most of the food grown on contaminated sites and do not use it for household consumption. The producers, however, said that they had no alternative, since agriculture was their main source of livelihood, food security, and household income. The farmers weigh the health risks against the benefits. One said: 'We have been eating food from these waste dump sites for several years, but we have never fallen ill. We have even educated our children up to University level using money from the sale of cocoyams grown in these areas.'

The farmers perceive these sites as unhealthy but also as fertile, free land available for farming and one of the few opportunities open to them to gain a minimum livelihood. They continue farming despite their awareness of the health risks. They are more aware of risks that have some visibility, such as cuts from sharp objects, sewage disposal with bad smells, and skin infections. However, they are less aware of the possible long-term health risks associated with farming on waste-disposal sites without adequate precautions.

Conclusions

Lessons learned and recommendations

Both men and women engage in urban farming in Kampala city, but a higher proportion of women grow food crops on rubbish tips than men. Women cultivate former rubbish dumps because these are cheap, accessible, and readily available to them. On the other hand, a higher proportion of the men cultivate in wastewater-fed wetlands, compared with the women.

A majority of the urban farmers sell some of their produce, and a higher proportion of these are men. While women constitute a larger proportion of those who consume all the food obtained from urban agriculture, they also sell to urban consumers within their neighbourhoods, hence improving the nutritional status of the urban poor and thereby playing a major role in feeding the (poor part of) the urban population in Kampala.

Women involved in urban agriculture are disadvantaged in many aspects, including access to resources, distribution of labour and benefits, and protection against health hazards.

Urban agriculture projects should target women farmers to improve their opportunities to access resources and their capacity to perform more efficiently with minimal exposure to health risks, and to empower them beyond reproductive roles. There is also a need to enhance their marketing facilities and efficiency for the sale of perishable vegetables and to minimize post-harvest losses.

Policies should aim to enable access to productive land for women farmers, for example by creating quotas for women in land allocation. Urban farmers in Kampala, especially women, are constrained by lack of formal access to land and security of tenure. Urban activities are neither planned, nor monitored, nor guided, and hence they put consumers and urban farmers, especially women, at risk of exposure to health hazards from biological and chemical contamination.

The study contributed knowledge and advanced understanding of chemical contamination in urban agriculture, complementing earlier studies on food security and nutrition. It increased public awareness regarding chemical contamination in urban agriculture, and changed the perceptions of urban communities and city authorities of the practice of growing food crops on rubbish tips and wetlands around the city. The study played a vital role in informing decision making regarding urban agriculture by facilitating the community-based review of urban agriculture ordinances in Kampala city. The research findings were used in the development of policy guidelines for Kampala City Council.

Urban agriculture is currently recognized as a legal activity in Kampala. There is a need for immediate implementation of the newly formulated guidelines and urban agriculture ordinances to address urban waste management and administration of land for farming, especially for women, in order to improve productivity and enhance food safety. Dissemination of knowledge of adequate crop selection and improvement, and of appropriate techniques such as composting and low-cost wastewater-treatment technologies would go a long way to improve the productivity, quality, and marketing of urban agriculture products.

Women should be encouraged to establish groups through which they could form partnerships with NGOs, development agencies, and action-oriented researchers. Since women comprise the majority of farmers who engage in urban agriculture, empowerment of this group is essential. Knowledge of public-health issues, high-yielding crops, farming safety, and nutrition, for example, would improve not only the quality and quantity of food produced but also the health of the farmers, their families, and consumers.

On a broader scale, the government of Uganda should incorporate urban agriculture into the Plan for Modernisation of Agriculture (PMA), to ensure more sustainable and improved food production, processing, and marketing at local and national levels. The government should also empower women through protected rights to ownership of land, which is an integral part of urban agricultural issues.

Urban agriculture projects should apply gender-sensitive methodologies and take into consideration the social-cultural values that favour men and often render women inferior within the household and the larger community; the projects should therefore specifically address the needs of women. This requires a good knowledge of the actual differences in contributions and problems of men and women in urban agriculture and the external influences such as culture, legislation, and policies or bylaws.

Differences in gender mainstreaming between urban and rural agriculture

Gender-mainstreaming initiatives in urban agriculture differ from those in rural agriculture, because the social values, structures, and institutions that create specific gender dynamics in rural agriculture are different from those in urban agriculture. In rural Uganda, ownership, control, and distribution of resources, and decision making regarding household budget are often a man's responsibility as head of household. The woman automatically takes on the reproductive roles defined by the society. The rural woman is also dependent on her husband for resources such as land, because land legally belongs to the man. The man often engages in incoming-generating activities such as marketing cash crops or livestock that empower him economically. The woman has no control over resources and the family budget, which renders her inferior in status and dependent on her husband.

In urban agriculture there is a diversity of cultural values, involving as it does people from different tribes, regions, and even countries, most of whom have no access to or right of ownership of resources such as land or irrigation water. Therefore, the traditional definition of gender roles within each group makes room for a new one in which the husband and wife contribute more equally to the livelihood of the household, either directly through digging, growing food, cooking, or selling produce or indirectly through paying for rent, tuition for children, or domestic needs. The women now contribute to the family budget, which demystifies the superiority of male over female.

References

Flynn, K. (1999) 'Urban Agriculture and Public Health: Risk Assessment and Prevention for Contamination and Zoonoses', Cities Feeding People programme initiative, IDRC, Ottawa (draft).

Hovorka, A. (1998) *Gender Resources for Urban Agriculture Research: Methodology, Directory and Annotated Bibliography*, Cities Feeding People Series Report 26, IDRC, Ottawa, Canada.

Kaneez, H. M. (1998) *Gender Capacity in Urban Agriculture: Case Studies from Harare (Zimbabwe), Kampala (Uganda) and Accra (Ghana)*, Gender and Sustainable Development Unit, IDRC, Ottawa, Canada.

Kiguli, J. (2004) 'Gender and access to land for urban agriculture in Kampala, Uganda', *Urban Agriculture Magazine 12, Gender and Urban Agriculture*: 34–35.

Maxwell, D. (1994) 'Internal Struggles over Resources, External Struggles for survival: Urban Women and Subsistence Household Production', paper presented to the 37th annual meeting of the African Studies Association, published by www.cityfarmer.org

Uganda Bureau of Statistics (UBOS) (2003) *Uganda Population and Housing Census 2002*, Kampala.

About the authors

Grace Nabulo is a Lecturer in the Department of Botany, Makerere University, Kampala, Uganda.

Juliet Kiguli is a Lecturer in the Department of Community Health and Behavioural Sciences, Makerere University School of Public Health, Kampala, Uganda.

Lilian N. Kiguli is a Graduate student at Rhodes University, Grahamstown, South Africa.

CHAPTER 6
Gender dynamics in the Musikavanhu urban agriculture movement, Harare, Zimbabwe

Percy Toriro

Abstract

This case study presents the gender dynamics of urban agriculture in the Musikavanhu Project in Harare. The project sought to understand the different roles and positions of men and women in urban agriculture in the Budiriro area, and the way in which these are changing over time. The majority of the producers in the area are women, males only assisting incidentally when they are not formally employed. While women play prominent roles in decision making regarding production and processing, men tend to take over when it comes to making decisions about the use of the income raised. The results of the study were used to facilitate gender sensitivity in the Musikavanhu project, in order to ensure that gender issues are taken into account when planning and implementing new activities, and to enhance gender equity in the areas where it is active. The positive impacts of these interventions are becoming visible as they influence local planning authorities' considerations of policy initiatives and actions related to urban agriculture in and around Harare.

Introduction

This case study presents the gender dynamics of urban agriculture in the Musikavanhu Project in Harare. The study was undertaken in the context of the RUAF–Cities Farming for the Future Project. Fieldwork took place in the Budiriro suburb of Harare in January–February 2008 as a continuation and update of a participatory field study done in 2004 (Mushamba and Mubvami, 2004).

Although commonly referred to as Musikavanhu *Project*, it is not a project in a proper sense but a loose assembly of a large number of community groups in the urban and peri-urban area. Each has 30–50 small plot-holders who have realized the advantages of working together in a group: advantages such as access to land, the economies of scale in securing inputs and marketing their products, and even protecting their crops from thieves. Musikavanhu currently has some 50,000 registered members (6,250 households, with every member of the household being a *de facto* member of the project) in Greater

Harare, and this is attracting attention from government institutions, NGOs, and FAO. The chairperson of Musikavanhu, Mr Zunde, estimates that there are 500 such small groups across Harare.

Background to the project

The Musikavanhu Project was founded in 1998 in the low-income suburb of Budiriro, located about 10 km west of the centre of Harare, on vacant municipal land which was used by sand miners to dig for pit sand sold for the construction of houses in neighbouring new settlements. Seeing their success in fighting the illegal sand miners, the municipality granted the members an annually renewable permit to use the reclaimed and adjacent lands for agriculture, attaching a number of conditions to protect the environment. However, the land remains reserved for future development.

Soon many other households joined, spurred on by the increasing unemployment initially caused by the economic structural adjustment programmes of the 1990s and later by the deep political and economic crisis in the country, which made farming a last resort for many poor urban families. These urban poor became involved in farming in order to maintain their food security, save money on food expenditures, and generate complementary income from regular sales. The households involved in the Musikavanhu Project consider their agricultural plots an important part of their livelihood support.

The Musikavanhu Project quickly spread to other low-income suburbs of Harare. By 2004 it was reported to have 20,000 members, and in 2008 the total reached 50,000. It is argued that the membership grew exponentially between 2002 and 2006 because prospective members thought they would use the project as a vehicle to be resettled on peri-urban land during the land-reform programme. All member households utilized vacant spaces left behind during the process of urban development, illegally or temporarily.

Musikavanhu Project's major contribution is in promoting the organization of the urban poor, who hitherto had farmed randomly and irregularly on any vacant land they could find in the city, and helping them to organize their farming activities. The platform offered by the Musikavanhu Project also ensures that members can effectively interact with other institutions in an informed way.

Musikavanhu has benefited from seed and fertilizer donations from commercial companies (who see this as a good marketing strategy); incidental training is provided by some extension workers of AREX (a government scheme), an agro-ecological NGO, and the staff of the commercial companies in relation to the hybrid seed and fertilizers provided. One organization also provided loans to buy farm implements. Although Musikavanhu has received donations from various organizations, its financial base is still weak, because it does not tax its members, apart from charging annual subscriptions.

Joining Musikavanhu Project has brought political and economic advantages to members, as well as providing much-needed unity of purpose when approaching the municipality to argue for preserving their land or adding more. Members of the project were conveniently allied to ZANU (PF), so that they could easily approach ZANU (PF) politicians for protection or get access to some offices, which might otherwise have been inaccessible. The platform offered by the Musikavanhu Project enabled members to collectively approach the input-support institutions such as SEEDCO and Zimbabwe Fertilizer Company to persuade them to set aside inputs specifically for them at a time when it was difficult for individuals to get inputs at an affordable price and on time.

Although 90 per cent of the farmers in the Musikavanhu Project are women, the project has no deliberate gender policy or mainstreaming strategy.

Farming in the Musikavanhu Project

Farming in the Musikavanhu Project is largely a seasonal activity, although some members grow throughout the year. Most of the members practise rain-fed cultivation during summer, when there is plenty of rain. Additionally in some places individual farmers have dug wells on their plots, while others practise wetland cultivation or cultivate areas very close to stream banks where irrigation is possible.

In most cases the farming activities are taking place on municipal land and largely have an informal (illegal) character. The field plots have an average size of about one acre (although, of the thousand members in the Chapter, 500 have been allocated 12.5 hectares of land, a measure that has been freely sanctioned by the local authority). The hoe is the standard tool. (In Zimbabwean culture, a hoe was part of dowry paid by the men for their women, indicating the place of women in the society: the hoe is predominantly used in tilling, weeding, and harvesting and seen as a symbol of womanhood.) Mixed cropping is practised on almost every piece of land. Maize is inter-grown with beans, pumpkins, and sweet potatoes, and sunflowers are grown to maximize production per unit of land. Produce is both for own consumption and for sale at markets and to neighbours.

Some of the members also have home gardens where they grow fruit trees, vegetables, herbs, and beans, mainly for home consumption with incidental sale of surpluses. During dry spells these home-based crops are irrigated using water from the Zimbabwe National Water Authority (ZINWA). However, the farmers say that supply is not reliable. Many households also rear poultry on their home plots, most keeping up to 20 birds at a time, largely for sale.

Members have also taken up other activities like mushroom production on the home plot (following some training carried out by FAO) and a fisheries project in Mufakose District. But such activities require investments that few can afford under current economic circumstances. The Project is even undertaking community projects like supplementary feeding schemes and recently established a school for training in urban agriculture in the nearby town of Marondera.

Table 6.1 Household incomes of 70 farming households in Budiriro area

Type of income	Sex of the earner	Annual income in Z$	% of total income
Urban agriculture	Predominantly female	5 billion	23.8%
Formal work	Predominantly male	264 billion	74%
Other (renting out rooms, vending)	Female /Male	7.8 billion	2.2%
Total income for 70 households interviewed		356.8 billion	100%

Source: 2008 study

This study found that most of the farmers in the Budiriro Chapter earn very little from farming, most of the foodstuffs grown being consumed at household level. Over three quarters (76.7 per cent) of the farmers earn below US$200 annually from farming and cannot afford to invest in the agricultural system because the land is not theirs and is not protected well. Households supplement their incomes with remittances, rents that they charge to lodgers who use part of their houses, and income from working relatives (see Table 6.1).

While the respondents could easily account for the sources of their income and the relative contribution of agriculture to their livelihoods, they had difficulty estimating the cash value of the vegetables produced, because of the absence of production and sales records.

Methodology of the gender study

The study was designed to enhance understanding of the main gender differences in urban agriculture in the Budiriro chapter (the study area). The study was not designed to explore detailed gender relations within the households, but to shed light on men's and women's positions and roles within urban agriculture activities, and to understand how the dynamics within this economic sector influence such position and roles. The study has taken into account methodologies that were used in similar studies (Hovorka, 1998 and 2006).

Primary data collection consisted of interviews with the owners of urban agriculture enterprises in Budiriro chapter of the Musikavanhu Project. This study site was large enough to generate an adequate sample size. Of the thousand members in the Budiriro Chapter, 500 have been allocated 12.5 hectares of land, freely sanctioned by the local authority, while the others farm on an informal basis on vacant public land. The 500 farmers on officially sanctioned land formed the sample population for this study. The study sample consisted of a total of 70 plot holders (or leasers, as the land belongs to the Municipality) of whom 31 were men (44.3 per cent) and 39 women (55.7 per cent). In the interest of equal gender representation, we purposely selected nearly equal numbers of men and women plot holders, even though women constitute 95 per cent of the project membership. Fieldwork took place in the Budiriro suburb of Harare from January 2008 to February 2008 as a continuation and update of a participatory field study done in 2004.

The study involved the following phases:

a. Secondary data collection; interviews with key informants, as well as review of academic papers and other documents.
b. Structured interviews (15 minutes each); a combination of short-answer and open-ended questions relating to:
   - net outcomes (type of agriculture production, gross earnings, and amount of foodstuffs produced, etc.);
   - socio-economic variables (gender, age, ethnicity, citizenship, income, access to capital, labour, natural resources, inputs, information, services, social networks, etc.);
   - locational variables (plot location, tenure system, cost of land, size, on/off plot production, process of acquisition, etc.);
   - environmental variables (water sources, soil type/quality, pests/ disease, climate, etc.).

These structured interviews generated a relatively rich data set, adequate for both quantitative and qualitative analysis.

Socio-economic characteristics of the farmers interviewed

The gender breakdown of farmers is different from that of plot-holders, indicating that it is predominantly women who engage in farming work, even in male-headed households. Most of the farmers interviewed were women (73.4 per cent), of whom 52.8 per cent were aged 50 years and above. Forty per cent of the farmers interviewed were widows and/or female heads of households. Persons under the age of 40 years are very few (16.6 per cent). Most of the farmers are literate, although very few have formal education (56.1 per cent have had no formal education). Of those who have Ordinary Level education, 52.8 per cent are females.

Most of the farmers (93.3 per cent) are of the Shona tribe. All of them are Christians.

Table 6.2 Summary of socio-economic characteristics of the farmers interviewed

Socio-economic characteristics	
Age average	48 years
%married	43%
%widowed	30%
Mean household size	6 members
Main type of farm enterprise	Family
Size of the land per family	1 acre
Annual average income from crops	US$ 215
Annual average amount of crops produced	1.5 tonnes

Source: 2008 study

Gender analysis of the Musikavanhu Project

Role in decision making

Decision making at the household level is a reflection of the distribution of power at that level (Feldstein Sims and Poats., 1989). Due to the fact that the men have not been very involved in agricultural activities, the women take most of the decisions on crop choice, cultivation practices, and marketing. However, decisions on financial investments in the household are dominated by men, who do not see agriculture as a priority. Since many married women are unaware of how much their husbands earn, it is difficult for them to obtain money for agricultural inputs or tools.

The participation of children in decision making is very low. The role of children is seen as taking instructions from adults and performing tasks set for them (to till, weed, harvest, and process farm produce).

Table 6.3 Decision making in farming households in Budiriro area

Decisions	Male	Male/female member jointly			Female	Comments
		Male dominates decision	Equal influence	Female dominates		
Inputs				X		Normally after consulting the male
Production					X	
Marketing					X	
Investments		X				
Reproduction	X					

Division of tasks

Women do the overwhelming majority of the farm work. The men may assist in the agricultural activities during the weekend or when they are not engaged in a formal job.

Typically, in the early morning, women first do daily household chores, such as cooking for the children, making sure that they are ready for school, and cleaning the house, before they go off to the fields for cultivation. After working in the field, women normally get back to their household chores of looking after the children, laundry, preparing food for the family, washing dishes, etc.

The history of the organization shows that men have always been sceptical about urban agricultural activities, and most of the women did not receive support from their husbands when they started. Men joined only after the results, in terms of harvest and income from selling produce, were demonstrated. But still, participation in cultivation has been largely left to women. Despite this obvious gender division of labour, respondents stated that there is no specific gender division of work, since tilling, planting,

Demonstration plot at Musikavanhu Urban Agriculture project in Harare
By Shingirayi Mushamba

weeding, harvesting, and processing are done by both women and (the few) men at work.

However, when it comes to reproductive tasks in the household, the situation is quite different. None of the men in the sample could accept doing work in the household, which they think of as 'women's duties', like cleaning dishes, cooking, scrubbing floors, cleaning children, doing the laundry, making fire, etc. Only under difficult circumstances and in exceptional situations (if the wife is away or sick) will men feel free to do 'women's duties'. However, male members do repair damaged property and electrical wiring and other 'hard' jobs. So of every ten tasks in the household, the male member attended to only three, and women shouldered the rest. However, none of those interviewed was able to explain how these different duties came about, although some pointed out that the Bible endorses such things.

Work in the fields takes place twice a day – in the morning and afternoon – which means that women are doing it on top of their daily household duties in the early morning, at lunchtime, and in the late afternoon and evening.

All interviewees concurred that gender roles are culturally prescribed and that it will take a long time to change them. The absence of a clear gender policy in the Project and the lack of emphasis on gender issues in the training workshops that are organized for farmers by supporting agencies worsen the situation.

Access to and control over resources

Access to land

Provided that cash is available for renting and the farmer has the right kinds of social relationship (for example, a long-standing relationship and good will, involving produce from farming), access to land is not a major constraint for either men or women in the peri-urban areas. However, since men dominate cash resources in the household and do not give high priority to agriculture, it might well be that some women willing to engage in agriculture in practice are not able to do so.

Within the Musikavanhu Project, all members have equal access to plots, which are annually allocated. However, due to growth in membership there is now a shortage of land and individual plots are getting smaller, while the Project is continually searching for new locations. There has been increased pressure on land because of the increased demand for agricultural plots, resulting in conflicts caused by new members – mainly male – seeking to farm on land belonging to women (adapted from Mushamba and Mubvami, 2004). The main problem for all members is land tenure, since the land is municipal property reserved for other uses.

Access to inputs

Agricultural inputs such as planting materials and agro-chemicals are normally purchased by male and female farmers from the market, although the lower financial status of female farmers, compared with their male counterparts, may be a problem. Married female farmers' control over buying inputs is restricted by the fact that they have to consult their husbands. As mentioned before, many men do not see agriculture as a priority and since they decide how much money will be allocated to agriculture and when it will be released, it is difficult for many married women to buy the required agricultural inputs or tools in a timely way.

The 6.6 per cent of respondents who said that they had no problems in securing and controlling their inputs were women with sound financial backgrounds, that is to say husbands with a regular income and who are supportive of the agricultural activities of their wives.

Access to credit and investments

Lack of access to credit is a major constraint for most farmers in Musikavanhu, and women tend to have more difficulties in accessing credit than men. Recently, the government has started a micro- credit scheme for women as a poverty-alleviation strategy.

Investment in fixed assets for urban agriculture among the interviewed farmers was much higher for the 31 men than for the 39 women interviewed, while in general the men's average overall investment in their agricultural enterprise was higher than the women's.

Access to family labour

As explained above, farming activities are mainly performed by the women, with men generally reluctant to participate. This means that many female farmers have to hire labour for heavier tasks such as tilling, clearing, ploughing, seedbed preparation, terracing, and the repair of terraces. Also, male-led farms employ more family labour, whereas female-led enterprises receive less labour from the household.

Access to information and services

Most of the members (87 per cent) indicated that their access to information and services was limited because of poorly developed extension services. The access of females to relevant information and services is further limited by problems of illiteracy and cultural factors. A large proportion (56.7 per cent) of the members has no education at all, or only a few years of primary education. This fact is not taken into account by most extension workers, who provide information in English (with a lot of jargon), rather than in the native language. Many husbands will not allow their wives to attend a training programme or meetings that would take them away from home for a whole day or more.

The above factors have prevented women from acquiring relevant skills and have inhibited their potential to take on leadership roles in the organization. As a consequence, while the level of enthusiasm regarding agricultural activities is high among women, their levels of knowledge and skills are relatively low (technical skills, organization and management, market information, relevant legal and policy issues). However, whenever such skills and information are imparted, uptake is very quick (as in the mushroom project, for example).

Impact of external factors on gender in urban agriculture

Generally agricultural yields in Musikavanhu are very low, despite the high percentage of the population that is engaged in agriculture. The basic constraints for the farmers are the limited access to land, capital, information, and support services. Lack of access to urban farmland and insecurity of tenure undermine the Musikavanhu Project members' capacity to make urban agriculture more profitable.

In Musikavanhu the farmers are fully aware that the land is not theirs and that the Council can take it away as and when it is needed. Most of the off-plot land is earmarked for future construction of schools and other urban development uses. Lack of secure tenure over the land negatively affects the capacity of the farmers to secure support for their farming enterprises.

Another constraint is the lack of attention to the specific needs of female farmers (training, input supply, research, etc).

Gender perspectives and strategies of the Musikavanhu Project

The Musikavanhu Project has no deliberate strategy of addressing gender issues. The Project leadership believes that because most of its members are women, gender concerns are automatically taken care of; and these leaders believe that they have done much to cater for women's needs. For most of the members, 'gender' means women's involvement in any form of activity. However, a deliberate gender strategy is required in order to ensure gender equity in the accessibility to and control over productive resources (land, credit, information and training, inputs) and in the distribution of benefits, as well as in redistribution of household chores, prevention of a 'double burden' for women, and a modernization of the farming enterprises led by women.

The following strategies are suggested to be applied by the Musikavanhu Project:

- Be aware of disparities of knowledge and preference among its members and take these into account when planning training and development actions.
- Provide gender-sensitivity training activities for its members, both husbands and wives, in order to make them aware of actual gender inequalities regarding division of labour, control of income and assets, participation in decision making at household and community levels, etc., to enhance their knowledge of existing legal rights of men and women, and to promote more equal participation of men and women in farming and household activities..
- Organize a centralized administration for easier co-ordination of training programmes, keeping records of who got what training, and controlling the costs/benefits of the services supplied by the Musikavanhu Project.
- Point out and correct gender imbalances in the services provided by organizations supporting the Project (inputs supply, extension services, distribution of land titles, provision of credit, etc.).

Municipal government can support the Musikavanhu Project in gender mainstreaming urban agriculture in the following ways:

- **Creating a conducive legal framework.** The creation of a facilitating legal framework for urban agriculture that includes gender issues and removes the structural hurdles to the development of sustainable small-scale farming enterprises. These hurdles include the conditions applied by the municipality to temporary use of municipal land by the Musikavanhu members, for example not allowing them to establish boreholes or any other agricultural infrastructure, obliging them to obtain a licence and pay taxes as soon as part of the harvest is sold, forbidding livestock keeping, etc. During Operation Murambatsvina in 2005, all on-plot activities that were not in accordance with the sanctioned plan (including simple irrigation facilities, poultry and mushroom sheds, etc.) were demolished by the government. There is a need for national and local government to recognize the role played by the urban farmers in mitigating urban food insecurity and the effects of the HIV/AIDS pandemic, and their contribution to local economic development. The support given by the government to the New Farmers through land reforms and AgriBank loans for mechanization needs to be extended to female urban farmers, in view of their crucial role in feeding so many people and their potential for local economic development. Government and public media should stop stigmatizing the urban women farmers and openly support them with political recognition and material support.

- **Raising awareness of gender and urban agriculture.** Local government, in co-operation with the Ministry of Gender, should organize gender-sensitivity courses for the members of the Musikavanhu Project and their husbands/wives. The municipality could also promote public recognition of the important role that the urban farmers play in enhancing local food security and income raising; this could be done, for example, by means of television and radio programmes. Also the common myths about urban agriculture (that the plots are breeding places for malaria mosquitoes and hiding places for thieves) could be tackled in this way.
- **Improving land tenure.** The government needs to review the existing land-tenure system and designate zones for permanent urban agriculture. Special attention should be given to giving women farmers not only access to land but also the ownership or leasehold of the farm land (rather than granting it to their husbands, as is the current practice).
- **Stimulating adequate support services.** The municipality should lobby national government and private organizations (NGOs, enterprises, research organizations) to enhance their assistance to female urban farmers. The government could stimulate such assistance by co-funding NGO programmes that provide training and technical assistance to female urban farmers, providing tax incentives for private enterprises that provide inputs and equipment on a cost-sharing basis to women farmers, making a risk-sharing arrangement with organizations that provide cheap credit to female urban farmers, and stimulating the provision of irrigation water.
- **Providing protection.** Theft of agricultural produce is a major problem for poor female urban producers, who cannot protect their fields at night themselves and cannot afford to pay watchmen. The local government could provide more protection until the Musikavanhu farmer groups are able to organize effective security services themselves.
- **Giving access to free medical support.** Most of the members of the Musikavanhu Project are providing home-based care for people affected by HIV/AIDS, and/or they look after orphans in addition to their farming activities. They need access to free medical treatment under the Primary Health Care Programme.

Conclusion

Women play an important role in feeding more than 50,000 people in poor neighbourhoods of Harare, including HIV/AIDS patients and orphans, and raising some additional income to sustain their livelihoods. However, they receive little support from their husbands in these activities, because working on the land to provide food for the family is traditionally considered to be the task of women.

Also government and media do not recognize the important role of women urban farmers, who are granted only temporary access to some municipal land, under such strict conditions that only traditional rain-fed crop production for self-consumption is possible, and the development of more productive and income-generating agriculture is blocked.

In view of these structural constraints, modernization of urban agriculture around Harare (for example, by facilitating irrigated year-round vegetable production, poultry rearing, mushroom and herbs cultivation) is not taking off. To succeed, institutional transformation at all levels, from farmers up to government, needs to take place. The self-esteem and leadership capacities of the female producers need to be enhanced, and the social movement of which they are members needs to be strengthened. The political and public valuation of the productive role of the female producers has to change – a process which will require a change in attitudes to urban agriculture and to gender relations at household and community levels.

References

Feldstein Sims, H. and Poats, S.V. (1989) *Working Together: Gender Analysis in Agriculture: Volume 1 (Case Studies)* and *Volume 2 (Teaching Notes)*, Kumarian Press, Connecticut.

Hovorka, A. J. (1998) *Gender Resources for Development Research and Programming in Urban Agriculture*. Cities Feeding People Series, Report No. 26, IDRC, Ottawa, Canada.

Hovorka, A. J. (2006) *Gender, commercial urban agriculture and urban food supply in Greater Gaborone, Botswana*. Available from http://www.idrc.ca/en/ev-85408-201-1-DO_TOPIC.html, IDRC, Canada.

Mushamba, S. and Mubvami, T. (2004) 'Mainstreaming gender in urban agriculture: A case study on Musivanhu project, Harare, Zimbabwe', in Proceedings of the RUAF–Urban Harvest Workshop 'Women Feeding Cities: Gender Mainstreaming in Urban Food Security and Food Security', Accra, Ghana.

About the author

Percy Toriro is an Urban agriculture and planning specialist for the RUAF Cities Farming for the Future Programme, Municipal Development Partnership (MDP), Harare, Zimbabwe.

CHAPTER 7

Key gender issues in urban livestock keeping and food security in Kisumu, Kenya

Zarina Ishani

Abstract

This case study was developed from a scoping study on interactions between gender relations and livestock keeping in Kisumu city. The focus of the scoping study was the improvement of gender-based division of labour, inequality between males and females in terms of power and resources, and gender biases in rights and entitlements to increased productivity, remuneration, and development of women livestock keepers. The study considered the following gender aspects: control and access to resources; roles in decision making; division of labour; knowledge; role of external factors; and the benefits and risks of livestock keeping in urban and peri-urban areas of Kisumu. The study showed that patriarchal norms are changing, especially in the face of HIV/AIDS. However, women's autonomy has not increased much in relation to decision making and the freedom to make choices.

Emanating from the study, a multi-stakeholder city forum on food security, livestock, and agriculture in Kisumu was established, and farmers' networks from the slum areas were formally set up in Nairobi and Kisumu. They continue to function as a major influence in enabling and regulating urban and peri-urban agriculture. Training activities for livestock keeping were undertaken to enhance the knowledge and skills of urban livestock keepers, which succeeded in increasing their capabilities and changing attitudes to gender relations.

Introduction

Background

The case study entitled 'Scoping Study on Interactions Between Gender Relations and Livestock Keeping in Kisumu' was first developed in 2004, based on a research project carried out in 2002. The study was implemented by Zarina Ishani and Kuria Gathuru of Mazingira Institute, funded by Natural Resources International Ltd. under the UK government's Department for International Development (DFID) Livestock Production Programme.

The study was presented at the Women Feeding Cities Workshop organized by ETC–RUAF, Urban Harvest, and International Water Management Institute (IWMI), held in Accra, Ghana, in September 2004. It was then revised and updated in January/February 2008 for inclusion as a chapter in this volume.

Study area

The study focused on livestock keeping in six poor areas of the city of Kisumu, Kenya. The city is located on the shores of Lake Victoria, which is the second-largest fresh-water lake in the world. The population of Kisumu is estimated to be 535,664 (Republic of Kenya, Ministry of Finance and Planning, 2002a), with population densities ranging from 5,771 to 14,484 persons per km^2.

According to the Kenya Population Census (1999), life expectancy at the time was 47.2 years for males and 50.7 for females in Kisumu (country average: 45 years). Total numbers of households were 123,341; of these, 43,169 (41.2 per cent) were female-headed. The Analytical Report on Gender Dimensions shows that 60.9 per cent of females in Nyanza Province had never attended school, compared with 39.1 per cent of men.

Levels of absolute poverty (rural and urban) stood at 53 per cent (Republic of Kenya, Ministry of Finance and Planning, 2002b). Contributions to household income in 2002 were recorded as follows: agriculture, 75 per cent; wage employment, 10 per cent; urban self-employment, 4 per cent; rural self-employment, 3 per cent. The remainder was classified as 'Others'. Seventy-one per cent of the female-headed households were living in owner-occupied houses in 1999 in Kisumu, compared with 48 per cent of male-headed households (Republic of Kenya, Ministry of Finance and Planning, 2002c).

Urban livestock practised in the study area

The majority of the 55 households interviewed kept, on average, chickens (11 per household), goats (six), pigs (five), ducks (four), and cattle (two). Other types of livestock kept in small numbers were sheep, turkeys, geese, pigeons, guinea fowl, and rabbits.

The most common livestock kept were goats, in both female-headed and male-headed households. Goats were numerous because they do not need a lot of care and do not present major health problems. The main reason given for keeping chickens was that they are affordable and can be slaughtered when visitors arrive. Local cattle were kept because they were four times cheaper than grade cattle. Sheep and ducks were not common, because the respondents did not like sheep or duck meat. Pigs, where found, were many, as they were kept for commercial purposes by a limited number of people. Turkeys and geese were kept for sale and as 'guard dogs'.

Female-headed households kept goats, chickens, ducks, cattle, and sheep (in that order). They did not choose to keep pigs, because the women found the workload to be heavy and considered pigs to be dirty animals. Other

A widowed woman showing off the cows she owns in Manyatta, one of the study areas in Kisumu
By Zarina Ishani

livestock not kept by female-headed households were turkeys, geese, guinea fowl, and pigeons. These were not traditionally preferred livestock; the male-headed households that kept them were doing so on an experimental basis.

Methodology of the study

The objective of the study was to acquire relevant information on the interactions between gender relations and livestock keeping in the city of Kisumu. Since this study was predominantly a gender study, the focus was not on women's issues but rather on the interactions between men and women. To understand and assess the contributions of women and men in development, and the impact of development on both, the information gathered examined who does what, the levels of resources available for both sexes, and benefits and deprivations imposed by society on both men and women.

The study was carried out in six informal settlements in Kisumu city: Manyatta, Bandani, Nyawita, Kibos, Nyalenda, and Migosi. Migosi is a low- to middle-income community, whereas all the others are low-income areas. The purpose of selecting Bandani was that it was the only Muslim settlement and it was necessary to obtain the views of women, especially on inheritance and access to and control of resources, which would perhaps give a different slant on the issue.

A questionnaire was used to gather information from 55 (39 male-headed, 16 female-headed) households keeping livestock in the urban and peri-urban areas of the city of Kisumu. A deliberate selection of female-headed and male-headed livestock-keeping households was made, which were identified with the help of the area Chief. From the list, a random selection of the households was done. The data were collected by interviewing male-headed and female-headed households. In the former, the section of the questionnaire with questions on women's access to, control over, and ownership of land and property (including livestock) was separately administered to the wife.

Focus-group discussions were carried out with 12 Nyalenda men and 12 Bandani women separately. A male and a female researcher both moderated each of the two discussions. In addition 16 key informants, representing various sectors, were interviewed.

The survey was conducted mostly in Kiswahili, as it was found that most of the respondents were more comfortable with speaking this language. The data collected were of both a quantitative and qualitative nature, with some open-ended questions also included in the questionnaire. The point of reference for the Kisumu survey was the household. For the purposes of this chapter, a household unit is defined as consisting of 'a house and the group of people who live in it, providing a central place for them to be fed and sheltered'. The Kisumu study considered a household as a dynamic institution, with different typologies, integrating multiple relationships and partnerships.

In this study, if there was a polygamous household the husband was assumed to be the household head – the reason being that in the Luo culture it is customary for the husband to be the decision maker in the family. There was one case in which the husband was bedridden. The wife said that she was the household head, as she was the main bread earner and also was the main decision maker. Where a household had a single woman with or without children as minors, it was considered to be a female-headed household. Where there were widows with adult married sons and their families living together, the researchers asked the family members whom they considered to be the household head. In all cases they were informed that the widow was the household head. Thus in such cases it was not presumed that the eldest male was the household head.

Limitations of the methodology

The main limitation was inadequate time to carry out an in-depth study. This meant that only simple techniques could be used. For instance, it was difficult to find out the exact number of hours spent by each household member on each activity. It was not possible to interview all the female members in an extended family, because most of the time they were not at home, and if they were in, there was not adequate time to go through a full interview with them. The focus-group discussions were fruitful. However, it was felt that it would have been better to interview men and women in one group from the same area in order to find out the similarities and contrasts in their perceptions.

Gender analysis of the local situation

The case study concentrated on six aspects of gender relations and interactions in urban and peri-urban livestock keeping: access to and control over resources; decision making; division of tasks; differences in knowledge and preferences; the impact of external factors on gender in urban agriculture; and the potentials and risks of urban agriculture.

In terms of age, sex, and marital status of the livestock keepers, more than half were over 45 years of age. Close to two thirds of the livestock keepers were married, about a fifth were widows, and more than a tenth were single women. Twenty-nine per cent of the respondents were female heads of households.

Access to and control over resources

The most important factor affecting women's equal right to own, control, and access resources (including property) in Kenya is culture. The dominant ethnic group in Kisumu is the Luo, and their customary laws are patrilineal. The last-born son inherits the property. Women do not own land or property. Upon divorce, or death of the husband, the woman loses all rights to her property. Traditionally, a widow is 'inherited' by one of her brothers-in-law, so that family property is not fragmented and the children are looked after. However, these days, this is a thinly veiled excuse for taking over the land and property of the woman, which should rightfully be hers.

Prior to the Kenyan Law of Succession (LSA) Act of 1981, each ethnic group had its customs related to inheritance. The goal of the LSA was to unify all inheritance laws. But it contains exemptions on inheritance of livestock and agricultural lands outside the municipality. These are governed by customary law. Fathers are under no obligation to provide for their daughters. Property rights are terminated for widows upon remarriage. Thus the statutory law can be circumvented, using the Succession Act.

It is useful to distinguish gender relations in the household with regard to different types of property: financial property, consumption property, and production property. Men in male-headed households own and control financial property. Concerning consumption property, of which livestock is an instance, men exercise authority on matters to do with large livestock, and women have authority on matters concerned with small livestock, such as chickens. Men also have a stronger say in practices concerning production property. In the case of livestock, which can also be classified as production property, the preferences of men prevail on the choice of type of livestock, questions of animal health, and the disposal of livestock (whether for consumption or for income).

It was found that inequality prevails in gender relations, irrespective of the type of property in question. Property, in essence, is a social construct. Social relations regulate property, and gender relations make up a part of social relations. Ownership can be defined as right of possession, that is, who owns the livestock, house, land or any other property. 'Control' here means the power to direct or determine the use of a resource – in this case, livestock. In Kisumu, the survey found that control over property was largely determined by whoever was the household head. Female heads held absolute control over the household property. This was especially true for widows who controlled land, houses, and other property, including livestock.

In female-headed households the women were free of the restraints imposed on their counterparts in male-headed households. They exercised their free will regarding financial property, consumption property, and production property (even when there were adult sons and their families living in the same compound). Sometimes the sons were consulted, but not the daughters.

The survey showed that although husbands predominantly own the livestock in male-headed households, in more than 10 per cent there was some joint ownership with the wives; and also about a tenth of the wives owned the livestock. The joint or sole ownership of livestock by wives in male-headed households does not mean that women exercise control over this asset, or that they have the authority to make decisions, such as disposing of the livestock, or any income accruing from livestock keeping.

Women in male-headed households had ownership and control over small livestock only. Even where the woman had bought the livestock, she neither owned nor controlled it: in such cases there was joint ownership and control.

Gender relations differed in terms of control. Livestock control was mostly in the hands of the male in male-headed households where the livestock was kept for production and financial reasons. Where the livestock was for consumption only, the male had control if there was large livestock involved, but their spouses had control in the case of small livestock.

Likewise in male-headed households, ownership and control over property (houses, land) were in the hands of the man. If the woman contributed more towards the household income, control was jointly shared. However, income earned and contribution to the total household income by the wife and children in male-headed households were not the main determining factors for ownership and control. Culture played a dominant role. One finding was that traditions are changing, with market liberalization and urbanization. If the Luo culture were to be completely adhered to, no cases of women owning property would have been found. Even in the case of widows, the sons would have had ownership and control over property.

In the case of other properties, again gender relations differ, mainly according to the prevailing culture. Land and houses are in the man's name, stemming from the cultural traditions of the Luo community. The Kisumu study found some deviations from the perceived norms dictated by Luo tradition. In male-headed households, access to and control of financial, production, and consumption property was usually the prerogative of males, but not in all cases. Where there was tangible contribution by the wife, say when the wife had bought the livestock, there was joint ownership and joint decision making on marketing and husbandry practices.

According to Luo tradition, inheritance of a widow and her property by a brother-in-law is part of their culture. The study showed that this was not so in all cases. The main reason is the prevalence of HIV/AIDS. The widow did in fact have some control over her property. Another result, contrary to other findings, was that the widowed daughter-in-law was usually found to be living in the same compound as the mother–in-law, instead of being thrown out, in accordance with tradition. The cases that we studied showed that the mother-in-law gave access to some of her livestock to the daughter-in-law, especially if there were grandchildren involved. However, further research in this area is required.

In the focus-group discussion with women in Bandani (predominantly Muslim), there were indications that norms are changing from patriarchal inheritance to the prescriptions of Islamic law. Islamic laws do not discriminate against women, and the recent trend is for parents to bequeath land and other property to their daughters. However, this could not be established conclusively, due to the terms of reference of the scoping study.

In female-headed households where adult male children live in the same compound, according to literature, ownership, access, and control are retained by the male children. In our study, it was found that this was not the case. Mothers had sole control of and access to the livestock. When it came to decision making on any aspect of livestock keeping, the sons were sometimes consulted, which shows that trends are changing and female household heads are not subservient, as presumed.

As for inheritance of property by male heirs (sons), the study was not able to ascertain categorically that male heirs inherit the property after the death of the father. Our cases showed that the widow inherited the property, including land, houses, and livestock.

Decision making

There was more collective decision making on husbandry practices in the male-headed households than in female-headed households. In the former, husband and wife jointly made decisions regarding production and marketing of livestock. The wife made all the decisions in the case of small livestock.

In female-headed households, the woman made all the decisions, even if there were adult sons and daughters living with her. Sometimes the sons were consulted, but not the daughters.

Table 7.1 Decision making on livestock-husbandry practices in male-headed households

	Manyatta	Bandani	Nyawita	Kibos	Nyalenda	Migosi	Total	Total%
Self	4	3	1	4	5	4	22	56
Spouse	0	1	0	2	0	1	4	10
Joint	1	1	2	0	6	3	13	33
Son	0	0	0	0	0	0	0	0
Daughter	0	0	0	0	0	0	0	0
Collective decision	0	0	0	1	0	0	1	2
Total number of households	5	5	3	7	11	8	39	100

Table 7.2 Decision making on livestock-husbandry practices in female-headed households

	Manyatta	Bandani	Nyawita	Kibos	Nyalenda	Migosi	Total	Total%
Self	4	2	2	2	4	1	15	94
Son	0	0	0	0	0	0	0	0
Daughter	0	0	0	0	0	0	0	0
Collective decision	1	0	0	0	0	0	1	6
Total number of households	5	2	2	2	4	1	16	100

Gender division of labour in the households

In all households, livestock keeping did not require much work. Local animals were kept and required little care, and either roamed freely or were taken out by hired labourers, who were usually men. The labourers also carried out most of the work of tending the animals. In male-headed households, men claimed that they shared the work of livestock keeping with their wives, but, when interviewed separately, the wives indicated that they themselves did most of the work.

The work that men were involved in was mainly to do with animal health. The routine work of animal care was left to the spouses. Children were not directly involved in livestock keeping, except for helping with waste disposal and egg collection. Girls were rarely involved.

In female-headed households both the mother and son(s) took care of animal health; in their absence, the other females in the household were involved. In

Table 7.3 Gender division of labour in the households

	Husband	Son	Wife	Daughter	Labourer	Total
Grazing	5	4	1	0	12	22
Getting feed	3	1	2	0	10	16
Purchasing feed	5	1	1	0	7	14
Cleaning	1	1	2	1	15	20
Milking	2	0	0	0	2	4
Treatment	2	1	1	0	16	20
Other	0	0	0	0	4	4
Total number of cases	18	8	7	1	66	100

the female-headed households the women were the ones who were involved in marketing of the livestock. The children were not involved, even when they were married and living in the compound. In the male-headed households, both husband and wife were involved in the marketing of the livestock, but the wife could not sell without the authority of the husband.

Products were not sold in large quantities. Fewer than a dozen eggs were sold (to neighbours in most cases). Where the production was commercial, the eggs were sold in the nearby market places. Milk was sold to neighbours, and the volumes sold were small.

Differences in knowledge and preferences

Knowledge and access to information were not specific to any one sex. Both men and women seemed to lack the appropriate information or knowledge about livestock husbandry practices. They were also not keeping any records pertaining to income and expenditure. Women felt that they should be getting training in livestock keeping, as they were around the house more than the men were. They were particularly interested in learning about basic treatments pertaining to animal health. One key finding was the need for training of the livestock keepers in basic husbandry practices, record keeping, environmental care, health – of both animals and humans – and sourcing properly formulated inputs for feed, among other matters.

Differences in benefits and disadvantages of urban agriculture for men and women

Main source of income

The scoping study revealed that the number of male-headed households who kept livestock purely for commercial purposes (33 per cent) was greater than for female-headed households. Another 33 per cent of the male-headed households kept livestock to supplement the family income in case of need, while 20 per cent held livestock for both subsistence and sale. The rest kept the livestock for consumption purposes during festive occasions or funerals.

Comparing the main source of income for male- and female-headed households, it was noticed that one third of the male-headed households depended on livestock, compared with only one eighth of the female-headed households. Almost half of the latter depended on rental income, compared with one tenth of the former. The reasons could be that the female heads of household were mostly elderly and, having been left the property by their husbands, could rent the houses out. Their adult children did not seem to take much interest in livestock, probably due to the fact that they did not have control over it. The male-headed households were comparatively young and kept livestock for commercial purposes, as the women in the household were responsible for taking care of the livestock.

Human health and environmental pollution

Livestock keeping has certain risks attached to it, due to poor husbandry practices, overcrowding in the slum areas, and lack of information and knowledge. Limited knowledge and information about environmental pollution and especially waste management was a general issue. Waste disposal was a major problem. However, the impact of mounds of waste piling up near the living quarters and wells could not be assessed. We surmised that this environment would have a more negative impact upon women than on men, as the women are mostly at home and the land area is small. Further research is required.

Impact of external factors on gender in urban agriculture

Unequal access to opportunities, assets, and the freedom to make decisions are the most important determinants of the level of poverty experienced by women. The main external factors influencing gender relations in urban agriculture are the following: traditional patriarchal norms related to land, property, and inheritance; the effect of HIV/AIDS; and legal, administrative, and regulatory barriers. These not only prevent women from contributing fully to the Kenyan economy, but also create an imbalance in the recognition of women's contribution to the agricultural sector, which forms an important element of the economy.

Gender and property

According to Luo custom, male heirs inherit property; but inheritance by male heirs was not clear in the urban setting. In the case of Muslim marriages, the wife was entitled to one eighth of the property. Considering livestock as property, in general 64 per cent of the female-headed households had bought the livestock, and 25 per cent had inherited it; 7 per cent was acquired as dowry, and 4 per cent was obtained as gifts.

Inheritance and purchase of livestock played equal roles in the access to livestock of female-headed households. This was surprising, because according to Luo tradition wives or daughters do not inherit – and yet the females had inherited the livestock, mostly from their husbands. There was also one case of a single woman who had inherited the livestock from her parents. This shows that the norms are changing and that widows do inherit. The daughter-in-law also had control of the property (the house) if she was widowed and living in the same compound as her mother-in-law. No cases of a widowed daughter–in-law owning livestock were found. One possible reason could be the HIV/AIDS scourge.

In male-headed households, 74 per cent had bought the livestock, and only 18 per cent had inherited it. These figures also contradict what has been commonly assumed about inheritance by male heirs.

HIV/AIDS

According to a World Bank report (2007),

> *HIV/AIDS has caused changes in land use, household labour, and financial standing because of loss of financial assets, higher cost of living, increased burden, particularly to grandparents, and disintegration of family ties. HIV/AIDS creates a 'missing' generation, distorting inheritance and transmission patterns to grandchildren; widows and their children can be vulnerable in terms of potential loss of land rights on the death of the male household head, with young widows being more vulnerable than old ones; and distress sales of land resulting from HIV/AIDS were rare as the value of family land as a safety net is recognised.*

For instance, possibly one third of all deaths in Kisumu are AIDS-related, mostly those of married sons. Widows were not 'inherited', as per traditional custom, by brothers-in-law, for fear of spreading the HIV infection. Where there was no wife inheritance, there was no property inheritance either. This conclusion, however, cannot be generalized, and further research is required. In female-headed households, the mother-in-law looked after the daughter-in-law and any grandchildren. This is contrary to Luo tradition. A widow who refuses to be inherited is usually thrown out of the household. But in the cases of our interviewees, this did not appear to be the trend. The reason given was: 'If I throw out my daughter-in-law, how will my grandchildren survive?' A bigger sample size would make it easier to draw out firm conclusions.

Kenyan land and by-laws

Kenya has 75 laws governing land, many of which are obsolete, while others conflict each other, supporting different land regimes within the same area. Thus, legal, administrative, and regulatory barriers specifically relating to insecure land rights limit women from making necessary investments in their land.

At the local authority level, the by-laws of Kisumu do not bar any person from keeping livestock in urban and peri-urban areas. However, most of the regulations are outdated because they were enacted between 1925 and 1951. Livestock keepers are expected to obtain permits for any livestock kept. None of the livestock keepers knew of this by-law. By-laws pertaining to the selling of milk (Milk and Dairies by-laws of 1951) stipulate that no one shall sell or process fresh milk in the municipality of Kisumu, unless such a person obtains a licence from the Dairy Board of Kenya (Cap 336, revised 1984). None of the farmers adheres to this by-law, with the result that milk is handled in an unhygienic manner.

Credit mechanisms

Women's limited land ownership restricts their access to formal financing mechanisms. Only two respondents had taken credit. The others had not, for several reasons: fear of defaulting, lack of information about credit organizations, high interest rates, and misuse of loans by family members.

Lending institutions specifically aiming to support livestock livelihoods were non-existent. However, most of the female respondents were members of informal groups that provide local rotating savings and credit systems ('merry-go-rounds'), and some men were members of welfare associations. The study found that when women contributed to household income, the balance of power shifted, as women now had more voice in (joint) decision making. Thus if women had the opportunity to increase their incomes by obtaining credit, then they would have more access, control, and ownership of livestock.

Strategies and tools used to incorporate gender in the project cycle

Strategies and tools used to incorporate gender in the diagnosis

From its inception, the study incorporated a gender perspective, with a focus on gender division of labour, gender differences in power relations and resources, and gender biases in rights and entitlements, with specific reference to livestock keepers. The household was considered as the starting point for the research, with deliberate selection of female-headed households and male-headed households. The questionnaire contained a section on women's issues which was administered to women only.

Focus-group discussions were constructed to ensure that there were women-only groups and men-only groups, and discussions were moderated by the researchers, using open-ended questions. This was deliberate, because it was assumed that women might not talk freely in front of their spouses, neighbours, or male relatives. The men-only group also was assumed to give more candid answers rather than the politically correct replies that they might have given in a mixed group. The women-only group was selected from a Muslim community, in order to find out whether they had different views on property and land compared with non-Muslim women.

Key-informant interviews were carried out to seek opinions of influential people in the communities, or those with whom the livestock keepers were in constant contact, to discover their gender biases and the extent of their knowledge.

Strategies and tools used to incorporate gender in policy influencing

Regional policy-oriented workshop

The scoping studies were followed by a workshop on 'Urban Livestock for Improved Livelihoods in sub-Saharan Africa' in March 2003. Cities represented were Nairobi, Kisumu, Addis Ababa, Kampala, and Dar es Salaam. Municipal and national government representatives, farmers and livestock keepers, civic and community organizations, international bodies, and donors participated in the workshop. The findings of the scoping studies, not only the Kisumu study but also others from the Eastern African region, were discussed. The studies show that while the urban poor rely considerably on urban agriculture and livestock, there is a dearth of access to knowledge and information related to urban agriculture and livestock. The workshop led to the establishment of City Focal Points in each of the participating cities. For each city, a Focal Point was formed. Mazingira Institute was selected as the Focal Point for Nairobi to develop the City Forum.

Establishment of a City Forum on Urban Food Security, Agriculture and Livestock

The 'Nairobi and Environs Food Security, Agriculture and Livestock Forum' (NEFSALF) was established in 2004 by Mazingira Institute. In Kisumu, the same multi-sectoral model was applied. Its strategy focuses on the desirable, acceptable, and feasible interactions among the community, government, and market sectors. This interaction is to enable and regulate agriculture and livestock-keeping practices employed by the urban poor to improve their well-being and earn a living. The Forum envisions creating a better way to enhance food security and sustainability of the many, rather than the few, in Nairobi and its environs, through urban crop production and livestock keeping.

Three meetings of the Forum are held annually. In the meetings policy issues identified in the scoping study have been discussed regularly, as a means to facilitate discussion of the main problems encountered in urban and peri-urban agriculture and livestock keeping, and to stimulate reflection and dialogue among the various stakeholders on possible measures to improve the actual situation.

NEFSALF organized a series of training activities, where 198 farmers, 92 of them women, have been trained in record keeping and gross-margin analysis; crop husbandry; livestock husbandry; group dynamics; and the production of organic fertilizer from waste. The criteria for selecting trainees include gender, type of urban farming activity being practised, location of activity, age, and languages spoken. Gender balance is a key concern in the training courses, as is maintaining a diverse range of participants, determined by age, type of farming, and location. Pre-training visits were made to the trainees' sites to ascertain what activities are going on, and post-training visits are made for assessment purposes. The courses were conducted in the form of lectures, site visits, and on-site demonstrations.

The courses have not only helped to improve the capabilities of the trainees in urban livestock keeping and agriculture but have also enhanced their understanding of gender roles and the importance of involving women in decision making. The committee of one group adapted rules to reflect gender equity and now has representatives of youth, women, and men. Other groups made considerable changes in the group composition, governance, and group activities. Some groups formed sub-sections to address the emerging need to keep dairy goats.

In terms of decision making and sharing of responsibilities, changes took place. For example, men now consult their spouses, and the women are given more control over livestock and access to bigger livestock, which previously was the domain of men. Men are now taking on responsibilities which traditionally they never used to undertake, such as marketing of livestock and livestock products.

The participants achieved a measure of self-esteem and confidence which did not exist before. They are now willing to approach research institutions, seek assistance from the government for veterinary services, and have a voice; and they are branching out into other urban agriculture-related activities such as production of yoghurt and compost making. The incomes of the farmers have increased, and they are reaching out to others to disseminate the information and knowledge gained through the Forum.

The courses have made the women more knowledgeable, not only about the opportunities that exist but also about the advantages of knowing their rights. Through the resulting networking they have been able to form groups to share their experiences and be proactive. The consequence of all this is that women have been given more choices; they are not as dependent on their spouses as before and have more freedom to do what they want with the income that they earn. Some women have bought land and/or livestock to increase their incomes.

Conclusions and recommendations

The scoping study showed that in some respects there is gender disparity, but it also demonstrated that norms are changing, particularly in relation to property rights and inheritance in the face of challenges posed by the HIV/AIDS pandemic. Knowledge and information are crucial for people living in informal settlements if they are to increase their well-being and gain a sense of dignity. The scoping study gave rise to a number of ideas which are being implemented by the Institute in collaboration with local stakeholders.

To enable and regulate urban and peri-urban livestock keeping, it is important to use a multi-stakeholder approach. Working with multi-stakeholders, however, is not an easy task. Links have to be made, and patience and persistence are required. The regime change in Kenya in 2002 opened new avenues for interaction with the government. However, the resources at the government's disposal are not adequate to enable it to reach out to all the communities. Both central and local government officials are willing to help, provided that adequate facilitation exists.

The Kisumu Forum has functioned well so far, but some challenges remain if sustainability is to be assured; a shortage of resources is one of them. The response from the market sector and the local authorities has not been as expected, although those who have been attending the Forums have found the interactions with the farmers very useful. Both women and men have become very creative after attending the forums, and they are more concerned about animal and human health than before. Women in particular have become the 'push' factors for change, and they now take the lead in the household to ensure that these changes do take place.

Suggested strategies to be applied in local policies and development projects on urban livestock and agriculture

- Mainstreaming of gender equity in the goal(s), objectives, activities, and outputs that a project aims to achieve is essential, in order to ensure that costs and benefits of an urban agriculture or livestock project, both economic and social, are distributed equitably among men and women.
- Gender equity, however, cannot be achieved at the project level if there is disparity in policies benefiting one sex and not the other. All policies, at the local level and at the central government level, have to be formulated in a gender-sensitive manner. Issues such as inheritance and succession, especially for women, should be of paramount importance in order for the whole household to benefit.

To ensure equitable division of labour and responsibilities in urban livestock

- Through better knowledge of husbandry practices, women's burden would diminish, leading to a higher standard of living and better mental and physical health. Mindsets of both women and men need to change, and this can be achieved through education, training, and networking not only on practices in livestock keeping but also on human rights.
- There is a need to establish the cost-and-benefit differentials existing between men and women; and the cost of women's reproductive and productive labour vis à vis that of men has to be taken into account at the household level and then at the national level. Currently, the contribution of urban agriculture to the overall Kenyan economy (GNP) has not been incorporated in the 'Strategy for Revitalisation of Agriculture (2004–2014)'. Urban and peri-urban agriculture needs to be recognized as a sector in the economy. All data accumulated for this purpose should be disaggregated by gender, so as to assist in the formulation and implementation of gender-responsive policies in urban agriculture.

To ensure equitable access to and control over productive resources

- All policies pertaining to customs, practices, and conflicting laws on inheritance and succession need to be revised. There should be specific laws allowing acquisition, access, and ownership of land and property (including livestock) by HIV/AIDS widows/widowers and orphans. Appropriate legal measures should be put in place to ensure co-ownership of land by spouses, so that men and women have equal rights, before, during, and after marriage. Matrimonial property should comprise all properties acquired, developed, and invested, including livestock.
- Access to justice should be made affordable, less time-consuming, and geographically accessible. Para-legal training and family courts could cut down on time and costs. When larger disputes pertaining to land and property are resolved amicably, urban agriculture will automatically benefit.
- Training for judges, magistrates, chiefs, and police in women's rights, including property rights, should be provided on a regular basis to ensure equity in decisions relating to land and property.

Improving women's access to finance for enhanced decision-making in households

- Women's access to finance for micro, small, and medium enterprises is negligible. According to a World Bank report (2007), women receive less than 10 per cent of the available credit. In terms of urban agriculture, including livestock keeping, no specific finance companies (public or private) exist. The government should encourage the provision of financing for women through local financial institutions, specifically for urban agriculture, and this may consequently improve women's decision-making power in households.

To enable both men and women to acquire relevant knowledge

- All stakeholders should work together to bring about a change in livestock-husbandry practices so that farmers acquire relevant knowledge to mitigate any potential health hazards, thus reducing conflict between the authorities and farmers.
- Knowledge of one's rights, particularly relating to land and property, should form part of the education and training offered to farmers.

To ensure equality in power relations

- Change in power relations can come about only if the first three objectives are successfully achieved. Attitudinal changes which render harmful and discriminatory cultural practices obsolete are slow to take root.

- Involving all members of the family in decision making and record keeping, particularly adult children, would ensure women's active participation in decision making.
- No strategy employed in isolation can be successful: all dimensions of gender inequality must be addressed. The government should include all affected stakeholders in its policy-making processes, to ensure that the policy products are favourable to both men and women.

References

Ishani, Z., Gathuru, K., and Lamba, D. (2002) 'Scoping Study on Interactions between Gender Relations and Livestock Keeping in Kisumu', unpublished report prepared for Natural Resources International Ltd.

Republic of Kenya, Central Bureau of Statistics, Ministry of Finance and Planning (2002a) *Kisumu District Development Plan, 2002–2008,* Government Printer, Kenya.

Republic of Kenya, Central Bureau of Statistics, Ministry of Finance and Planning (2002b) *Kenya Population and Housing Census. Analytical Report on Housing Conditions and Household Amenities.* Volume X, Government Printer, Kenya.

Republic of Kenya, Central Bureau of Statistics, Ministry of Finance and Planning (2002c) *Kenya Population and Housing Census. Kenya Analytical Report on Gender Dimensions.* Volume XI, Government Printer, Kenya.

The World Bank (2007) *Gender and Economic Growth in Kenya,* World Bank, Washington.

About the author

Zarina Ishani is Programme Officer for the Mazingira Institute, Kenya.

CHAPTER 8

Urban agriculture, poverty alleviation, and gender in Villa María del Triunfo, Peru

Noemí Soto, Gunther Merzthal, Maribel Ordoñez and Milagros Touzet

Abstract

This chapter describes gender mainstreaming in the multi-stakeholder action-planning process (MPAP) in the district of Villa María del Triunfo, which began in 2005. It reflects the inclusion of a gender perspective in every stage of the process developed, and the project's contribution to promoting gender equity in urban agriculture.

The multi-stakeholder action-planning process was conceived in the city as a dynamic process of local development, planning, and implementation of public policies and strategic actions in urban agriculture, strengthening the capacities of local stakeholders. This process developed participatory methodologies and inclusive strategies which contribute to reducing inequalities and inequities affecting women. Urban agriculture represents an opportunity for productive activities for women and a decisive contribution to food security for their families.

As part of the process, an Urban Agriculture Forum was created in the city, providing a space for dialogue with local stakeholders concerning the planning and implementation of public policies and strategic actions related to urban agriculture. The Forum has formulated a Strategic Agenda 2007–2011 for urban agriculture. As a result, 570 urban farmers are organized and registered in the Urban Producers Network, and 86.5 per cent of this group are women.

The main references used in preparation of this chapter were the case study 'Urban Agriculture and Gender in Villa María del Triunfo', conducted in 2004, and an update conducted in 2008. In both cases we used participatory methodologies which make visible the division of labour in urban agriculture, as well as access to and control over resources and decision making.

Introduction

The information presented in this case study is based on the results of a participatory analysis of urban agriculture and gender undertaken in 2005 in co-operation between the Municipality of María del Triunfo and the RUAF–

Cities Farming for the Future (regionally co-ordinated by IPES – Promocion para el Desarrollo). In 2007 this diagnosis was upgraded by a special study of gender and urban agriculture in Villa María del Triunfo.

Villa María: the local context

Villa María del Triunfo is one of the 43 municipalities that make up the city of Lima. It forms part of the poverty belt that surrounds the city, which is located in the desert area of the central Peruvian coast where it hardly ever rains (but the air is humid, which regularly creates fog and a monthly precipitation ranging between 5 and 30 mm). A major feature of the city is its rugged geography, with hills ranging from 200 to 1,000 m above sea level and slopes ranging between 7 and 43 degrees, creating harsh living conditions for much of the population.

Villa María has over 350,000 inhabitants, of whom 75 per cent are under 39 years of age. Of its population, 57.3 per cent live in poverty, and 22.2 per cent in extreme poverty. Chronic malnutrition affects 23 per cent of children under the age of eight, and 25 of every 1,000 inhabitants suffer from tuberculosis. The official unemployment rate is about 8 per cent (but it should be noted that the official statistics consider persons engaged in street vending as employed, although this is a precarious and unstable activity). Of the employed population, 77 per cent are involved in commerce (formal and informal), 5 per cent in industrial activities, and 18 per cent in services (DESCO et al., 2005).

Villa María is divided into seven consolidated zones which include in total 292 neighbourhoods and settlements, 60 of which are informal and without any basic sanitary facilities. One of the main characteristics of Villa María is that it is a city under constant construction. At present it has over 73,500 houses, most of which are made of durable materials (bricks and cement) and 24 per cent of other materials (cardboard, wood, stone).

The response of the municipal government to the poverty and precarious livelihoods in the city is limited by its scarce resources: municipal revenue totals US$6.6 million ($27 per capita), of which the largest part must be spent in accordance with national government policy.

Urban agriculture in Villa María del Triunfo

In Villa María del Triunfo, urban agriculture is understood as the agricultural activities (crops and livestock production, production of inputs, and processing and marketing of products) in the intra- and peri-urban areas of the municipality which make use of local resources (labour, land, water, solid and liquid wastes, etc.) in order to generate food products for household consumption as well as for sale in the market.

In Villa María many resource-poor urban households are involved in some kind of urban agriculture. The total number of such households is unknown, but 570 have joined the recently established Urban Producers Network. Agricultural activities are conducted in family gardens, gardens belonging to community kitchens, community gardens, and school gardens. These gardens may be located on any of the following types of land:

- **Private land:** backyards and vacant private plots (often partially developed and abandoned).
- **Communal areas:** areas that have been recognized or earmarked by the municipality as communal land for use as green or sports areas.
- **Public land:** land owned by the municipality that is made available for the development of community gardens and soup-kitchen gardens. This includes land where construction is restricted, such as land lying beneath high-voltage power lines and along main roads.
- **Institutional land:** land owned by institutions that may be used by the institutions for their own gardens to generate income or food for their staff, or for educational purposes (school gardens) or land that is made available to the community for gardening (community gardens).

A mapping exercise identified the availability in Villa María del Triunfo of about 175.4 acres of vacant land with some potential for urban agriculture.

The activities of the urban producers in Villa María include the following:

- **Local production of inputs for urban agriculture.** Animal wastes (manure from chickens, guinea pigs, and other small animals) along with vegetable wastes are composted locally and reused in the gardens. No chemical inputs are used. The municipality has a greenhouse for the production of seed and seedlings, but its production is not sufficient to meet the local demand.
- **Agricultural production.** About 83 per cent of the urban farmers produce vegetables, 45 per cent fruits, 31 per cent aromatic plants, 18 per cent ornamental plants, and 3 per cent fodder. About 52 per cent of the urban farmers raise animals, especially smaller animals (chickens, hens, ducks, quails, turkeys, etc.).
- **Processing and marketing the produce.** About 20 per cent of the farmers occasionally process their products, for the most part producing jellies, juices, cakes, and other items.
- **Commercialization of the produce.** About 20 per cent of the farmers sell some of their production, for the most part locally, either fresh or processed.

Characteristics of the urban producers in Villa María del Triunfo[1]

The Urban Producers Network of Villa María del Triunfo has registered 570 families as members. Of the 2,850 persons involved, 86 per cent are women. Only 17 per cent of the families engaged in urban agriculture have a household income above the minimum wage ($152 per month), meaning that 83 per cent of the families engaged in urban farming in Villa María have an income per person equal to or less than $1 per day. The female-headed families have lower incomes than the families led by men. For these poor and female-headed households, urban agriculture is an opportunity to enhance the family's food security, to save on food expenditures, and, for a minority, to generate an additional income. Most of the women in Villa María used to dedicate their time mainly to domestic activities and participation in the *comedores populares* (government-subsidized community kitchens). They were involved only sporadically in income-raising activities such as selling hand-produced dresses and ornaments.

Of the farming households, 49 per cent do not have access to water and drainage services, while 23 per cent do not have electricity.

The case-study update in 2008 showed that most of the members of the Urban Producers Network (and especially the women) are also members of other programmes; for example, 72 per cent of the members participate in the 'Vaso de Leche' programme[2] (a government-sponsored feeding programme for children under 13 years and elderly people); and 44 per cent participate in the 'Comedores Populares' programme and, to a lesser degree, in the Association of parents in schools.

The above indicates both the great need of these households, and the culturally determined distribution of tasks within them. Women's membership of the programmes mentioned can be understood as an extension of the reproductive tasks that are mainly seen as a woman's duty, whereas men participate more than the women in neighbourhood organizations that address issues of housing, services, and infrastructure.

Eleven per cent of the members, all of them women, have never received any kind of formal education, and 31 per cent of the female producers have

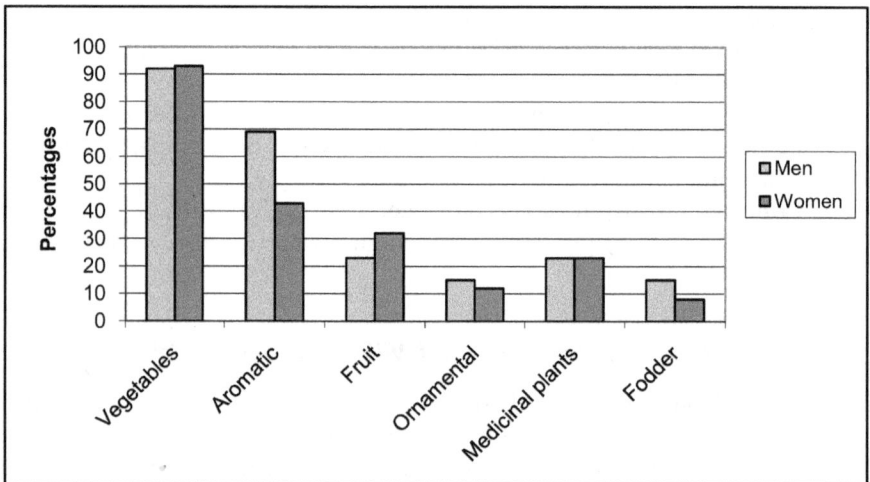

Figure 8.1 Crop cultivation by male and female producers

only elementary education, compared with 15 per cent of men; women are also under-represented among those who have been educated beyond the elementary stage. This clearly indicates unequal access to education for men and women.

Most of the female urban farmers are between 30 and 50 years old, while most of the men are more than 40 years old, with a notable proportion who are more than 60 years old. This is probably due to the fact that younger men tend to dedicate themselves to other work outside the home.

Most of the urban farmers are married or cohabiting[3] (77 per cent of the men and 76 per cent of the women). While most single men do not have children, most of the single women are head of a household (over 13 per cent).

Only 18 per cent of the urban producers in Villa María were born in Lima; the rest were born in other provinces of the country, mainly in the highlands (85 per cent of the men and 64 per cent of the women), indicating the strong migration wave that reached Villa María and the city of Lima three decades ago in response to the marginalization and exclusion that affected the country's provinces at that time. In addition, Peru suffered from nearly 20 years of terrorist violence that had its focus in the country's interior, leaving 69,000 people dead, mostly Quechua-speakers in the mountains and forests in the centre and south.

The development of Villa María's policy on urban agriculture

The local government of Villa María began to support the development of urban agriculture in 2001. Before that date, urban agriculture was largely invisible; no data on such activities were available, and no regulatory framework or public policies regarding urban agriculture were in place. In 2002 the local government created the 'Municipal Programme on Urban Agriculture', and in 2004 the programme became a sub-unit of the Economic Development Unit.

Due to the lack of information on urban agricultural producers, the initial supporting activities organized by the municipality were not conceived of in a strategic way, and they responded insufficiently to the priority issues of the different categories of urban producer; as a result, the activities had limited impacts. Moreover, severe restrictions in financial and human resources limited efforts to meet the needs of the urban farmers. Several local organizations had initiated projects in Villa María's neighbourhoods, but without co-ordination among themselves or with the municipality. Some of these initiatives were unsustainable and/or had low impacts.

In 2005 the local authorities of Villa María, supported by the RUAF–Cities Farming for the Future Programme (RUAF–CFF), initiated a process to enhance the effectiveness and sustainability of the efforts to promote urban agriculture, and to optimize the use of the scarce human and financial resources of the municipality and of other local actors involved. The objective of this process was the participatory formulation and implementation of a strategic action plan for urban agriculture.

A multi-stakeholder platform on urban agriculture was established (City Forum on Urban Agriculture), including 21 organizations and institutions (the municipality, NGOs, community-based organizations, universities, representatives of the urban producers, etc.). The main objective of the Forum is to promote urban agriculture at the city level and to co-ordinate the related development activities undertaken by its members. The Forum, with the help of RUAF–CFF, elaborated a 'Strategic Action Plan' for the development of urban agriculture in Villa María. The plan contains the city's vision for urban agriculture and its expected contributions to the realization of the City Development Plan, the principles and strategic objectives for the promotion and consolidation of urban agriculture in the city, and the main actions / projects to realize these objectives. This plan concretizes the City Policy on Urban Agriculture and forms the main instrument for its implementation.

In August 2007, the plan was approved by the municipality, together with Ordinance No. 021-2007/MVMT, which recognizes urban agriculture as an ongoing and legitimate activity in the city and defines the principles and strategic objectives of its promotion (social inclusion and participation, the promotion of equity, justice, and solidarity, and conservation of natural resources). When the City Development Plan of Villa María was updated in 2007, urban agriculture was included as one of the four priorities for action in the field of economic development.

Also a Network of Urban Producers was established which integrates 570 households involved in urban agriculture. The Urban Producers Network is now representing the urban producers' interests in the City Forum.

After the elections in January 2007, the new municipal authorities endorsed their commitment to the realization of the Strategic Agenda on Urban Agriculture and maintained the Urban Agriculture Sub-Unit in the municipality, indicating the importance that the urban agriculture movement has gained in Villa María in the past few years,

The integration of gender in the multi-stakeholder action-planning process

The multi-stakeholder action-planning process was developed in the city of Villa María del Triunfo from August 2005 to December 2007. The main objective of the process was the formulation and implementation of an inclusive and equitable public policy and a strategic action plan to strengthen urban agriculture in the city, with the active participation of all stakeholders. The following actions were taken to ensure that the policy and action plan would be inclusive and equitable.

In the preparatory phase

During selection of the members of the Local Co-ordination Team (LCT), which would facilitate the planning process, the gender sensitivity of the candidates was taken into account as well as the representation of women. The LCT includes four representatives of local government departments and two professional staff of IPES, together forming a gender-balanced team of three women and three men. Two women assumed responsibility for the day-to-day co-ordination of the LCT (one from the municipality and one from IPES). In this way, attention to gender issues throughout the process and the active participation of women in decision making were ensured.

The gender sensitivity of the staff who would play a role in the diagnostic phase was enhanced by giving proper attention to gender issues in urban agriculture during the training course on urban agriculture and multi-stakeholder planning that was organized by IPES/RUAF–CFF. During the training the conceptual framework for the multi-stakeholder action planning (with a clear 'gender and development' focus) and the integration of gender in each phase of the planning process were discussed and adopted by all involved. Also during this training, the capacity of the staff to work with gender-sensitive diagnostic methodologies and tools was enhanced.

In the diagnostic stage

The participatory diagnosis of the existing situation of urban agriculture in Villa María del Triunfo was undertaken with a clear gender perspective and the application of participatory and gender-sensitive methodologies and tools.

During the stakeholder analysis, an inventory was taken of the main stakeholders in urban agriculture and their actual and potential roles in promoting urban agriculture. Attention was given to the gender sensitivity of these organizations and the extent to which they represent the interests of women – and how they do so. It became clear that there are important differences in the organizational linkages maintained by men and women and in the services offered by these organizations to women and men respectively. This was taken into account when selecting organizations to be invited to participate in the action-planning process.

In the analysis of the actual urban production systems, participatory tools were applied, to ensure that (a) men and women could independently voice their specific problems and interests; (b) the data collected were gender-disaggregated; and (c) insight was developed into the local gender division of labour, access to resources, roles of men and women in decision making at household and community levels, etc. This also enabled men and women to recognize the existing inequities in the distribution of benefits and access to decision making, as well as the differences in the daily workload of men and women in urban agriculture. It made them aware of the low value accorded to domestic work.

The gender-specific analysis of problems encountered and their perspectives on the development of urban agriculture in homogeneous groups of men and women resulted in the identification of gender-specific problems, interests,

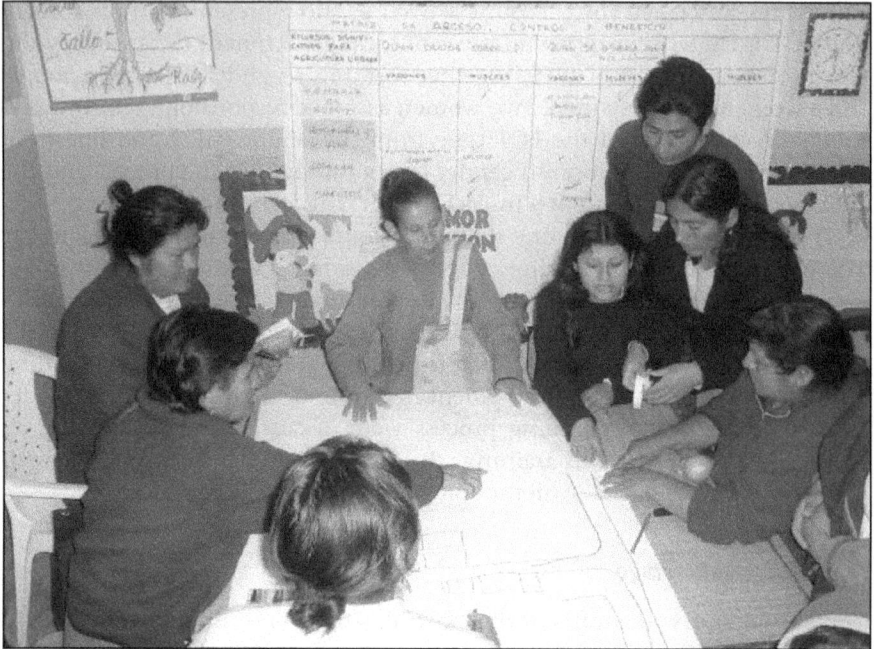

Women producers analysing access to and control over resources in Villa María del Triunfo
By IPES

and priorities, next to others that were common for both groups. Such gender-specific differences referred to aspects of production as well as to processing and marketing.

During the mapping of available open spaces (the participatory identification of areas in the city that are vacant and have potential for urban agriculture) attention was given to the gender dimension. This allowed recognition of differences in men's and women's knowledge about their surroundings, with women displaying less knowledge than men.

The review of actual policies, norms, and regulations related to urban agriculture included gender as one of the six elements of the analysis, trying to identify differential impacts of the actual legal framework for men and women.

In the planning stage

This phase of the multi-stakeholder process had two stages:

  a. The establishment and strengthening of the Villa María Multi-stakeholder Platform on Urban Agriculture, constituting an independent forum in

which the problems and proposed solutions to various issues related to urban agriculture in Villa María are discussed between the various stakeholders, and developed actions are co-ordinated.
b. The formulation of a Policy and Strategic Action Plan on Urban Agriculture.

When establishing the multi-stakeholder platform, gender was taken into account in the following ways.

- Including in the Forum organizations that are more gender-sensitive and involve women.
- Including in the constitution of the Forum the statement that it is 'inclusive and equitable, because it seeks to involve all stakeholders, without exclusion or discrimination of any kind, with a focus on equity and gender perspective'.
- Ensuring a gender balance in the committee that co-ordinates the Forum and in the working committees that were formed to elaborate various components of the Action Plan.
- Enhancing the gender sensitivity of the members of these committees by giving training in participatory and gender-sensitive action planning, social inclusion, and gender equity.
- Assisting the urban producers (a large proportion of whom are women) to organize themselves to be better able to present their interests in the Forum and its working committees. During the diagnosis phase the urban producers were contacted and workshops were conducted to support them in electing their leaders (five women and two men) and in defining the objectives and operating principles of the Network, including 'to develop urban agriculture with equal opportunities for men and women' and 'to promote equitable relationships between men and women'. During these workshops the farmers' representatives for the Forum were selected, and these subsequently received training in gender-equity issues and citizen participation, among other topics.

When formulating the Strategic Agenda, the integration of gender was facilitated as follows.

- Building a collective vision of the desired development of urban agriculture in Villa María del Triunfo, including a clear gender perspective as one of the five principles of the Strategic Agenda on Urban Agriculture. This principle has to be taken into account when planning specific projects and actions, and each project plan should clarify how it will benefit men and women.
- Selecting appropriate meeting places and times, providing child-care facilities, devising working methods that encourage women to speak in public, and adopting gender-inclusive language which does not reaffirm conventional stereotypes.

- Presenting the main results of the diagnosis in a gender-specific way to the working committees to encourage them to develop policies and actions that take into account women's specific needs and interests and contribute to achieving inclusion and equity in the local society

The formulation of the plan was worked out by the members of the Forum on Urban Agriculture in about three months, resulting in a Strategic Agenda, with six strategic objectives, each with a number of priority actions and projects and a set of success indicators, including the differential impacts of the implementation of the Strategic Agenda on men and women.

The effects of the MPAP process on gender and urban agriculture in Villa María del Triunfo

In 2008 a study was undertaken to improve understanding of the changes brought about by the participatory and gender-sensitive planning process in Villa María del Triunfo described above: to what extent have changes occurred in the gender-differentiated division of labour, in access to and control over resources, and in participation in decision-making? And have the views of male and female urban agriculture producers changed since 2004?

The study team consisted of one representative of the Municipal Agricultural Unit, three representatives of IPES, seven representatives of the Urban Agriculture Network (five women and two men; one from each of the seven zones of Villa María), and three support staff (students from the area).

Techniques and tools were selected in direct relation to the main issues indicated in the objective of the study: the gender differentiation in the division of labour, access and decision making, mobility, views, knowledge, interests, preferences, and needs, as well as the influence of external factors (age, education level, place of origin, family composition, economic level, community participation). Some of these techniques were applied during individual semi-structured interviews (for example, gender differentiation in division of labour, access to resources, and decision making), while others were applied during a workshop (for example, gender-differentiated views on key problems and opportunities in urban agriculture).

As in 2004, 20 per cent of the members of the Urban Producers Network were included in the survey. The following criteria were taken into account when selecting the respondents:

- distribution of the network members over the seven formal zones of the municipality and the various neighbourhoods and settlements within these zones;
- distribution by sex;
- distribution by type of garden (family gardens, community gardens, and institutional gardens).

This resulted in a sample of 114 persons, of whom 86 per cent were women and 14 per cent men. Fifty per cent of the respondents are home gardeners, 44 per cent are involved in community gardens, and 6 per cent are participating in institutional gardens.

The interviews were performed by the students, who were trained beforehand. One member of the Urban Producers Network accompanied the students during the interviews in each zone. The interviews were performed in a simple and easy language.

The workshop was organized in co-ordination with the Urban Producers Network and convened by their zonal co-ordinators. The workshop was attended by 22 people, of whom 17 were women and five men. The president of the Urban Producers Network gave the introduction, and one of the facilitators presented the objectives and the agenda of the workshop. In order to achieve information disaggregated by gender, the discussions were held in homogeneous groups (two composed of women and one of men).

Division of work in the farming households

The overall picture in 2008 is that male and female producers in general perform similar activities/roles, with men tending to engage more in raising animals and processing the products, and especially the commercialization of the products. However, women also show an increasing interest in these activities, are already strongly involved in the selection and packaging of crops to be sold (since they know the criteria of the mainly female buyers better), and ask for more training in these fields.

In comparison with the situation before the start of MPAP process, one can observe that the following shifts have taken place in the division of labour in urban agriculture. In 2004, 83 per cent of the female farmers performed activities on their own, and only 10 per cent of them were supported by their husbands and 7 per cent by their sons or daughters. Now the situation has considerably changed. Currently, only 29 per cent receive no support in their urban agriculture activities; 40 per cent are assisted by their husbands and 31 per cent by their sons and daughters, which indicates an increase in family commitment to urban agriculture.

Agricultural production

In general, in crop-production activities men and women perform the same tasks, with a slightly stronger tendency for the women to predominate in the harvesting of the vegetables. Both male and female producers tend to produce a diversity of crops, with the main emphasis on vegetables. However, we notice that male producers are more active in the production of aromatic plants than the female producers. On average the male producers spend 10 hours per week on urban agriculture activities, compared with the female producers' average of eight hours. This difference is explained by the extra demands of women's reproductive role in the family. It is interesting to note that male producers receive ample support from their sons in crop-production activities, and to a lesser extent from their daughters, while female producers are mainly supported by their daughters.

Livestock keeping

The raising of animals is mostly done at home, and most animal-husbandry activities are done by both men and women, with women slightly more involved in providing water to the animals, most probably since the women are more often at home. Male producers are slightly more likely to undertake the rearing of animals (77 compared with 64 per cent). Both men and women prioritize raising poultry (54 per cent of males and 52 per cent of females), while guinea pigs are kept more by men (54 per cent of the men) than by women (21 per cent of the women). Ducks are raised by 23 per cent of the men and 16 per cent of the women, and rabbits are raised by 15 per cent of the men and 4 per cent of the women. In addition, 8 per cent of the men raise pigs, and 4 per cent of the women raise quails.

Processing

Processing of agricultural products is undertaken by 31 per cent of the male producers and only 13 per cent of the women. However, we also find that the majority of men who are processing agricultural products are supported by women at home (wife and daughters). The men and women involved in processing activities in general both perform nearly all the tasks, but women are more involved than men in the selection of the products that will be processed (90 per cent compared with 50 per cent) and in the packaging of the processed products (30 per cent compared with 0 per cent). Males involved in processing activities on average spend some six hours per week in these activities, while the women involved indicate that they spend on average some four hours per week on similar tasks.

Marketing

In 2004, the focus was mainly on production for home consumption, and only some women showed interest in selling part of their produce. Nowadays men spend more time than the female producers each week in the commercialization of the products, in order to earn extra income. In most households commercialization of urban agriculture products is mainly undertaken by men (62 per cent) and much less so by women (33 per cent). Most of the products are sold to neighbours and relatives and in local markets, fairs, and soup kitchens. Male producers spend on average eight hours per week on marketing activities, while women spend only some four hours per week on similar activities. This reflects the fact that the male producers tend to be more market-oriented than the female producers, who tend to emphasize the home consumption of the products. However, women also show an increasing interest in these activities, are already strongly involved in the selection and packaging of crops to be sold, and ask for more training in these fields.

The support received by male and female producers from sons or daughters in the marketing activities is more or less equal, with men receiving more support from their sons (63 per cent of men and 48 per cent of women) and women more from their daughters (28 per cent of women and 13 per cent of men).

Access to resources and decision making by gender

The person who manages the garden, either male or female, is usually the one who provides the resources that are needed for production (seeds, fertilizers, etc.). However, various male respondents mention that they also have access to the resources of their wives, a fact that is hardly mentioned by the female producers.

The gender-differentiated decision making in respect of various agricultural activities (see Table 8.1) shows the same pattern (taking into account that 86 per cent of the gardens are run by women): the person who is in charge of the garden makes most of the decisions, and the figures indicate a growth in female leadership since 2005. The women estimate their role in decision making substantially higher than the men rate it, which seems to indicate that not all men yet have accepted that women take their own decisions regarding their gardening activities.

Table 8.1 Gender differentiation in decision making on agricultural activities in Villa María

Who decides about:	Answers by the female producers (%)				Answers by the male producers (%)			
	Men	Women	Both	Others	Men	Women	Both	Others
What crops to grow	11	71	13	5	62	15	23	0
Which crops to process	20	60	10	10	25	50	25	0
What products to sell	16	76	8	0	38	37	25	0
Which products / animals to consume	6	81	13	0	34	33	33	0
Which products / animals to sell	5	81	10	4	27	45	10	4
How to use income raised from agriculture	9	80	4	7	33	44	23	0

Mobility of male and female urban producers

Table 8.2 shows the degrees to which male and female producers travel away from their homes to engage in agricultural production. The table indicates that most of the movements of both men and women are within their own zone. Women leave their zone more often for production activities, while men do so more often for the acquisition or production of inputs.

Table 8.3 shows to what extent men or women have freedom of movement to undertake agricultural activities, indicating that men and women in general either inform each other or just take off, and only to a minor extent have to negotiate or ask permission to leave the home to farm. Some women hide from their husband when they are going to sell some products.

In 2004 we noticed that women's mobility was limited to a small local area. In 2007 we found that women move within a much wider area within the district and play a co-ordinating role at a communal organizational level.

Table 8.2 Agriculture-related mobility of male and female producers

	Production		Processing		Marketing		Acquiring / producing inputs	
	Men	Women	Men	Women	Men	Women	Men	Women
Inside the zone	100	61	100	100	74	75	62	75
Outside the zone	0	39	0	0	26	25	38	25
Total	100	100	100	100	100	100	100	100

Table 8.3 Freedom of movement of male and female producers

	For production		For marketing		For processing		For input acquisition & production	
	Men	Women	Men	Women	Men	Women	Men	Women
Ask permission	15	8	0	7	0	5	15	8
Negotiate to leave	38	8	23	11	0	3	0	9
Announce that they are leaving	0	32	26	36	43	40	38	11
Leave secretly	0	0	0	16	0	0	0	0
Just leave	57	52	51	30	57	52	46	52
Total (%)	100	100	100	100	100	100	100	100

Views of male and female producers on the strengths and weaknesses of urban agriculture

In separate groups, male and female members of the Urban Producers Network identified and prioritized strengths and weaknesses in the urban agriculture situation. Table 8.4 shows that male and female producers agree on various strengths that they have gained since 2005 (organization, technical

Table 8.4 Views of male and female producers on strengths and weaknesses of urban agriculture in Villa María

| | Strengths | | | Weaknesses | |
Both	Specific Women	Specific Men	Both	Specific Women	Specific Men
'We are organized now' 'We know how to produce' 'There are now institutions that support us' (the UA Forum members; the City Strategic Agenda on urban agriculture)	'There is unity and commitment' 'We produce healthy products without chemicals'	'There's land available for planting' 'We know to whom to sell our products'	Lack of water Low agricultural quality of the available land Lack of capital to buy agricultural inputs, seeds, and fertilizer Need for more training and technical assistance for more people	Special request for more technical assistance in: ecological pest control and processing and marketing	Poor knowledge of how to plan production properly and maintain a continuous supply Weak market integration

knowledge, recognition by and support received from institutions) as well as the weaknesses that remain (lack of irrigation water, poor land quality, need for more training and technical assistance). However, there are also some specific differences between men and women. The women stressed organic 'healthy' production and the need for more assistance regarding pest management, processing, and marketing. The men stressed access to land (which remains predominantly a 'male' issue), the need to improve production planning, and links with the market.

Conclusions of the case study

Over the past three years, urban agriculture in the district of Villa María has been developed in two respects:

- *Qualitatively,* since before the multi-stakeholder planning process urban agriculture was a poorly recognized activity, and now it is a line of public policy and support; before, the urban farmers were poorly organized, and now they have their own organization; before, the support organizations hardly knew each other, and now they participate in the Forum and co-ordinate their activities; before, women's role in urban agriculture was undervalued and their participation in decision making was restricted, and now their leadership is widely recognized.

- *Quantitatively*, because the number of urban farmers has increased, as has the size of the urban area in use by agricultural producers.

The Network of Urban Producers is an active member of the Multi-stakeholder Forum on Urban Agriculture in Villa María del Triunfo and is actively promoting the strategic interests of women producers in this Forum. One result is that the City Strategic Action Plan on Urban Agriculture includes gender equity as one of its key values and includes among its main strategies high-priority attention to women producers, providing them with technical and social training, and promoting women to leadership positions in the community gardens.

Drawing on the analysis of the information presented in this chapter, we can conclude that the gender-sensitive action-planning process for urban agriculture in Villa María del Triunfo has contributed to the empowerment of women; has improved their self-esteem, leadership, and capacity building; and has increased the independence and freedom of mobility of the women involved. In the past two years, women's participation in urban agriculture has increased; currently 86 per cent of all farmers are women, running home gardens as well as community or institutional gardens. It is important to note that for these women urban agriculture is not resulting in an overload of activities but rather constitutes a means by which to build their personal development and their capacity for social interaction and organization, and to overcome conditions of devaluation, subordination, and exclusion.

However, some inequities in urban agriculture still remain, like the fact that agricultural activities performed by men – which tend to be market-oriented and result in cash income – are still more highly valued in the local social context than those performed by women, which are more oriented to home consumption and barter and are often seen by the community as an extension of the household duties of women. However, female producers at present are engaging more in market-oriented gardening and income generation, and further changes in gender relations may be expected.

Although the participation of women in community roles has grown in recent years, women are still better represented in community organizations that relate to reproductive activities, while men are better represented in community organizations associated with the management of the territory, infrastructure development, and the political sphere.

Notes

1. The data in this paragraph are taken from the survey implemented by the authors in 2008 in the updated case study of Villa María del Triunfo
2. The 'Vaso de Leche' (Glass of Milk) programme was initiated by the left-wing city government of Lima in 1984 and later extended to the whole country, with state funding. The programme benefits some 4.3 million people, of whom 1.8 million live in Lima, including children between 0 and 13 years, pregnant and lactating mothers, older people, and TB

patients. The *comedores populares* (community kitchens) were created in the late 1970s as a survival strategy of the poor in the slums of the larger cities of Peru. Women provide the required labour without payment. Actually there are some 15,000 *comedores populares* in the country, many supported by churches and NGOs and (about 20 per cent) by the state. Many others are run by the women without any external financial support. In both programmes the women were the ones who took the initiative to organize themselves. The groups attend to problems of nutrition and food security, as well as social and gender problems such as domestic violence. In Villa María del Triunfo, various *comedores populares* have taken up urban agriculture. At the moment, some 40 *comedores populares* belong to the Urban Producers Network. Part of their production is used in the community kitchens and part is used by the participating women to feed their families.

3. For this study, we use 'cohabiting' to refer to couples whose relationship is not legalized.

References

DESCO, FOVIDA, IPES and SEDES (2005) *Concerted Plan of the Local Economic Development*, Municipality Villa María del Triunfo, Lima, Peru.

About the authors

Noemí Soto is Adviser to the Urban Agriculture Projects, IPES–Promoción del Desarrollo Sostenible, Lima, Peru.

Gunther Merzthal is Regional Co-ordinator at the RUAF Cities Farming for the Future Programme, IPES, Lima, Peru.

Maribel Ordoñez is Adviser to the Urban Agriculture Projects, IPES, Lima, Peru.

Milagros Touzet is Adviser to the Urban Agriculture Projects, IPES, Lima, Peru.

CHAPTER 9

Gender perspectives in organic waste recycling for urban agriculture in Nairobi, Kenya

Kuria Gathuru, Mary Njenga, Nancy Karanja and Patrick Munyao

Abstract

This case study presents strategies to identify and address gender issues in the project cycle of community-based compost production and briquette-making initiatives in Nairobi, Kenya. Community-based waste management and composting activities were studied in Nairobi, using a semi-structured questionnaire in a research project on organic waste management for urban and peri-urban agriculture, implemented by Urban Harvest and partners in 2003–2004. Gender issues within waste management and composting groups were documented through gender-focused group discussions, guided by a checklist, and also through interviews with key informants. As a follow-up to the organic waste-management research project, Soweto Youth in Action initiated a briquette-making action-research project in partnership with Urban Harvest and Kenya Green Towns Partnership Association in February 2007. Two baseline surveys (one on potential sources of raw materials and another on market opportunities) were carried out, using semi-structured questionnaires. Gender-responsive training courses in group development and governance (including issues of leadership, conflict resolutions, networking and advocacy, and project management) and fuel-briquette production and marketing were conducted, and a business plan and marketing brand for the fuel-briquette initiative were developed.

Introduction

The study area

Nairobi is located in southern Kenya, 500 km from the coast at an elevation of 1,670 m above sea level. The city covers an area of 700 km². Mean annual temperature is 17°C, while the mean daily maximum and minimum are 23°C and 12°C respectively. Mean annual rainfall ranges from about 800 mm to about 1,050 mm, depending on altitude (Situma, 1992). Most of it falls in two distinct seasons: the long rains from mid-March to June, and the short rains from mid-October to early December (Hide et al., 2001).

The current population of Nairobi is estimated at 3 million people, with an annual growth of 4.5 per cent (Ministry of Planning and National Development, 2003). Sixty per cent of Nairobi's population live in very low-income informal settlements; this group of urban poor is projected to increase to 65 per cent by 2015 if the trend continues. Unemployment rates are estimated at 14.5 per cent for males and 25 per cent for females (Ministry of Planning and National Development, 2003). It is estimated that over 1,740 tonnes of solid waste are generated daily, of which 60 to 70 per cent is organic (JICA, 1997).

Kahawa Soweto, the centre of the briquette project which is examined in this chapter, is located 21 km north-east of Nairobi city centre in Kahawa location, Kasarani division, Nairobi North district. It is classified as an informal settlement by the local authorities. According to the 1999 National Population census, there were 1,000 households within the village, occupying 700 dwelling units and constituting a population of about 8,000 people. A majority of the residents were former workers on a sisal farm, but they settled in the area after the business closed down, and the population now includes the second and third generations of the original workforce. The village has high unemployment levels, coupled with a 15 per cent HIV/AIDS prevalence. Kahawa Soweto has a village committee that oversees the running of the village, a slum-upgrading committee linked to work by UN-Habitat, and a community policing group (six women and 24 men) that is paid a fee for guarding slum-upgrading structures such as the public toilets and social hall. All the three organs of village governance are recognized by the government of Kenya.

Despite the legal restrictions on urban agriculture, Nairobi has always hosted many urban farmers, from Maasai pastoralists in the 1800s to zero-grazing units (with dairy cattle confined to a stall and fed by a cut-and-carry fodder system) and kitchen gardens and distance gardens that now dot the landscape, occupying road and railway reserves. In the mid-1980s, 20 per cent of Nairobi households were growing crops within the city limits, and 17 per cent of the households kept livestock in the urban areas (Lee-Smith et al., 1987). In the 1990s, the number of households involved in urban agriculture in Nairobi rose to 30 per cent (Foeken and Mwangi, 2000). In 1998, there were about 24,000 dairy cattle in Nairobi, worth roughly US$13 million, producing annually about 42 million litres of milk worth about $11 million (priced at $0.3/litre) (Ayaga et al., 2005). Estimates indicate that 50,000 bags of maize and 15,000 bags of beans are produced in Nairobi annually (Ministry of Agriculture, 2002).

Methodology of the study

This case study is based on two research projects. The first is entitled 'Management of Organic Waste and Livestock Manure for Enhancing

Agricultural Productivity in Urban and Peri–urban Nairobi'. It was carried out in 2003–2004 by Urban Harvest in partnership with International Livestock Research Institute (ILRI), World Agroforestry Centre (ICRAF), Kenya Agricultural Research Institute (KARI), and Kenya Green Towns Partnership Association (KGTPA). Fourteen community-based compost-production initiatives were identified and studied. Gendered focus-group discussions and interviews with key informants were held and guided by a gender-responsive checklist and a semi-structured questionnaire, involving 155 male and 87 female respondents. Composting techniques were documented, nutrient movements were mapped, links between compost producers and urban agriculture producers were studied, and challenges in compost production and marketing were identified. Also gender issues within waste management and composting groups were documented through gender-focused group discussions guided by a checklist and also through interviewing key informants.

In 2005, a stakeholder feedback workshop was conducted, in which Urban Harvest and partners presented the outcomes of the waste-management project to a variety of stakeholders.

As a follow-up to the organic waste-management research project, an action-research project entitled 'Enhancing Livelihoods of the Urban Youth through Recycling of Organic Waste for Energy Briquette Making' was implemented in 2007–2008 at Kahawa Soweto informal settlement by Soweto Youth in Action (SOYIA), in partnership with Urban Harvest and Kenya Green Towns Partnership Association, the University of Nairobi, and Terra Nuova. The overall objective of this project was to enhance income generation, food security, and urban environmental management through community-based energy-briquette production. The briquette-making project involved seven young women and 13 young men, members of SOYIA group. Eight thousand people living in Kahawa Soweto village and about 200,000 from other neighbourhoods gained a cheap and convenient source of high-quality fuel. The initiative would also be beneficial to the charcoal dealers and waste-paper collectors through the sale of waste paper and charcoal dust.

Two baseline surveys were carried out, using semi-structured questionnaires: one on potential sources of raw materials and another on market opportunities, including customers' willingness to pay and perceptions of fuel briquettes among households, institutions, eating places, and charcoal sellers in low-, medium-, and high-income areas. The resources study involved 160 households (22 male and 137 female respondents) and 100 persons from institutions and enterprises (61 male and 39 female). Six members of SOYIA, four males and two females, were trained in research skills and participated in the survey as enumerators. In the market a survey was carried out to establish the perception of customers, including willingness to pay, and to identify potential traders among individuals and supermarkets. This survey involved 26 male and 24 female respondents.

Two training courses were organized for five female and 12 male members of the SOYIA youth group, who were trained (1) in community organization and group development and governance (including issues of leadership, conflict resolutions, networking, advocacy, and project management) and (2) in fuel-briquette production and marketing, including environmental management aspects. The training courses applied a gender-responsive and participatory learning approach (see Njenga et al., 2008). Also a business plan and marketing brand for the fuel-briquette initiative were developed with the SOYIA members through participatory planning and budgeting activities. Over 300 members of the village, including men, women, youth, and school children, were introduced to the idea of briquettes as an alternative source of fuel, through meetings, demonstrations, and household visits.

Some members of the composting groups and the briquette-making group also received training on urban crop production, livestock keeping, and waste management and re-use, organized by the Nairobi and Environs Food Security Agriculture and Livestock Forum (NEFSALF), involving resource persons from the Ministry of Agriculture, Ministry of Livestock and Fisheries Development, and Kenya Green Towns Partnership Association and Urban Harvest.

In addition, partnerships were developed with organizations and individuals with ample experience in fuel-briquette making, including Terra Nuova, an Italian NGO, two private individuals also working in briquette technology, and the University of Nairobi. These partners assisted in the development of training materials, sourcing of appropriate briquette-making machines, and improving the quality of the briquettes.

Analysis of gender relations in compost production, fuel-briquette making, and marketing by urban community-based organizations (CBOs)

Access to and control of resources in compost and fuel-briquette making

Results from the diagnostic survey of potential sources of organic waste for use as raw materials in briquette making showed that women (69 per cent) and youth (20 per cent) play a significant role in the re-use of household wastes. Organic kitchen waste was re-used for feeding livestock and compost making, while waste paper was sold to recyclers. The survey also revealed that in the majority of households, especially in middle-income settlements, men contributed more money for purchasing fuel than women, while in informal settlements in half of the households women contributed money.

Lack of composting space was the main drawback to compost making by community groups, because the existing activities were taking place on rented or temporally leased public land. There was one group that had been officially allocated a plot by the Nairobi City Council. Supplies of water, which is a major ingredient in both briquette and compost making, were a major constraint at most of the places where recycling activities are taking place. When SOYIA youth groups started their waste-recovery activities in early 2002, the Kenya Railways Corporation had allowed them to use some space near the railway line for waste sorting and compost making, but they were not allowed to construct on the site. This constrained the group, because they had to shift the tools and the compost to their living quarters, which had very limited free space. Compost production was also frustrated by lack of markets, and so the group lost interest. Recently, the group has started an initiative involving energy-briquette making at a new site that was allocated to them by the Kahawa Soweto Slum upgrading committee. Here they have constructed an office, briquette production shed, and a store. SOYIA group also had discussions with the community policing group, who offered to provide security for the briquette-making enterprise at a subsidized fee. The project team played a key role during these negotiations between the youth group and the slum-upgrading committee and the community policing group. SOYIA youth group now has a good working relationship with these other local development groups, with whom in the past their relationships were full of mistrust and conflict.

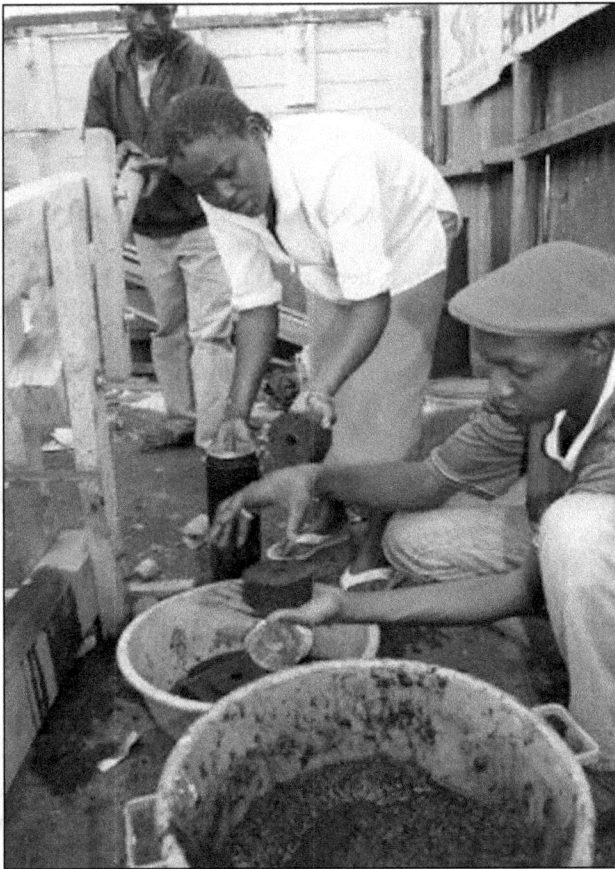

SOYIA youth group preparing briquettes
By Mary Njenga

As for the sourcing of organic materials for making compost, most of the groups obtain the waste materials for free, but they do incur transport costs. In briquette production, the SOYIA group use sawdust, which they purchase, while charcoal dust and paper are collected at no cost from the Kahawa Soweto village and environs by all group members, irrespective of sex (members are obliged to deliver a certain volume of organic materials). Urban Harvest, jointly with International Livestock Research Institute (ILRI), is exploring ways for the SOYIA group to access the large amounts of waste paper that are generated by the institutions based on the ILRI campus. Income realized from sales of compost and fuel briquettes were shared among the group members, following guidelines set by the group.

SOYIA members' involvement in the two research projects enhanced their skills in research and development work, and many of its members are now being hired by other development organizations such as World Vision, Rainbow (a faith-based organization), and Farmers Choice (a factory that processes pig products). However, female SOYIA members cannot participate, because they do not have the required educational qualifications, while most of their male counterparts did receive basic education.

Gender composition and decision making in compost and fuel-briquette making

Of the 14 community-based organizations studied in the first research project, 11 were making compost and three were practising waste management, as well as crop production or livestock keeping. An analysis of group composition and decision making showed that gender and age were major determinants of what took place in the groups. Six of the studied groups had more males than females, three had more females than males, one had equal numbers of males and females, two had females alone, and two had males alone. Eight groups had a mixture of both youth and elderly people, with their ages ranging from 25 to 71, with the exception of one group which had 32 children below 15 years of age. There were six youth groups, four of which had both male and female members, and two of which had male members only (Njenga et al., 2004).

Internal conflicts were reported in six out of the 14 CBOs mainly originating from gender and age differences and relating to duty allocation and management of finances. Most conflicts occurred in groups with members of both sexes, while age difference as a cause of conflict was noted only in groups with male members. Three types of group which had low degrees of internal conflict are (a) youth groups, whether mixed or of one sex only; (b) groups with females of different ages (elderly and young); and (c) mixed groups where women remain silent and assume all the work, while the men only share in the benefits.

SOYIA youth group, involved in fuel-briquette making, had an umbrella committee that consisted of a male chairperson and a female vice-chairperson;

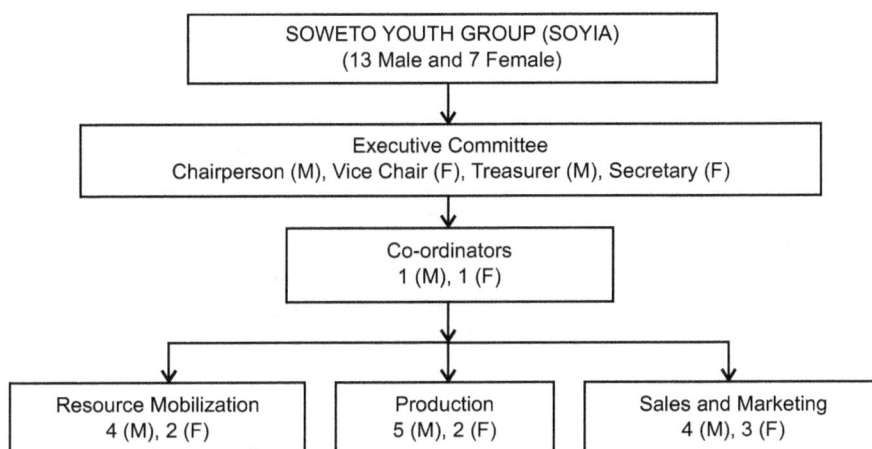

Figure 9.1 SOYIA Youth Group decision-making and management structure

both the treasurer and the secretary were men. During the community organizational development and institutional strengthening (CODIS) training course, the youth group developed a governance structure for the management of the briquette-making project as part of the practical session. They also established various sub-committees for resource mobilization, production, and sales and marketing. The selection of the members was based on the skills and capabilities of each member for effective and efficient project management. The three sub-committees were co-ordinated by two directors, all under the executive committee of SOYIA youth group, as illustrated in Figure 9.1. The duties and decisions of each committee are based on agreed set rules and regulations (Njenga et al., 2008).

SOYIA members belong to development networks such as Pamoja Trust and Nairobi and Environs Food Security Agriculture and Livestock Forum (NEFSALF) and Muungano wa Wanavijiji (a human-rights advocacy group). Through their participation in these activities they are able to contribute to the local and national agricultural and environmental policy-development processes. The youth group is also working closely with the village committees at Soweto and directly giving ideas on security, waste management, and slum upgrading.

Division of tasks/labour

Composting involves a number of activities, including collection and sorting of waste, preparation of waste heaps through systematic layering of different waste materials, periodic turning of waste heaps for aeration, sourcing water, and application to compost stacks. Once ready, the compost is sieved, packed, stored, and sold.

Allocation of duties among the composting members was done on an *ad hoc* basis, determined by the availability of individual members on their activity days. However, sharing of manual labour and leadership roles was observed in the youth groups: five out of the six had planned schedules and duty rosters. An example was the Tuff Gong Garbage Recycling youth group, which had a clear duty roster; the women had been allocated the tasks that were considered 'less dirty', such as fetching water from a tap, while the men were involved in sorting, turning, and sieving the compost. Most of the women did not wear protective clothing, unlike their men counterparts, who had overalls and gumboots.

Illustrations of the gender division of tasks/labour in compost and fuel-briquette making initiatives are presented in Boxes 9.1 and 9.2.

Box 9.1 Division of labour in compost production, packaging and marketing by City Park Environmental Group

City Park Environmental Group is located at the City Park Asian Market in Parklands, about 5 km south of Nairobi. The composting group was started in 1993 with a membership of four men and eight women and the objectives of generating income, supporting destitute children, and cleaning the market. The challenges of uncollected market waste led to the formation of the composting group, which was supported by UN–Habitat Nairobi office, whose staff trained the group. Asian Foundation constructed the shaded area where waste sorting and composting takes place and also assists the group in advertising for the compost. The group is guided by a constitution which governs its management and sharing of roles/tasks, decision making, and income.

At the beginning of the compost-making initiative, females were allocated the role of sorting the waste and transferring it to the shade, where males heaped it. The men also turned and watered the heaps on a weekly basis until maturity. Sieving and packing, including storage, were done by the whole group. One male member whose stall in the market was closest to the store was allocated the duties of selling the compost and keeping the store keys. The cash obtained from the sales was handed to a female cashier for banking. At the start, income accrued from the sale of compost was shared equally among the members at the end of the year, irrespective of their participation. However, during the study there was a lack of male participation in the manual work of heaping the materials, turning, and watering them. The male members preferred to do record keeping, which prompted conflicts among the group members. There were complaints that the men were reported to attend meetings whenever visitors with potential funding appeared. In 2002 four males and the chairperson, a woman, took some compost to the Kenya Agricultural Show, where they made some sales but failed to remit the cash for banking according to the agreed procedures. This brought further conflict among group members, and the group went into a decline as a result.

Only eight women, almost half of whom are over the age of 65, remained active. This affected the productivity of the composting activity, since they are unable to perform the heavy duties that are crucial to production of good-quality compost: for instance collecting different types of organic material, watering and turning it regularly, and packaging it. The women no longer follow composting techniques that they had learned but have reverted to their local know-how. They argue that the techniques are labour-intensive and time-consuming. The store keys are kept by one of the elderly women, who opens the store twice a week and sells the compost; the money is shared among the women without being banked. The group had not kept records of quantities of compost produced or the income generated in the last three years.

> **Box 9.2 Division of labour in briquette making by Soyia youth group**
>
> The SOYIA youth group chose to embark on fuel-briquette making because the marketing of compost had run into difficulties. The briquette-making process involves sourcing the raw materials i.e. charcoal dust, sawdust, and waste paper; preparing materials, such as shredding the waste paper and soaking it for two–three nights, and sorting charcoal and sawdust to remove impurities; mixing materials and binding them with the soaked waste paper; measuring the materials and putting them into the PVC cylinder; pressing the briquettes to dry on racks; packaging them; and marketing them. These tasks of resource mobilization, production, and sales and marketing are carried out under the three sub-committees that were formed during training sessions offered to the group. The sub-committees give instructions to the group members and keep records, while the actual processing is done by all members, irrespective of gender. Members' daily participation in these activities depends on each individual's availability: young women have been more involved than the young men, because the latter are involved in other casual jobs. To address gender inequalities in the sharing of income, the group has set guidelines which specify that income sharing should be based on each member's participation in production and selling of the briquettes. Records of the numbers of briquettes produced and income realized from their sales are kept by the respective sub-committees.

Differences in knowledge and preferences

The SOYIA group's inadequate skills of governance and group cohesion had been identified as threats to its sustainability, so members were offered a training course in community organizational development. Women were more active and consistent in attending to group activities than the men, because the women have a better understanding of the importance of the collective approach to sharing ideas and overcoming burdens. The SOYIA members defined 'gender' as the roles played by men and women, the rights of young and old, inequalities in opportunities, involvement of men and women, the issue of employed and unemployed males and females, and discrimination against and exploitation of males and females.

During the formation of sub-committees to manage the briquette-making project, members voluntarily joined specific committees, depending on their individual skills and preferences. For instance, those who liked being involved in business preferred to be on the sales and marketing committee, while those who liked being involved in making products chose to be in the processing committee. When leadership in each of the sub-committee was analysed, it was found that members elected people who had been involved in other development works in the village, such as projects involving non-government organizations. These women commanded respect and trust from the local/village leaders, and through frequent participation in the village advocacy meetings they had acquired skills and were being elected to leadership positions within the villages, as well as on the SOYIA group management committee.

In case of the compost-making groups, the low level of education and communication skills of women and elderly people limited their involvement in leadership activities. Composting groups were made up of young and elderly persons. The young men, due to the common belief that they are flexible and agile and the fact that they are more educated, represented the groups in meetings and workshops, which resulted in unequal empowerment within the composting groups. An example is the City Park Environmental Group, whose male members were young and educated and thus better placed to get involved in marketing and public relations activities, while women, most of whom were illiterate and elderly, were left to do all the difficult compost-processing activities. The group, though located in an ideal compost-making facility, collapsed because the roles and responsibilities were not allocated in a participatory manner, which created mistrust due to lack of transparency (see Box 9.1).

The training courses organized by Nairobi and Environs Food Security, Agriculture and Livestock Forum (NEFSALF) have enabled the organic waste recycling groups to overcome such problems and to involve women in the representation to NGOs, government departments, and research institutions. In the case of SOYIA, the women are more articulate in their management and decision making, and this is because of their better advocacy skills acquired over time through training and participation in other groups' neighbourhood meetings and in leadership roles of their groups.

Traditional compost making produces bad smells, which result in negative attitudes among neighbours; complaints are presented to city enforcement officers, who are frequently in conflict with community-based organizations like the SOYIA group. Handling of solid waste without proper protective clothing such as gumboots and gloves may result in physical injuries. This is made worse if organic waste or naturally decomposed compost is sourced from rubbish tips where waste products from industries, hospitals, and markets are dumped together. This results in contamination of compost with heavy metals, which may degrade the health of the soil.

The gender strategies of the project

The following strategies were applied in order to encourage gender awareness in the two projects.

Inclusion of gender-equity concerns in the project objectives

The objectives of the organic waste management and the briquette-making projects were formulated with gender equity in mind, and efforts were made to obtain a clear understanding of how men and women were involved in the composting and briquette-making activities, and to ensure that gender was taken into account in all stages of the project.

Inclusion of gender in the diagnosis

Gender-sensitive checklists were applied in order to obtain gender-disaggregated data and to develop understanding of key gender issues such as access to land, participation in decision making, and the division of labour.

Inclusion of gender in the design of the interventions

The SOYIA youth group, which was one of the 14 CBOs involved in the organic waste-management research project, was selected to represent all other groups in a pilot project on making fuel-briquettes from urban organic wastes, based on their previous involvement in waste recovery, the cohesiveness of the group, and their location in an informal poor suburb. Male and female members were invited to participate in the prioritization of possible initiatives. The options presented were crop production, compost making, and fuel-briquette making. The briquette making ranked highest, based on its potential to contribute to a clean neighbourhood, income generation, and self-employment for young people. Leaders of the SOYIA youth group directly participated in the action planning undertaken by Urban Harvest and Kenya Green Towns Partnership Association. In this process, gender-responsive activities and desired outputs were defined, in addition to related monitoring indicators (see below).

Attention to gender in the implementation stage

During the implementation, male and female members of the various sub-committees participated in the day-to-day planning and budgeting of project activities. As indicated above, initially men tended to dominate the internal and external roles in the groups. During planning meetings, women from the SOYIA group expressed fears that the male members would frustrate their efforts. These concerns were addressed in the community development and leadership training provided by the project team, during which resource mobilization, production, and marketing sub-committees were formed according to skills and capabilities, irrespective of sex. Guidelines were also set to govern income sharing, whereby a member will get 25 per cent if involved in production and another 25 per cent if involved in selling, and 50 per cent of the price of each briquette is retained in the group's account. To enhance community support through purchase of the fuel briquettes, a community sensitization and project launch meeting was held in the second month of the project at the Kahawa Soweto village (Njenga et al., 2008). During this meeting, men and women in the village were requested to support the initiative, particularly by purchasing the fuel briquettes. Terra Nuova participated in identification of gender-friendly briquette pressers and locally assembled paper shredders. Participatory testing of the cooking qualities of the energy briquettes was conducted at the village in January 2008, when men, women, youth, and children were involved in preparing a traditional mixture of maize and beans called *Githeri*. One male participant commented: 'When my wife asks for money to buy the briquettes, at least it will be something that I know and have seen cook so well without smoke and very fast'. Due to this open forum, the village committees have developed interest in the activity and have been creating awareness of the briquettes, as well as referring interested visitors to the project site.

Defining gender-sensitive project indicators

Gender-sensitive indicators were selected, including the numbers of men and women participating in meetings and training sessions; the numbers of issues raised by men and women during such meetings; the numbers of men and women occupying leadership roles in the groups; the involvement of men and women in the production and selling activities; and the production and sales realized by them. Also changes in the behaviour of men and women were observed qualitatively, for example, men's views on women's involvement in leadership.

Monitoring and evaluation activities in the fuel briquette-making project helped the group and project team to track performance of the project in order to remain focused on the set objectives, including the gender focus. Resource mobilization, production, and sale and marketing sub-committees kept records of their activities. The three sub-committees hold regular meetings to evaluate progress against set milestones, and to plan and budget for activities, guided by the minutes taken during group meetings. Some of the research-team members also participate in these meetings, and they also provided technical support during site visits two or three times per month.

Inclusion of women in networking and policy influencing activities

Most of the waste-composting groups and the SOYIA youth group are affiliated to the Nairobi and Environs Food Security, Agriculture and Livestock Forum (NEFSALF). The forum has a large and diverse membership which includes government ministries and departments, NGOs, development partners, and research organizations. The forum presents an environment for the composting groups and the urban producers to discuss policy issues informally with policy makers in face-to-face dialogue. As a result of this, a national policy platform on urban agriculture has been formed which will review the national urban agriculture policies. The forum is also used by researchers to disseminate research findings and technology dissemination. As indicated above, the project partners systematically promoted the inclusion of more women in the Forum and in other external contacts with support organizations and government departments, among other means by training members in networking and advocacy skills.

Main conclusions and recommendations

Principal lessons learned

- Gender and age differences in composting groups are likely to lead to internal conflicts about labour division and financial matters.
- Young people were willing to share responsibilities and benefits equally between male and female members.
- Access to and control of publicly owned resources requires community-based approaches and advocacy, while roles and regulation need to be set within the groups to ensure equal sharing of benefits, tasks, and decision making between male and female members.
- Well-structured participatory training sessions based on needs assessment can play a significant role in identifying and addressing gender issues in research and development.
- Community training in advocacy and networking in urban agriculture interventions may enhance women's participation in policy-development processes.
- Involvement of men and women beneficiaries in project development and implementation builds beneficiaries' skills in both technology and project management, and hence improves the sustainability of the project results.
- Consideration of gender issues in all phases of research and development projects is important for enhanced gender-responsive impact. Gender inequalities in sharing benefits and decision making, if not addressed, may adversely affect the success of the project and /or result in gender inequalities.
- The study of the 14 compost-making groups identified gender inequalities in participation and benefit sharing. This knowledge was used in developing the briquette-making project with the SOIYA youth group. However, little was done to address these issues among the other 13 studied groups. Research projects should create room for follow-up studies with actions to address gender inequalities identified in the study; otherwise the status quo remains unchanged, despite the gender analysis being carried out.

Differences between gender mainstreaming in urban agriculture and rural agriculture

Mireri (2007) argues that although men 'officially appear' as the owners of the farms, women are responsible for the success of urban agriculture. African culture bestows ownership of land and land resources on men, even though men's contribution to farm production is limited. Women undertake day-to-day management of the farms and invest financial capital in them. According to Smit (1996), urban farming provides jobs disproportionately for women, youth, and the elderly. It requires co-operation and partnership and so creates communities and reconnects urban people with nature. Urban agriculture produces three to 15 times as much per hectare as common rural methods. It is more organic and sustainable, because urban wastes – which are 70 per cent organic – are more abundant than rural waste, while the urban farmer's labour-intensive methods use less land and water per unit of production than industrial agriculture (Lee-Smith et al. 1987; Karanja et al. forthcoming).

Mainstreaming of gender issues in urban agriculture appears to be easier due to the 'softening' of cultural attitudes, especially towards women farmers whose cultivation of certain crops and keeping of certain livestock would have been restricted in rural settings. However, vulnerability to new forms of urban violence is common, and there are frequent conflicts among women and youth working on old garbage dumps, for example over wastewater supplies and space to display their agricultural products. Social and economic exclusion of urban farmers, mainly women, children, and youth, make it difficult for them to participate in making decisions and influencing policy. The rural farmer is 'protected' by social ties, while urban farmers have to fight for both physical and institutional space.

References

Ayaga, G., Kibata, G., Lee-Smith, D., Njenga, M., and Rege, R. (2005) *Policy Prospects for Urban and Peri-Urban Agricultura in Kenya*, Policy Dialogue Series No. 2, Urban Harvest–International Potato Centre, Lima, Peru.

Foeken, D. and Mwangi, A. (2000) 'Increasing food security through urban farming in Nairobi', in N. Bakker, M. Dubbeling, S. Gundel, U. Sabel-Koschella and H. de Zeeuw, *Growing Cities, Growing Food Urban Agriculture on the Policy Agenda. A Reader on Urban Agriculture*, German Foundation for International Development (DSE), Feldafing, Germany.

Hide, J. M., Kimani, J., and Thuo, J. K. (2001) 'Informal Irrigation in the Peri-Urban Zone of Nairobi, Kenya. An Analysis of Farmer Activity and Productivity', Report OD/TN 104.

JICA (1997) *Master Plan Study of Nairobi*.

Karanja, N., Njenga, M., Gathuru, K., and Karanja, A. (forthcoming) 'Crop–livestock–waste interactions in Nakuru: a scoping study', in G. Prain, N. Karanja, and D. Lee-Smith (eds.) *Urban and Peri-Urban Agriculture in Sub-Saharan Africa – 20 years on: Case Studies and Perspectives*.

Lee-Smith, D., Manundu, M., Davinder, L., and Gathuru, P.K. (1987) 'Urban Food Production and the Cooking Fuel Situation in Urban Kenya', Mazingira Institute, Nairobi, Kenya.

Ministry of Agriculture (2002) *Annual Report*.

Ministry of Planning and National Development (2003) *Economic Survey*.

Mireri, C. (2007) 'Credit and Investment in Urban Agriculture in Nairobi City, Kenya', Dept of Environmental Planning & Management Kenyatta University, Nairobi.

Njenga, M., Romney, D., Lee-Smith, D., Gathuru, K., Kimani, S., Frost, W. and Carsan, S. (2004) 'Management of Organic Waste and Livestock Manure for Enhancing Agricultural Productivity in Urban and Peri–Urban Nairobi', Project Report supported by World Bank.

Njenga, M. M., Karanja, N. K., Gathuru, K., Malii, J., Munyao, P., Mwasi, B., Awika, H., Mugwanja, A., and Muthoni, M. (2008) 'Enhancing Livelihoods of the Urban Youth through Recycling of Organic Waste for Energy Briquette Making at Kahawa Soweto Informal Settlement, Nairobi', Project Report supported by IDRC.

Situma, F.D.P. (1992) 'The environmental problems in the city of Nairobi, Kenya', *African Urban Quarterly 7*.

Smit, J., Rattu, A., and Nasr, J. (1996) *Urban Agriculture: Food, Jobs and Sustainable Cities*, UNDP, New York.

About the authors

Kuria Gathuru is National Co-ordinator at the Kenya Green Towns Partnership Association, Nairobi, Kenya.

Mary Njenga is Research Officer for Urban Harvest, International Potato Center (CIP), Nairobi, Kenya.

Nancy Karanja is Regional Co-ordinator SSA for CIP–Urban Harvest, Nairobi, Kenya.

Patrick Munyao is Research Assistant for CIP–Urban Harvest, Nairobi, Kenya.

CHAPTER 10

Urban agriculture as a strategy to promote equality of opportunities and rights for men and women in Rosario, Argentina

Mariana Ponce and Lucrecia Donoso

Abstract

The Urban Agriculture Programme of the Municipality of Rosario was developed in response to high levels of unemployment, especially among poor women, due to economic and social unrest dating back to the 1980s. This case study reflects on the results of the programme (2001–2004), which aimed to enhance the incomes, quality of life, and equal opportunities of men and women, through urban agriculture production, processing, and marketing activities. Strategies applied in the programme included the installation of community gardens on vacant municipal land, the establishment of a network of producers, the establishment of agro-industries, and the enhancing of links between producers and consumers. The chapter elaborates on the background of the programme and the process of integrating gender in these activities; it concludes with the eventual effects of the programme and the lessons learned in terms of gender.

Introduction

This chapter presents the results of an analysis of the lessons learned from the Urban Agriculture Programme (UAP) of the Municipality of Rosario concerning urban agriculture as a strategy to achieve equal opportunities and rights for women and men. This analysis was conducted in 2007, involving representatives of the UAP and the Women's Services (both belonging to the Department of Social Advancement of the Municipality of Rosario) and the Urban Agriculture Producers Network of Rosario (*Red de Huerteros y Huerteras*). We would like to thank the following authors for their contributions: Vanesa Calvin, Andrea Mazzucca, Graciela Veliz, and Analia Santa Cruz, technicians from the Urban Agriculture Programme, and Sandra Tolsa from Women's Services.

Background of Rosario's Urban Agriculture Programme

Rosario is the third most populated city in Argentina. Since the 1980s, the systematic application of neo-liberal policies has caused a downsizing of the major local industries and the disappearance of many small and medium-sized businesses which had stimulated the local economy and were crucial providers of employment. On top of this, migration from rural areas to the city increased, owing to the implementation of an agricultural policy that stimulated a 'modernization' of agriculture, introducing technological packages that enhanced mechanization, created high dependency on external supplies, promoted mono-cropping, and expelled manual labour. This led to large groups of urban people outside the formal labour market establishing homes in informal settlements surrounding the urban centres, including Rosario. The sharp social and economic crisis that hit Argentina in 2001 further worsened the situation, increasing poverty by more than 60 per cent.

It was during this profound crisis that the urban poor, and especially women, engaged extensively in urban agriculture as a strategy to overcome the emergency. The women began to occupy vacant public and private land in order to cultivate food to meet their families' needs. Some of these women had a mainly domestic role before they got involved in urban agriculture, whereas others had worked in various (often precarious) jobs away from the home, mainly in the informal sector without social security or medical coverage (working, for example, as housekeepers, hospital aids, and street food vendors).

In response to this situation, the Municipality of Rosario (Department of Social Advancement) developed the Urban Agriculture Programme, with the goal of responding to the needs of the unemployed in a productive manner. In 2001, the Women's Services Department initiated the Plan of Equal Opportunities for Men and Women (2001–2004), and within the framework of this plan the Urban Agriculture Programme was launched. The Department of Social Advancement signed an agreement with the Centre of Studies of Agro-Ecological Production (CEPAR, an NGO with vast experience in developing urban irrigated land) and the National Programme Pro-Huerta (garden) Food Security Programme, led by the National Institute of Agricultural Technologies (INTA), which constituted the Urban Agriculture Programme as an alliance between civil society and local and national government.

Objectives of the Urban Agriculture Programme of Rosario

The goals of the Urban Agriculture Programme (UAP) are to promote the social integration of poor and vulnerable urban families, to enhance their incomes, and to improve their quality of life through group and individual agricultural production, processing, and marketing. The programme also seeks to promote equal opportunities for men and women, applying a gender-sensitive approach.

Vegetables produced by urban farmers in Rosario
By Hans Peter Reinders

The following priorities were established between the Municipality and the responsible organizations of the UAP:

- To make municipal vacant land available to poor or vulnerable urban households (women, youth, and the elderly) and assist them to put these lands into production.
- To introduce production methods that will be easy for poor urban dwellers to adopt and will provide good results without creating dependency on external resources.
- To produce nutritious food in order to improve the diet of indigent families.
- To establish a system of direct marketing from producers to consumers at strategic locations in the city, with attractive presentation of the products.
- To promote the sustainability of the UAP by institutionalizing urban agriculture as a public policy.
- To promote equal opportunities for men and women.

Strategies and methodology of the Urban Agriculture Programme

The UAP applied the following strategies.

- Installing community gardens on vacant municipal land (often after cleaning up informal waste dumps) and training households in the production of vegetables and medicinal and aromatic plants.

- Establishing an independent network of producers (*Red de Huerteros y Huerteras*).
- Organizing groups that produce the tools, fence poles, bio-fertilizers, and compost required by the producers.
- Subsidizing these groups' purchases of seeds and seedlings, wires and posts for fences, tanks and hoses for irrigation, and tools.
- Establishing two Social Agro-industries (SUAs) to add value to the products obtained from urban agricultural activities.
- Assisting producers and groups in the commercialization of their products, including training to prepare the products for sale and marketing; the creation of weekly markets for organic vegetables and processed food products (sweets, baked goods, jams, wine, medicinal and aromatic plants, and natural cosmetics); and transport of their products to the market.
- Promoting producer–consumer linkages through the public media and the organization of practical demonstrations and meetings with consumers. These actions also result in further strengthening of the confidence and self-esteem of the producers.
- Enhancing co-ordination of activities among municipal departments and with other public institutions, universities, NGOs, and neighbourhood organizations, favouring the optimization of resources and achieving a more integrated development approach.
- Incorporating agriculture into the city's territorial planning.

The approach applied by the Urban Agriculture Programme is characterized by working methods that favour participatory learning, derived from community education, such as interactive workshops, exchange meetings, and discussion groups, which allow for exchange, dialogue, and collective construction of knowledge among the participants. Capacity building to develop group organization and leadership skills was continuous. Equal participation of men and women in all project activities was promoted, in addition to applying a gender-sensitive approach to daily practices (such as encouraging male and female producers to accompany each other in all tasks), enabling the development of trust and new relationships. The short- and medium-term plans were flexible and took into account the specific social context of each district where the activities were taking place. This participative and horizontal form of project organization was quite innovative in government circles.

The results of the Urban Agriculture Programme

Urban agriculture proved to be a valid strategy for the reduction of poverty and for social integration, bringing together and involving different city sectors. The following impacts were achieved.

- Access to land for the urban poor has been improved, mainly by the creation of five 'garden-parks' in various city zones (new multi-functional spaces which incorporate the production of food in the use of public spaces) and by cleaning up informal waste dumps on vacant municipal land (which also improves local living conditions).
- Vulnerable families (especially women) feel more valued and recognized and better included in the local society and economy.
- A network of some 500 groups has been created, involving more than 2,500 producers of organic vegetables, of whom 1,000 are women, producing vegetables for approximately 12,500 people.
- Each productive group generates a monthly revenue equivalent to between US$100 and $500 per member, a substantial sum compared with an average monthly income of $90 for the urban poor.
- In addition, families have made important savings on food expenditures.
- Six marketplaces have been established, as well as a home delivery scheme for bagged vegetables.
- Two social urban agro-industries have been established: one for processing vegetables and one for producing natural cosmetics, aromatics, and medicines; 300 small bakeries have been installed; and the development of community and school cafeterias has been stimulated.
- The local government has formulated a public policy on urban agriculture which includes new norms and regulations pertaining to urban agriculture.
- The gender approach adopted in the UAP helped to strengthen the integration of gender in the public policies of the local government (concerning urban agriculture and other sectors).

The gender perspective in the Urban Agriculture Programme

The gender-mainstreaming process

A technical interdisciplinary team was formed, including staff of the Women's Services, the Urban Agriculture Programme, and the Services of Employment and Social Entrepreneurship, to design an action plan to make the urban producers more gender-conscious and to help them to contribute positively to the modification of asymmetrical gender relations. The plan included the following activities.

Surveys of the roles and functions of male and female producers

In early 2003 an extensive study was made in 13 gardens, investigating the roles that women (and men) fulfil as producers (productive role), in the domestic field (reproductive role), and in their organizations and in community services (community role). In later years such surveys will be repeated to document the changes in roles and functions that men and women fulfil in production, processing, and marketing activities.

Photographic report

Based on the results of this survey, a photographic report was produced on the Female Producers of Rosario, their role in urban agriculture, and their rights regarding the possession of productive land.

Training in new forms of female leadership

Participatory workshops were held district by district, to enhance women's potential to act as group co-ordinators and to improve their technical skills. Three hundred women were trained.

'First Encounter of Female Producers'

The Urban Agriculture Programme and the Women's Services of the Department of Social Advancement of the city of Rosario jointly organized in August 2003 the 'First Encounter of Female Producers' with the motto *'Constructing new forms of leadership; towards secure possession of fertile land'*. The purpose of this meeting was to acknowledge the important role of female producers in Rosario and give them institutional recognition. During the encounter, workshops were organized to enhance the self-esteem and confidence of the women so that they would be encouraged to participate actively in the organization of the gardens. Group dynamics were applied to reflect on *'How I am'* in personal matters, as well as in the role of producer and with regard to the ownership of land. Also the capacity of the women gardeners to act as co-ordinators of training workshops was developed. The workshop enhanced female leadership and capacity to effectively exercise their political, social, and economic rights, to democratize the garden groups, and to secure their co-ownership of the land that they work. The encounter also identified the demand of the female producers for training in marketing of the products and equal distribution of the profits.

'Second Encounter of Female Producers'

The second encounter was organized to consolidate the democratization of the garden groups and the new forms of female leadership, as well as to provide training in marketing and distribution of benefits.

Gender mainstreaming in the Network

Gender was integrated in the construction and consolidation of the Urban Agriculture Producers Network of Rosario. The above-mentioned activities enabled the female producers to achieve the following:

- Reflect on actual conditions and significance of domestic, productive, and community roles.
- Analyse the roles and functions of male and female members of the garden groups.
- Analyse difficulties encountered in achieving access to and possession of fertile land, and in the marketing of products and distribution of benefits.
- Initiate a process of gaining equality between men and women by modification of some aspects of their daily life: personal, family, and work.
- Strengthen their role in the community by creating opportunities to assume leadership in the various programme activities in the public arena (group leadership, training workshops, marketing activities, group exchanges).

The strengths and weaknesses of gender mainstreaming in the Urban Agriculture Programme

The achievements

It has been clearly established that women's role in the management and operation of the community gardens has gradually grown, and that they at present make a major contribution to ensuring the continuity of the garden activity:

- 70 per cent of the group leaders are women.
- 100 per cent of the women participate in marketing activities and consider it a positive experience.
- 49 per cent of women producers manage the income derived from the gardens.
- 44 per cent of them have received training and considered it a very good experience.
- 93 per cent consider work in the garden as a job.
- 92 per cent feel that the garden work has improved their family's nutrition.

Through their roles as leaders in training, marketing, and community activities, starting with their work in the garden and participation in group meetings, the women have discovered and strengthened their capacities and knowledge, have enhanced their self-esteem, and have increased their presence in the public arena (in social and cultural terms).

The producers organize the work of the groups, taking gender into account. The groups reflect on the asymmetric relationships of power and the unequal allocation of roles that are socially established between men and women, and they search for ways to address such inequalities. As a result, the tasks are distributed and performed equally by men and women. The producers share activities of training, production, marketing, consumption, and processing of the products and jointly agree on the tasks in order to develop and share the same workspace with equal responsibility and ability.

The productive role of the female producers in the generation of income and in the making of decisions concerning the use of income generated has been strengthened. Earning their own income has generated, in some cases, greater levels of independence. Greater participation in income generation and in the management of the garden groups has contributed to women's growth and has strengthened their roles in personal, family, and social spheres. It has enabled them to acquire a fundamental role in the construction and consolidation of the Producers' Network, thus further enhancing their social status.

The limitations

The Urban Agriculture Programme, although it promotes equality of opportunities for men and women, is problematic in that for some female producers who are heads of their household their active participation in production, marketing, and organizational activities overloads them with work, and they have little chance of sharing the burden of labour and responsibilities with others. The expansion of the spaces for action and participation by the female producers requires new forms of organization in the domestic and public arenas in order to avoid the imposition of extra work and responsibilities.

On the other hand, female producers sometimes find themselves in a situation where their work in the garden identifies them even more strongly with their reproductive role, in that women are socially ascribed the task of feeding the family. In such cases, more emphasis needs to be given to the income-generation and cost-saving aspects of women's gardening activities, rather than the nutritional and food-security value of those activities. Although the Programme incorporates a gender-sensitive approach, it needs to identify more affirmative actions, in order to consolidate egalitarian practices between men and women, as much within the gardening groups as in the wider community. It will also need to assist women to widen their areas of action.

This year, the Urban Agriculture Programme will review its lines of action regarding gender equity and will set up a new series of training workshops.

Although the local government has made some progress towards mainstreaming gender issues when designing public policies and programmes, there is still scope for further efforts to address the strategic and practical needs of women. During the implementation phase it was noted that the political and institutional process takes a long time, which may lead to demoralization and a loss of trust on the part of the poor and vulnerable participating producers. Harmonizing the political process and the needs of the producers requires close attention from the start.

Enabling secure access to vacant land in the city is another crucial factor in the success of urban agriculture. Vacant municipal land that is not suitable for construction (for example, land under power lines, on flood plains, and in earthquake zones) and derelict open spaces (illegal or formal rubbish dumps, former industrial areas, etc.) had been identified as the most appropriate for urban agriculture in Rosario. The creation of community gardens in public parks, developed in Rosario as another valuable option for the development of urban agriculture with the urban poor, constitutes an example to be followed.

Although urban agriculture is in a process of consolidation as local-government public policy, it is not included in policy or planning at provincial or national level. Since 2007 this process has been undergoing revision, as the provincial authorities are showing themselves to be in favour of the change.

Lessons learned

The integration of gender-sensitive approaches in the Urban Agriculture Programme (UAP) has contributed to an enriched vision of development at both local and policy levels. The search for better living conditions for poor and vulnerable people requires development planning that is rooted at the micro level and takes the specific needs and interests of women into account, applying their knowledge and experience to the realization and management of the development activities.

The UAP has demonstrated that urban agriculture can form an important strategy for poverty alleviation and a significant opportunity both to promote a process of equitable local development, including men and women on an equal basis in production, processing, marketing, and organization-management activities, and also to ensure equal access to resources and income for men and women. Urban agriculture is an activity appropriate for women, given that it allows them to organize themselves in a manner that fits in with the demands of the multiple functions that they fulfil in the household. However, to make a positive impact on gender relations, various affirmative actions have to be undertaken: the actual roles and contributions of women in agriculture, in the household, and in the community have to be acknowledged, men's and women's awareness of gender issues has to be raised, women have to be trained in leadership skills, and opportunities have to be created for them to assume leadership roles in training, group management, production, and marketing.

The opportunity to generate an income through urban agricultural production or related processing or input-producing activities allows the women to gain recognition and independence within the family and community spheres. Assuming roles as producers and generators of income also allows women to call into question the actual division of roles in the domestic sphere and re-negotiate the organization of domestic chores between men, women, and children in the household. It is important to foster encounters between female producers (workshops, exchange visits, etc.), and to facilitate exchange and reflection on crucial themes (such as leadership, ownership of land, distribution of benefits, and redistribution of household chores). Gardening is related to women's strategic interests, such that the construction of their identity as women and producers is supported. Strengthening the role of female producers in the garden groups also contributes to strengthening their role and recognition in the community as persons capable of acting and making decisions in their own right.

About the authors

Mariana Ponce is Gender adviser at the Centro de Estudios de Producciones Agroecológicas (CEPAR), Rosario, Argentina.

Lucrecia Donoso is Psychologist at the Women's Services Unit, Social Development Department, Municipality of Rosario, Argentina.

CHAPTER 11

The role of women-led micro-farming activities in combating HIV/AIDS in Nakuru, Kenya

Mary Njenga, Nancy Karanja, Kuria Gathuru, Samwel Mbugua, Naomi Fedha and Bernard Ngoda

Abstract

The case study is based on an action-research project entitled 'Combating HIV/ AIDS in Urban Communities through Food and Nutrition Security: The Role of Women-led Micro-livestock Enterprises and Horticultural Production', which is being implemented in Nakuru, Kenya. Nakuru is the fourth largest town in Kenya, with a multi-ethnic composition of about 320,000 residents. Crops are cultivated in people's compounds, along roads and railways, and under power lines; livestock are kept in compounds or on vacant land, while others are free-range. The main objective of the project was to improve the food and nutrition security of households in HIV/AIDS-affected communities in Nakuru Township, in order to contribute to mitigation of the impact of HIV/AIDS on the livelihoods of households. The case study describes how gender was incorporated in diagnosis, design, planning, implementation, monitoring and evaluation, and policy-influencing phases of the project.

A baseline survey was conducted in 11 out of 15 wards of Nakuru municipality, where 85 male-headed and 70 female-headed households were interviewed. Results of the diagnostic study were used to design two interventions: a vegetable intervention currently involving six male and 44 female representatives of households, of whom 40 are also participating in a dairy-goat project. Households that actively participated in the vegetable project were selected to receive the dairy goats. Produce from both interventions are for domestic consumption, and any surplus is sold. Beneficiaries received basic training in vegetable production and dairy-goat husbandry techniques including feeding, housing, and detection of basic diseases; and they were assisted with initial inputs such as seeds, pesticides, land preparation, and irrigation networks.

Introduction

Nakuru

Nakuru Municipality is a mid-sized town, the fourth largest town in Kenya. It has a multi-ethnic composition due to its location at the crossroads of many of Kenya's ethnic groups. It is located in the heart of the Great Rift Valley between latitude 0°10′ and 0°20′ South and Longitude 36° and 36°10′ East, at a distance of 160 km north-west of the capital city Nairobi (Foeken and Owour, 2000). According to the latest population census, the population of Nakuru stood at 239,000 people in 1999, with a growth rate of 4 per cent. The main economic activities in Nakuru are commerce, industry, tourism, agriculture, and tertiary services (MCN, 1999).

It is estimated that the poverty level in Nakuru is 65 per cent of the population, with the poverty line defined as a per capita income of less than one US dollar a day (MPND Kenya, 2003). Sub-Saharan Africa continues to bear the brunt of the HIV/AIDS pandemic; the region is home to almost two thirds (63 per cent) of the global HIV-positive population (UNAIDS, 2006). Almost two thirds of people living with HIV/AIDS are women, who also have to live with the related social stigma. In many countries moral judgements are passed upon HIV-positive women, much more than on HIV-positive men, regardless of how they contracted the disease (ILO, 2003). HIV/AIDS continues to be a significant problem for Nakuru, where it is estimated that one in four adults is infected with the virus (FHI, 2000 in Anderson, 2007).

The action-research project described in this case study was carried out in 11 out of 15 wards in Nakuru Municipality: namely Kaptembo, Shabab, Rhonda, Shauri Yako, Langa Langa, Lakeview, Bondeni, Kivumbini, Menengai, and Nakuru East.

Urban agriculture in Nakuru

Foeken and Owour (2000) established that 35 per cent of Nakuru households were engaged in urban farming, 27 per cent of whom were growing crops, while 20 per cent kept livestock. In 1998, it was estimated that about 5,200 acres of land were cultivated within the built-up urban area. The most common crops grown were maize, kale (*sukuma wiki*), beans, onions, spinach, tomatoes, Irish potatoes, cowpeas, bananas, and a local vegetable commonly referred to as *sageti* (Foeken and Owuor, 2000). In the built-up areas, crops were being cultivated not only in people's compounds but also along roads and railways, under power lines, and on every piece of vacant land (Ibid. 2000). There were about 160,000 head of poultry, 25,000 head of cattle, 3,000 goats, 3,500 sheep, and 1,500 pigs within the municipality (Ibid. 2000). Livestock are kept in residential compounds, and free-range grazing is also a common sight as animals roam around in open spaces and streets.

The action-research project

The main objective of the project is to improve the food and nutrition security of HIV/AIDS-affected households in Nakuru Township, in order to mitigate the impact of the pandemic on people's livelihoods. The project commenced in May 2006 and is expected to end in April 2009.

The project includes a diagnostic study and two urban agriculture interventions. The projects are implemented through Badili Mawazo Self Help Group (BMSHG), an HIV/AIDS psycho-social and welfare development group for People Living with HIV/AIDS (PLWHA). Badili Mawazo has 200 clients, who came together as a group to fight stigma, isolation, and loss of livelihoods caused by their HIV status. The Presbyterian Church of East Africa (PCEA)–Nakuru West Parish offered land at no cost for the production of vegetables and fodder, and hosted one cluster of 15 beneficiaries for the dairy-goat project. The church also offered spiritual support and space at the church compound for people to hold their Friday support-group meetings. The project received technical support from the Urban Harvest programme of the International Potato Center (CIP) and several other organizations (universities, national government, Catholic dioceses of Nakuru, and others). In addition to contributing technical support, the Ministry of Agriculture offered land at no cost for the production of vegetables and fodder.

The diagnostic study: livelihood and nutrition survey

A cross-sectional study design, involving both qualitative and quantitative measurement, was used. A total of 85 male-headed and 70 female-headed households were interviewed. These households were drawn from all the HIV/AIDS-affected households with a child aged between two to five years known to three main HIV/AIDS support organizations working in Nakuru, namely Catholic Diocese of Nakuru (Love and Hope Centre), International Community for the Relief of Suffering and Starvation (ICROSS), and Family Health International (FHI). Livelihoods status was determined by administration of a gender-responsive questionnaire based on the Sustainable Livelihoods Approach framework (Chambers and Conway, 1991), as well as outcome measures of age-specific mortality and child illness. Nutrition status was determined by anthropometric measures (weight, height, mid–upper arm circumference, and triceps skin-fold measures) and an interactive 24-hour dietary recall; food-insecurity status was determined by the household dietary-diversity scale and household food-insecurity access status tools. A potential health-risks assessment was done for the vegetables and dairy-goat interventions, which included identification of potential impacts on human health (hazards and benefits), opportunities for health promotion and disease prevention through policy and practice implementation, and opportunities for monitoring and evaluation. The survey was used to advise the project on potentially good urban agriculture interventions.

The majority of households in the study live below the Kenyan urban poverty level of US$31 per month, and female-headed households had less income than male-headed households (Andersen, 2007). Seventy-three per cent of the households were found to be severely food-insecure (Mbugua and Andersen, 2007). The female heads on average were older (41.69 yrs) than male heads (38.89 yrs). Elderly female heads of households may experience greater-than-average labour problems on their plots, and the resulting food shortages may be made worse if they have to take care of children, most of whom are grandsons and granddaughters.

The baseline survey also revealed that only 35 per cent of the HIV/AIDS-affected households had access to land. Female-headed households had less access to land (23 per cent) than male-headed households (45 per cent) (Andersen, 2007). These findings are in accordance with the results of the study of crop–livestock–waste interaction by the Urban Harvest Programme which also found that land ownership was higher for men than for women-headed households (Karanja et al., forthcoming). This could be explained by Kenyan traditions which dictate that women do not inherit land, a fact which increases their vulnerability to poverty.

The vegetable-production scheme

Eighty households were identified for participation in the vegetable project and then divided into two categories: those with their own farming space and those without. Those without space were clustered into four groups, based on their geographical location, and allocated communally managed plots that were rented or donated by the project partners. Inputs in the form of seed, fertilizer, manure, and tools, together with labour for initial land preparation, were supplied to the four group farms and to the individual farmers working their own plots. Before introduction of the vegetables, the beneficiaries were trained in vegetable production, utilization, and marketing.

In the vegetable-production groups, the work is divided among the members according to a duty roster. Land preparation is done using tractors and hired casual labourers. The group members carry out planting and application of manure, and casual labourers are occasionally hired to assist in weeding, due to periodic illnesses of the members. Use of short-handed hoes may cause backache and members are advised to use long-handed hoes. Pest and disease control is carried out using both traditional methods and pesticides. Pesticides are used only when absolutely necessary; otherwise, Integrated Pest Management practices are emphasized. The Manyani plot, formerly used as a rubbish tip, was strewn with pieces of broken glasses and batteries, which the members are encouraged to collect. The Friday support-group meetings are used as a place for discussing any issues or observations pertaining to potential health risks, and remedial measures are then taken.

Vegetables are harvested by the members, who also decide what should be used for household consumption and what for sale. All the Badili Mawazo

Women working in an irrigated vegetable garden in Nakuru
By Samwel Mbugua

members have access to the vegetables being produced on the four farms, but at variable prices which are based on members' involvement in production activities: active members working daily on the farm were allowed to take home *0.25 kg free; all other members paid 3 KSH per bunch, while non-members paid 5 KSH).* Female members particularly are buying vegetables from the group for resale in kiosks to their neighbours. The groups also sell vegetables to NGOs for preparing lunch when they are conducting training and workshops on HIV/AIDS advocacy, treatment, and prevention. A small part of the income generated from vegetables is put towards the group's savings account for the maintenance of the irrigation equipment.

At the beginning of the vegetable project, some people were reluctant to get involved in farm activities because they had become used to the dependency syndrome created by the handouts that many HIV/AIDS projects give to affected people. HIV/AIDS-affected people also have a tendency to favour initiatives that have immediate benefits, on account of their uncertainty and desperation. The other factor that was observed to affect participation in the farm activities was social and cultural diversity among beneficiaries. These challenges were addressed as described below (see under 'Strategies and tools used to incorporate gender in the project implementation'). However, the number of active members now stands at six men and 44 women.

The dairy scheme

For the dairy-goats rearing scheme four men and 36 women household representatives were selected, based on their degree of commitment shown in the vegetable-production scheme during the first six months. The 40 households were clustered according to their geographical location into two groups of 15 members and two groups of five members. A goat house was constructed in each of the four clusters, and each participating household received one Kenyan-Toggenburg goat, which they would repay in kind by donating the first female kid to another project member. Sweet potato vines (*Ipomea batata*) and napier grass (*Pennisetum cladistenum)* were established as fodder banks in three farms.

All men and women beneficiaries are involved in all husbandry activities, which include feeding, milking, and weighing. Health care, such as immunizations and attendance at births, is provided by a veterinary doctor, whose services are paid for. Feed is transported on bicycles to farms without adequate fodder banks by male members, who are compensated for this service.

Results of the two interventions

The two interventions have increased self-esteem and hope among the participating households through their involvement in social-economic activities. For instance, during one meeting a woman member said: 'I now have something to do on a daily basis and going to the farms, in addition to getting food, has also given me an opportunity to meet and chat with my fellow HIV/AIDS affected people'. Another one commented: 'I am now not hopeless, worthless and rejected but energetic. I am now able to effectively participate in other social groups, feeling good like any other human being.'

Sub-committees, elected by the farm members and including men and women, manage the activities in each of the farms of the four vegetable and dairy-goat groups. Also the involvement of Badili Mawazo members in the workshops with various institutions has contributed to their self-esteem and has given them a chance to contribute to policy development in urban agriculture.

The participation of men and women in the project has helped them to share their knowledge and skills in vegetable production and dairy-goat rearing. Women have a lot of experience in tending to vegetables, including production of traditional African vegetable seeds, while men know more about milking and the reproductive health of the dairy goats. The knowledge and skills of both men and women in the vegetable production and dairy-goat rearing have been complemented through training provided by the project. With the skills they have learned, the farmers are now able to implement environmentally sound practices such as reuse of goat manure as organic fertilizer in the vegetable and seed-production plots. The participating households have also learned the nutritional value of vegetables and dairy-goat milk. Both men and women who are directly involved in the interventions are now able to keep records for monitoring and evaluating the performance of the projects.

The increased supply of vegetables for the household meals and the increased use of goat milk as a substitute by HIV-positive breastfeeding women and have helped to enhance the nutrition and health of the members of HIV/AIDS-affected households.

Strategies applied to incorporate gender into the project

The following strategies were applied to ensure attention to gender issues in the project.

In the diagnosis

The review of available secondary data and discussions with affected community-based organizations and NGOs working with people living with HIV/AIDS showed that women are most vulnerable to the effects of the HIV/AIDS pandemic and most likely to be infected and affected. It also showed that most organizations involved in HIV/AIDS programmes focus on advocacy and care giving, but pay little attention to enhancing access to food and nutrition, although this is critical for mitigating the effects of the HIV/AIDS pandemic. Accordingly, this project sought to empower women in HIV/AIDS-affected households through vegetable production and livestock rearing activities, with the hypothesis that the benefits would trickle down to the whole household.

All data gathered during the diagnostic survey were gender-specified (derived from both male and female heads of households). Also a gender-responsive semi-structured questionnaire was applied which paid special attention to crucial issues like access to land, income, and food security among the male- and female-headed households, which formed the basis for the design of the interventions component of the project.

Male and female enumerators were trained to conduct the interviews in Kiswahili, which is the national language which many communities understand.

In the project design

At the inception of the project in 2006, a stakeholder workshop was held in Nakuru, bringing together 25 women and 16 men from self-help groups, various government ministries, academic partners, NGOs, and Urban Harvest. The workshop defined priority interventions for enhancing income and food and nutrition security among HIV/AIDS-affected households and it established that support for safe infant feeding to prevent mother-to-child transmission had to be strengthened. The workshop also identified gender-responsive strategies for implementing the same objectives.

After the diagnostic study, vegetable and dairy-goat interventions were chosen as the most appropriate interventions focusing on HIV/AIDS-affected households with at least one child aged 2–5 years.

Male and female direct beneficiaries of the vegetable intervention were selected after discussions between the project team and Badili Mawazo community leaders regarding the ways to promote equal benefit sharing and participation by men and women. For instance, members (mainly female heads of household) without land for vegetable production or goat rearing were grouped together and given access to land rented or borrowed on their behalf by the project.

Other concerns addressed during the project-planning stage to ensure gender equity included the following:

- Interventions were identified that would not require intensive labour.
- The combination of a dairy-goat project and a vegetable-production project (with largely the same households) was chosen in order to meet the needs of both men and women.
- A training-needs assessment was carried out among men and women beneficiaries for the two projects, and the results informed the preparation of the training sessions.
- All farming activities were planned to be managed by sub-committees in each group farm, applying gender-responsive rules and regulations.
- Technical support and co-ordination roles in the project were shared among the male and female members of the project team.

During project implementation

To ensure participation of women in the project implementation, all households were given equal access to starter inputs, while a continued supply of vegetable seed was ensured by training the members in seed production.

All activities were carried out by male and female participants, following a flexible duty roster on each farm, constructed to ensure an equitable division of labour between men and women, and managed through a participatory elected sub-committee, including both men and women.

Equitable benefit-sharing strategies based on member participation were developed by the direct beneficiaries, as discussed above. The 36 women and four men raising dairy goats have all been empowered with skills to manage the goats. Although livestock rearing is believed to be a male domain, the women have been keen to own the goats and have managed them with minimal supervision.

The social-cultural diversity among the beneficiaries in both vegetable and dairy-goat interventions affected participation, particularly in provision of labour. For instance, those with an agricultural background took good care of the vegetables, while those from a fishing background were reluctant to do weeding and harvesting. To address these challenges, a Community

Organisational Development and Institutional Strengthening (CODIS) course (Gathuru et al., 2007) was conducted by Urban Harvest in conjunction with Kenya Green Towns Partnership Association. During the course, beneficiaries were taught the importance of working as a group and the need for everybody's participation for the good of the whole group, in addition to individual benefit sharing. The CODIS training course addressed topics such as group formation and development, personality, leadership, conflict resolution, division of labour, networking and advocacy, and project management.

The active participation of men and women was noted to be affected by their dairy calendars and health status. Duty rosters and meeting schedules were developed in a participatory manner and were reviewed from time to time. To cope with the fact that women could not transport animal feed from one farm to another, the project purchased bicycles to be used by the men, who received compensation in recognition of the fact that they had to put aside their own daily chores in order to perform this service for the group.

The diverse communities' differing preferences for traditional or exotic vegetables were addressed in training sessions which emphasized the higher nutritional value of traditional African vegetables.

Handout-dependence syndrome was common among the beneficiaries, and it affected their participation in activities that require a lot of attention, such as goat rearing. Awareness raising and counselling sessions were held, particularly during the Friday meetings and training sessions, to improve beneficiaries' attitudes towards vegetable production and goat keeping as sustainable means of improving the social, natural, and financial capital of men and women, especially people affected by HIV/AIDS.

Whenever necessary, members were allowed to use vernacular language in meetings and training sessions in order to prevent members with low literacy levels being intimidated by other people who were fluent in Kiswahili and/or English.

In monitoring and evaluation

All the monitoring information is disaggregated by gender: for instance, the numbers of men and women trained, the number of beneficiaries by gender per household for both projects, members' involvement in farm activities by gender, time, and date. Records are maintained on the performance of dairy goats belonging to men and women (on feeding, weight gain/loss, and health). The yields of vegetables, milk, and income produced per (male- or female-led) household are recorded.

In scaling up and influencing policy

In August 2007 a workshop was held in Nakuru during which the results of the baseline livelihood and nutrition survey were discussed with policy makers from the Ministry of Agriculture and the Municipal Council. Urban agriculture was presented as a productive sector that needs to be planned for and supported with extension services. In September 2007 a second workshop was held, involving Badili Mawazo members and stakeholders from NGOs, government departments, researchers, trainers, policy makers, and representatives of the private sector to discuss the potential benefits of the two projects and provide an opportunity for Badili Mawazo to share their agricultural policy-related concerns. In this way the project has influenced other support groups to consider urban agriculture as a viable initiative for people living with HIV/AIDS and to give special attention to female-headed households. As a result, among others, UNGA (Ltd.) Kenya Company has provided the Badili Mawazo members with layer chicks.

Through testimonies by male and female project members regarding their HIV/AIDS status, more people in the Presbyterian Church of East Africa have declared their HIV/AIDS status and consequently can be assisted by the church and the project.

Conclusions

Lessons learned about gender mainstreaming in urban agriculture

- Acceptability, viability, and impacts of agricultural interventions among men and women depend on the beneficiary's involvement in their identification and prioritization.
- The social and cultural background of the potential beneficiaries of urban agricultural projects should be well understood, because it affects their participation.
- Gender needs to be addressed in agricultural projects throughout the project cycle.
- Gender issues arising during project implementation need to be addressed even if not planned for, otherwise the intended outcomes and impact will not be realized. Some research organizations may shy away from addressing gender issues, viewing it as a task for NGOs than for research organizations. However, it is more efficient for these issues to be addressed directly by the research organization, rather than seeking to involve an NGO partner to do so.
- Various gender issues emerging during implementation can be addressed through capacity building: for instance, through courses in community organizational development and institutional strengthening (CODIS).
- Access to land, although crucial to the success of any urban agriculture intervention and especially for poor female-headed households, is a big challenge in urban areas. There is a need for timely identification of project partners / institutions who could play a role in securing access to land; and a need to allocate money and time for the identification of available vacant land.

- Agricultural interventions with disadvantaged groups such as people living with HIV/AIDS require broad partnerships with agencies including faith-based organizations that can offer them spiritual support.
- Equal benefit sharing, division of labour, and participation in decision making in urban farming groups is best addressed through rules and regulations set by the beneficiaries themselves, taking into consideration their challenges and opportunities.
- Previous studies, for instance in Kisumu by Ishani (2002) and by Urban Harvest in Nakuru (Gathuru et al., 2005; Karanja et al., forthcoming) have indicated that decisions about livestock are normally made by men. Hence there is a need to ensure that women's right to own dairy goats is ensured, particularly in households where the direct beneficiary is female. Agreements and/or contracts could be developed between the Badili Mawazo group and the beneficiaries in order to protect the right of women to own the dairy goats, as well as protecting the household members from losing the animals in case of death of the direct beneficiaries. The agreements and/or contracts will also ensure appropriate use of the dairy goats and avoid the chances of their being used for unintended purposes such as slaughtering for meat and as security for loans.
- Research and development organizations targeting disadvantaged groups such as those affected by HIV/AIDS should adopt sustainable approaches, such as household empowerment through skills and training and provision of opportunities other than food donations which create a dependency syndrome while promoting a feeling of helplessness. Agricultural projects provide HIV/AIDS-infected people with an opportunity to grow their own food and generate an income, and also give them a chance to fulfil their strategic needs of socializing and supporting each other.

Gender mainstreaming in urban vs rural agriculture

- There is more social and cultural diversity among urban farmers than rural farmers, which increases the challenge of introducing gender equity into project management. However, the social and cultural beliefs are weaker in urban areas than in rural areas, and discussions of gender issues such as access to income and land are more open in urban areas.
- Urban producers have more opportunities for interaction and networking and thus better changes of gender mainstreaming, due to the availability of better infrastructure, especially communication facilities, and a greater presence of institutions.
- As in rural agriculture, women face greater challenges than men when aspiring to access and own land and livestock. In rural areas the traditional land tenure and inheritance system gives women no chance to inherit land, while in urban areas the competition for open spaces is fierce, and

women are disadvantaged in this 'survival of the fittest' scenario in which masculinity prevails. The insecurity of the roadsides and riverbeds where urban agriculture is taking place is another disadvantage for women.
- Urban agriculture is more commercialized and hence provides better opportunities for the empowerment of women.

References

Andersen, N. (2007) 'Assessment of Sustainable Livelihoods, Foods Security and Illness in HIV/AIDS Affected Households in Nakuru, Kenya', Research Paper, University of Toronto, Canada.

Chambers, R. and Conway, G.R. (1991) *Sustainable Rural Livelihoods: Practical Concepts for the 21st Century*, Institute of Development Studies Discussion Paper, IDS, Brighton.

Foeken, D. and Owuor, S. (2000) *Urban Farmers in Nakuru, Kenya*, ASC Working Paper No. 45, Africa Studies Centre, Leiden, and Centre for Urban Research, University of Nairobi, Kenya.

Gathuru, K., Njenga, M., Karanja, N., and Githe, E. (2005) *Community-Based Research and Development Centre Urban Agriculture and Waste Management Nakuru*, supported by International Development Research Centre (IDRC). Project Report, IDRC, Canada.

Gathuru, K., Njenga, M., and Karanja, N. (2007) 'Community Based Organisational Development and Institutional Strengthening', draft training manual.

ILO (2003) *Socio-economic Impact of HIV/AIDS on People Living with HIV/AIDS and their Families*, New Delhi, Delhi Network of Positive People, Manipur Network of People Living with HIV/AIDS, Network of Maharashtra by People Living with HIV/AIDS, Positive Women's Network of Southern India, International Labour Organisation, New Delhi.

Ishani, Z., Gathuru, K., and Lamba, D. (2002) 'Scoping Study on Interactions between Gender Relations and Livestock Keeping in Kisumu', unpublished report prepared for Natural Resources International Ltd.

Karanja, N., Njenga, M., Gathuru, K., and Karanja, A. (forthcoming) 'Crop–Livestock–Waste Interactions in Nakuru: A Scoping Study', in Prain, G., Karanja, N. and Lee-Smith, D. (eds.) *Urban and Peri-Urban Agriculture in Sub-Saharan Africa – 20 Years On: Case Studies and Perspectives*, Chapter 10.

Mbugua, S. and Andersen, N. (2007) 'Sustainable Environments and Health Through Urban Agriculture (SEHTUA)', Nakuru Project, feedback workshop flyer.

MCN (1999) *Strategic Nakuru Structure Plan. Action Plan for Sustainable Urban Development of Nakuru Town and its Environs*, Municipal Council of Nakuru, Nakuru, Kenya.

Ministry of Planning and National Development (MPND) (2003) *Economic Survey*, MPND, Kenya.

UNAIDS (2006) Report on the Global AIDS Epidemic, UNAIDS, Geneva.

About the authors

Kuria Gathuru is National Coordinator for the Kenya Green Towns Partnership Association, Nairobi, Kenya.

Mary Njenga is Research Officer for the Urban Harvest, International Potato Center (CIP), Nairobi, Kenya.

Nancy Karanja is Regional Coordinator SSA at the CIP-Urban Harvest, Nairobi, Kenya.

Patrick Munyao is Research Assistant at the CIP-Urban Harvest, Nairobi, Kenya.

CHAPTER 12

Gender dynamics of fruit and vegetable production and processing in peri-urban Magdalena, Sonora, Mexico

Stephanie Buechler

Abstract

The gender dynamics of fruit and vegetable production and processing in San Ignacio, a peri-urban area of Magdalena, Mexico, are the focus of this applied research project. The cities of Magdalena and Nogales, in close proximity to the community studied, influenced the volume of water available for agriculture. Water scarcity and climate change are negatively affecting fruit production. The reduction in fruit production is in turn affecting the small fruit-processing businesses run by women and men. Some of these women and men have their own fruit orchards and vegetable plots, some purchase the fruit and vegetables they process, and some both produce and purchase the fruit and vegetables. The involvement of women and men in processing is influenced by gender. Women are more involved than men in canning fruit, making sauces, and pickling vegetables. They sell these items to larger processing businesses within their community and/or to vendors in cities located nearby, or to the state capital. Women also use these products to maintain important social networks. Men dominate quince-jelly production and sale. The major production constraint reported by women was the rising costs of inputs such as sugar and glass jars, whereas men tended to emphasize the high cost of labour and lack of capital to produce sufficient quantities of the jelly. Both women and men expressed their concern about increasing water scarcity and rising temperatures. Women were especially interested in learning new ways of processing the fruits and vegetables in order to be able to offer more unique products. Recommendations are made to improve the natural-resource base in this community and the production and processing of these fruits and vegetables, particularly for those activities performed by women. Women are more vulnerable than men to water scarcity and climate-change effects on these agricultural processing businesses, owing to gender inequities that include prevailing lower wages for women in alternative employment.

Introduction

This case study of the gendered production and processing of fruit and vegetables in San Ignacio, a peri-urban community of the city of Magdalena in Mexico, is based on a research project which examines the effect of water scarcity and climate change on gendered agriculture-based livelihood strategies. Fruits and vegetables are preserved by several different methods in San Ignacio, and men and women are involved in different phases of the processing. With some products, women direct and dominate production; with other products, men dominate production. Women and men also use the products for different purposes. The research was conducted by Dr Stephanie Buechler of the Bureau of Applied Research in Anthropology at the University of Arizona. The study was funded by the Resource Centre on Urban Agriculture and Food Security (RUAF) Foundation and a Magellan Circle Fellowship Award from the University of Arizona. The project was carried out between October 2007 and April 2008.

San Ignacio – the study area

The peri-urban community of San Ignacio, with a population of 720 people (INEGI, 2006b), is located about six kilometres from the city of Magdalena (which had a population of approximately 30,000 people in 2005) and 30 km from the city of Santa Ana (with a population of 14,538 in 2005).

The study area is located in the northern state of Sonora, approximately 75 km to the south of the 350-mile border that this state shares with the state of Arizona in the United States. The entire stretch of the US–Mexico border area is gaining population, and some have projected that this area will become the largest urbanized area in North America or even in the world (Weaver, 2001: 110).

The close proximity to the border has affected agriculture-based livelihoods in this area in a variety of ways. Most agriculture-dependent households in the study site have at least one immediate or extended family member residing in the US. In 2007, Mexicans working in the US sent back an estimated US$24 billion to their families (Malkin, 2008). In the first quarter of 2008, however, remittances from Mexican migrants in the US to Mexico dropped 2.9 per cent from the first quarter of 2007, representing the first major decline since the Mexican Central Bank began tracking the transfers in 1995. This decline is mainly attributed to fear of job loss on the part of legal and illegal migrants, to fewer jobs for migrants, and to lower real incomes as a result of inflation (Preston, 2008). Due to stricter border controls, it is now less common than it was a year ago to have family members work in the US during the week and return to their community at the weekend. Instead, family members must engage in permanent migration to the US or seek employment in other Mexican cities. In addition to obtaining remittance income from migrant relatives, direct economic benefits from the proximity of the US are derived from the sale of crops to truckers, who take the produce directly to the US. The fruit and vegetable preserves are occasionally sold to vendors on the border, who sell them in the US; most of these vendors sell the goods in Nogales, Arizona.

The community of San Ignacio is situated approximately 780 m above sea level (INEGI, 2006a) in the Sonora (semi-) desert, which has an average rainfall of approximately 330 mm per year. Temperatures in recent years have been above normal and erratic, with unpredictable, heavy rainfall which makes the area vulnerable to erosion (Vásquez-León and Bracamonte, 2005).

Increasing competition for water from urban centres, industry, and agriculture is putting pressure on water resources in the area (Magaña and Conde, 2000). Water from the Los Alisos basin/River Magdalena is piped to supply the nearby (75 km) twin cities of Nogales (on both sides of the Mexican–US border) with a combined population of 300,000 (INEGI, 2006a and US Census Bureau, 2007). The domestic water supply of these cities reduces the availability of groundwater and surface water for agriculture. The cropped area in the state diminished by 40 per cent from 1996 to 2004 (Bracamonte et al., 2007: 54). Water for irrigation in the peri-urban community in this case study comes from springs channelled into irrigation canals, from wells, and from municipal domestic supply (groundwater).

Dynamic agricultural production in San Ignacio

Agricultural production has experienced vast change in the area. Until approximately 55 years ago cotton was produced, but cotton blight caused farmers to switch to producing mainly wheat. A widespread pest attack approximately 40 years ago caused the closure of the surrounding wheat-grinding mills and another shift to vegetable production. Gradually, fruit trees were planted in response to fluctuating vegetable prices. Orchards are now common in many of the towns surrounding the city of Magdalena. However, the cropped area is becoming smaller as a result of internal and international migration, water scarcity, climate change, and off-farm employment in Magdalena and other surrounding cities. Today, fruits, olives, vegetables, flowers (for cut flowers), alfalfa, and grains are cultivated. Flowers and trees are also grown in small nurseries and sold in pots. Ranching is another agricultural livelihood activity in this community. The varieties of fruit that are grown are changing. At present, mainly quince, peaches, persimmons, pears, and citrus fruits such as oranges, pomelo, grapefruit, lemons, and various types of orange are produced. Plum and apricot production once predominated; however, these fruits have almost completely disappeared. Fig and olive trees have also become much rarer in the area. The producers in the area attributed this change in cropping pattern to water scarcity, particularly evident in the last seven years, to warmer average temperatures, and to nematode infections of the tree roots. Olive trees were infected with a white fly. Community members with tree nurseries noted that an increased number of small trees are dying because of higher temperatures. Other factors involved in the change in cropping patterns include the abandoning of farming for other occupations in urban areas in Mexico or in the US. This has a negative impact on labour availability for farming.

Fruits are canned and also made into jams or preserves. Quince jelly, a kind of sweet that is sliced before being consumed, is also produced. A wide variety of vegetables are preserved by pickling and canning methods. Chiles, *nopales* (edible cactus), green and red onions, radishes, cilantro, cabbage, lettuce, spinach, and *quelite blanco*, a vegetable with small waxy leaves and stems, are the principal vegetables produced. Chiles, nopales, and onions are the main vegetables that are pickled and canned.

Fruits and vegetables as well as canned fruit, jams, and jelly are sold to a wide variety of buyers. The main marketing channels for vegetables include buyers who come with their trucks to the fields; municipal markets in Hermosillo, the capital city of Sonora state, with a population in 2005 of 707,838 people (INEGI, 2006a), located about 200 km from the community; and the municipal market in the city of Nogales on the border with the US. For fruits and fruit products, the main marketing outlets are local vendors with stands next to the highway in and near Magdalena and Santa Ana (30 km away), who often come to the homes of canned-goods producers; individuals who come to the producers' homes or fields (such as religious pilgrims walking to area mission churches in the month of October); community members; urban grocery chain stores; smaller urban grocery stores; and urban municipal markets in

Woman producing canned fruits
By Stephanie Buechler

Hermosillo and Nogales. Bakeries in Hermosillo also purchase the quince jam for a particular type of pastry named *coyotes*. In the case of some contracts, for example between smaller grocery stores and quince-jelly producers, sugar is exchanged for quince jelly, so the producers do not need to have large sums of cash for this important ingredient.

Study methodologies

The methodologies used in this study included field observations, participant observation, semi-structured, in-depth qualitative interviews, and informal conversations with female and male adults living in the peri-urban community studied. Observations were also made in the nearby cities where the produce and processed products were sold. For example, visits were made to supermarkets, smaller grocery stores, and highway fruit stands in Magdalena, to learn about the effect of the city on the marketing of fruit and vegetable preserves. A snowball sampling method was used to select most interviewees for the study. The author also selected interviewees via field observations of women and men working in orchards. Sixteen women and ten men were interviewed in depth, and follow-up interviews took place with many of them. Each interviewee sold fruits and/or vegetables and fruit and vegetable preserves; most grew part or all of the produce that they preserved. Statistics on the state's agriculture sector and water resources were also collected.

Semi-structured interviews were framed as conversations: the interviewer included common points of reference, including experiences of producing canned fruit, pickled vegetables, jam, and olives, and experiences of mothering and the juggling of family and work. Women were interviewed alone, without the presence of men. This helped to foster an atmosphere in which women felt freer to speak without being interrupted or overshadowed by male household members. In the interviews with women, questions were asked to probe the way in which women's work (including agricultural work) interacts with the work of other household members. Women and men were asked separately about gender divisions of labour and any flexibility in these divisions.

The interviews and follow-up interviews were multi-seasonal and multi-local, in order to gain a better sense of the changes in agriculture-based livelihoods according to the month of the year. Interviews were also conducted at different times of day. Women's 'triple day' became very apparent as a result. In the evening women were still cooking dinner, washing dishes, taking care of children, and preparing for the next day. Interviews took place at a range of locations, such as the home and the fields. This gave a much better indication of the normally wide gamut of agricultural and other activities such as food processing in which the interviewee engages. It also provided insight into the activities of other household members and an opportunity for future interviews with those family members. Transect walks along waterways and participatory mapping of water sources were other methodologies used for the study.

Gender analysis of the local urban agriculture situation

Access to labour

Women and men are employed in both agricultural and non-agricultural activities. Due to the lack of public transportation to the surrounding areas, women's ability to obtain employment in the city is limited. This is particularly the case if they have small children, which means that they cannot come and go with ease. The landless and women and men with small landholdings work in a variety of on- and off-farm employment in the fruit orchards, vegetable fields, and cattle ranches, and in the production of canned goods, quince jelly, and cheese. They also work as managers and workers in nearby foreign-owned greenhouses which produce tomatoes and cucumbers. Men's non-farm employment includes work in copper mines and other types of mine, usually entailing a two-hour commute each way every day in a company van to a mining area within Sonora. Women and men find work also in construction jobs and in grocery stores, auto-repair shops, and other local businesses in Magdalena. Some young women work in *maquiladoras* (assembly plants) in Magdalena which provide transportation to and from the plants, but one was rumoured to be closing soon. Others make the long daily commute to Nogales; however, they must have access to private transport to Magdalena in order to catch a public bus from Magdalena to Nogales. Many who go to Nogales are women who work as housekeepers and nannies. Many households have members who are living and working in the US, but remittance levels are being threatened by a poor economy over the border and stricter immigration enforcement. Women in Mexico in general earn less than men for comparable work and they advance much less quickly, thus contributing to Mexico's low ranking (109 out of 128 countries surveyed) with respect to gender equality in economic participation and opportunity (Hausmann et al., 2007: 4; 9). This helps to explain why, even if women or their daughters in San Ignacio are employed in the formal sector, additional sources of income and food are usually necessary to sustain their households.

Access to land and control over resources

Agricultural land in the community is mostly in the hands of men. This is mainly due to a patriarchal and patrilineal system of land ownership whereby land is passed down between generations from father to sons. The plots of land that are farmed (many are left unfarmed due to absentee landowners) average 3 ha in size; fruit orchards vary greatly in size, ranging from 200 m² to 8 ha. Some of the fruit orchards are located in the back patio or *solar* near the house. Others are located at a distance from the house in areas with agricultural fields. Vegetable plots are frequently rented in a share-cropping arrangement in which the landowner invests in the groundwater-pumping costs and the plants, while the share-cropper pays for a tractor and driver to prepare the land, and for the labour to plant the crops. The share-cropper and landowner invest equal sums in the costs of pesticide, fertilizer, and harvesting (by field labourers). After expenses are deducted, half of the profits go to the landowner and half goes to the renter. Tree and flower nurseries are frequently located in the back patio.

Canned-fruit producers, as well as jelly producers who do not grow sufficient quantities or who do not have any orchards, purchase the fruit from growers in San Ignacio or from neighbouring communities. Arrangements vary; either the fruit is purchased, or the canned-fruit or jelly producer goes to the orchard with his/her own labourers and picks the fruit. In the latter case, the fruit is then purchased at a discounted rate from the owner of the orchard. Men and women producers of canned fruit and jelly often put their own as well as family labour into picking the fruit. If family members are hired as labourers, they are often paid by the bucket or bag of fruit. Men and women are paid the same rate for each bucket or bag. One woman proudly claimed that she was able to pick more than most men; she said that she picked rapidly because she had 'six hungry children to feed' with the income that she earned from canning and jelly production.

Gender division of tasks in processing and marketing of fruits and vegetables

Quince jelly and jam production

Men dominate the production of quince jelly (*cajeta de membrillo*) and quince jam (*jalea de membrillo*), and women tend to control canned-goods production (see Table 12.1). Although women and men initially reported a strict division of labour in the production of canned fruit and vegetable and quince jelly, in practice there is substantial sharing of tasks. In fact, one woman explained that the main reason why she engaged in canned-fruit production and jelly production with her husband and children was so that they could all work together. Even when sons and daughters graduated and then continued their studies in another city, or lived and worked in another city, or commuted from the community to the city to work, they often helped their parents and other family members in the production of these goods during peak times in the year. It was very evident that this type of production fostered a sense of family unity.

In quince jelly and jam production, women and men wash the quince and cut it. Children and elderly family members also often share the tasks of washing and cutting the quince. The tasks most consistently divided by gender in quince jelly production are the stirring of the hot quince jelly in the huge open copper pots over the wood-burning fire or stove, which is considered to be heavy work, and the cutting of the cooled jelly into blocks that are later individually wrapped. Men mainly perform those tasks. However, workers are hired for a daily rate to help with the heavy work of stirring. Most of the workers hired for this particular part of the production process are men. Women and men wrap the blocks of quince jelly in plastic and place them in boxes for sale. Mainly female workers are hired to do the wrapping and packing. In the case of one quince jelly producer family, the son-in-law pours the jelly into moulds to create quince jelly shaped into designs for sale to customers who buy this product for special occasions.

The quince jam is produced by boiling the quince seeds that remain after the fruit is used for the quince jelly. These seeds are ground, then boiled with sugar. Men and women share in most of these tasks except for the grinding, which the men tend to do. Hired labourers help at different stages of the production process, again following gendered divisions of labour, so that only men, for example, are hired to do the grinding. Whenever possible, relatives such as nephews or nieces are hired. Household members are not paid, however, for their work, and it is frequently daughters-in-law who help in quince jelly and jam production.

Canned-goods production

Women control the production of canned goods (see Table 12.1). They boil and stir the fruit and vegetables for the canned fruit and pickled vegetables. Women and their sons and daughters and/or their daughters-in-law wash and generally prepare the fruit. Unlike the case of male-dominated quince jelly and jam production, very little or no labour is hired. Women mainly perform the tasks related to the boiling process, which include adding the sugar and stirring it with the fruit, as well as continuously skimming the froth off the top and discarding it. They cook the chillis that go into the pickled vegetables, wash and cut the vegetables, and add the vinegar. They perform these tasks in their kitchens on the same stove that they use for cooking their family's food. They are also the primary vendors of these products.

Marketing

Women display greater knowledge of customer preferences in canned goods. For example, women rotate the peaches and quince in the jar so that the part where the stem originates faces inward. They undertake this labour-intensive activity because customers have told them that this makes the fruit look more appealing in the jar. They also painstakingly skim the froth and impurities from the top as they boil the fruit, for visual appeal.

The locus of women's marketing is more circumscribed than men's; women mainly use their community and family networks to sell their products, whereas men tend to have more far-flung networks, garnered from their participation in employment in Magdalena as well as larger cities in Sonora. It is also easier for men than for women to leave their homes to travel to cities that are farther away, and they are thus more able to take their produce and processed products to markets and businesses in these more distant areas. Women are responsible for performing a far greater number of household tasks than men, including child care, cooking, cleaning, and washing clothes, which

Table 12.1 Gender division of labour in the peri-urban study area

Socio-economic activity	Females (♀)			Male (♂)			Locus
	Child	Adult	Elder	Child	Adult	Elder	
1. Production of goods and services							
a) Fruit							
Irrigation		+			+++	+	Within the field
Harvesting	+	++		+	+++	+	Within the field
Transport	+	++		+	+++		Field to transport vehicle
Sale in the market (vendors different from farmers/producers)		+			+++		Municipal market; roadside stand; to businesses
Sale from home		+++	+		+++	+	Within home
b) Vegetables							
Irrigation					+++		Market to home
Harvesting		++			+++		Within the field
Transport		++			+++		Field to transport vehicle
Sale from the field		+			+++		Within the field
Sale in the market		+			+++		Municipal market; roadside stand; to businesses
c) Canned fruit							
Fruit preparation	+	+++	++	+	+++	++	Within home
Boiling of fruit with sugar		+++	+		+		In kitchen
Skimming off impurities		+++					In kitchen
Sterilizing jars		+++	++		+		In kitchen
Placing fruit and syrup in jars		+++	+		+		In kitchen
Sale in market		+			+++		Municipal market; roadside stand; to businesses
Sale from home	++	+++	++		+		Within home
d) Jam		++			+++		
Fruit preparation (and grinding seeds in case of quince jam)	+	+++	++	+	+++	++	Within home
Boiling of fruit or seeds		+	+		+++		In kitchen
Boiling of fruit or seeds with sugar		+++			++		In kitchen
Sterilizing jars		+++	++		+		In kitchen
Placing jam in jars		+++	+		+		In kitchen
Marketing of jam		+			+++		In municipal markets; bakeries; roadside stands; to businesses
Sale from home	++	+++	++		+		From home
e) Quince Jelly							
Fruit preparation	++	+++	+++	+	+++	+++	Within home or garden
Stirring of thick fruit and sugar mixture in large copper cauldron		+			+++		Outside in garden or in separate shed
Pouring of jelly into pans					+++		In home kitchen
Cutting jelly into rectangular bars		+			+++		In home kitchen
Wrapping jelly in plastic		+++	++		+		In home kitchen
Marketing of jelly (producers sell to vendors; they don't sell it themselves)		+			+++		In municipal markets; roadside stands; to businesses
Marketing of jelly		+++	+				From home

Socio-economic activity	Females (♀)			Male (♂)			Locus
	Child	Adult	Elder	Child	Adult	Elder	
f) Pickled vegetables or sauces							
Vegetable preparation	+	+++	++				In home kitchen
Cooking vegetables		+++	++				In home kitchen
Sterilizing jars		++			++		In home kitchen
Placing vegetables in vinegar	+	+++	++				In home kitchen
Marketing of pickled vegetables		++			+++		In municipal market; roadside stand; to businesses
Marketing of pickled vegetables	+	+++	++				From home
2. Social reproduction and maintenance of human resources							
Child care	++	+++	++	+			Within home
Care of sick children	+	+++					Taking children to hospital
Care of the elderly		+++		+			Within home
Care of the elderly or hospital		+++		+			Taking the elderly to clinic
Household management (cooking, cleaning, washing)	++	+++	+				Within home
Collecting water during dry summer months	++	+++	+	+	++		To springs and other well
3. Community management							
Participation in local institutions/ associations							
Ejido meetings *		+	++		+++	++	In community hall
Community meetings of *comuneros* (those who do not have *ejido* rights)		+	++		+++	++	In community hall
Women's church group	+	+++	+++				In community hall or church

* The **ejido** system is a process whereby the government promotes the use of communal land shared by the people of the community.
A child is a girl or boy below the age of 16 years. An elderly person is a person 60 years old and above.
+++ indicates that frequency is high
++ indicates that frequency is medium
+ indicates that frequency is low

constrains their ability to take the time to travel to distant markets. However, men take products produced by women to market, which aids women who have older sons, husbands, or other adult male household members.

Social dimension of canned-fruit production

Canned-fruit production supports household food security, gift exchanges, and income generation. Women in particular are responsible for household food security. Fruit is retained for snacks and desserts, and fruit juices are made from the fruit for family consumption. Canned fruits are retained for special family occasions and holidays. They are also given as gifts to maintain reciprocal relationships, particularly in kin and fictive kin networks, or are presented to important social figures such as the community priest. For example, one woman explained: 'I cannot give birthday gifts because of the cost, so I bring along a can of my peaches when I attend a birthday party'. Another woman, who makes canned fruit with the help of several extended-family members and whose husband produces quince jelly, explained her methodical approach to gift giving and to retaining food for home consumption:

> *I keep two boxes of canned peach jars and four boxes of quince jelly bars to give as gifts. I also keep one box of canned peaches and a few pieces of quince jelly, but these are for the house [her household members]. I keep some fruit to make fruit juice for my family and also to eat as snacks* (Field interviews, November–April 2008, author's translation).

These exchanges are important to women, because they depend on mutual aid arrangements in the form of reciprocal gifts and labour inputs in agriculture and small-scale enterprises. Women and men invest in their fruit and vegetable processing businesses and also use the income from these products for household expenses. The woman whose words are quoted above produced a total of 70 boxes or 840 jars of canned fruit in 2007 (20 boxes were produced with fruit from her own orchard, and the remaining 50 boxes were produced with fruit that she purchased). She and her husband produced about 450 boxes of quince jelly, or about 23,400 bars of the jelly in 2007. Women interviewees consistently maintained that one of the principal ways in which they invested their money was to pay for their children's education. As one woman explained: 'Men work in the orchards, but women kill themselves doing these things [canning fruits and vegetables and helping with quince jelly production] so that their children can get an education' (field interviews by author, November–April 2008, author's translation).

Constraints on production

Fruit producers mentioned numerous constraints on production. These problems include the length of time that elapses between tree planting and fruit production. Trees must also be replanted after a certain number of years. Quince trees, for example, take three years to begin to bear fruit and must be replanted every 15–20 years. There are also fewer orchards, a fact which producers attribute to owners taking land out of orchard production due to water scarcity and higher temperatures, the sale of orchards for ranching activities, or the abandonment of the land by migrant families.

Another major problem is increasing water scarcity. This causes many related problems, such as greater susceptibility of plants and trees to pests, lower production, decreased life-span of the trees, and, frequently, the need to reduce the cropped area. The response of one woman when asked why she did not replant the old peach and quince trees that had to be cut down in her orchard was: *'ya no hay agua'*, 'There is no water any more'. The canal that helped to irrigate the community for more than one hundred years no longer has water, and the well owned by her neighbour, on which she and her family depend, has less and less water. Women depend to a much greater degree than men on fruit produced in their home gardens. Women irrigate and generally care for these orchards. The orchards near their homes enable them to combine farming, child care, and household tasks more easily. These are the orchards where water scarcity is currently most severe; most are not located within easy access of the springs used to irrigate the fields farther from the homes. Instead, these home orchards depend primarily on municipal water (obtained from a community well), which is insufficient even for household use in the dry season. As the locus of production moves farther from the home, it is likely that women will have less control over access to produce for their canned goods and for household consumption. Costs of production are also likely to increase, as more of the produce for the preserves is purchased.

Women and men who canned fruits and vegetables mentioned the obstacles that they faced in their efforts to achieve a decent and steady income from the sale of their products. A major problem cited by the women is the lack of affordable jars in Mexico. A woman in the community sells jars purchased by her son, who lives in the US. These are considered to be expensive. Therefore, when possible the female producers often travel to the US to purchase the jars in large quantities in chain stores such as Walmart. However, border customs regulations restrict the number of boxes of jars that can be taken across the border into Mexico without duties. Women reported that they also made use of social networks to obtain glass jars (either used or new jars), usually from relatives in the US. Men producers tended to emphasize the problem of the high cost of labour and the significant amounts of capital necessary to produce quince jelly in large enough quantities to earn sufficient profit. These men noted that it was not normally possible to borrow money, due to the lack of available sources of credit in the community. A bank in Magdalena extends credit, but male fruit producers and processors complained that 'one pays double because of the high interest payments'. Some of the male farmers with fields and orchards irrigated by the springs operated a rotating credit fund in order to obtain more capital to purchase inputs, but this fund was temporarily suspended in 2008 by the government when a few of the farmers (reportedly the better-off farmers) did not repay their loans. Women operated with less capital than men and did not attempt to obtain credit. Both men and women stated that there was a need for a municipal market in Magdalena where they could sell their fruits, vegetables, and derivative products. Producers must instead go to Hermosillo, approximately three hours' journey away, to sell their crops to vendors in the municipal market. As mentioned, men are more able than women to make this trip, for reasons related to women's greater burden of household responsibilities as well as to greater restrictions on women's movements away from the watchful eyes of their community members.

Both women and men also mentioned other barriers. These included costs of production, competition with other producers, and difficulties in obtaining legal permits required for labels. These barriers included the rising costs of sugar, gas, and wood for cooking, and rising labour costs. Labour costs are significantly higher on the border than elsewhere in Mexico. Female and male producers mentioned that there was a lot of competition for their products from other producers in the community. Another limitation is the lack of knowledge, the cost, and the bureaucracy involved in obtaining a government permit from the local office of the Ministry of Finance in order to be able to label their products and sell them directly to supermarkets. Most must sell their products therefore to intermediaries who are able to sell the products with a label to large supermarkets. One larger quince jelly producer in the community grows his own quince and purchases some quince fruit to make jelly. He also purchases additional quince jelly, which he markets under his label. However, there were no women who had this label. As one woman stated, even though her daughter works in Magdalena in *Hacienda* (the Ministry of Finance, which administers labels) she does not know what obtaining a label for her canned products and her quince jelly would entail in terms of bureaucratic procedures, paperwork, and cost.

Another constraint for producers of both jelly and canned goods is the limited demand for these products. This is caused by the generally low purchasing power of most customers – a state of affairs that is becoming worse with the recession in the US, which affects the level of migrant remittances sent from the US to Mexico. In a recession, non-essential food items such as the ones produced are either eliminated from the list of goods purchased by households or are bought for special occasions only. Restrictions on border crossings and long queues at border crossings have also limited the number of customers. Relatives, friends, and tourists used to come more frequently to the community and buy the products. In addition, canned products and quince jelly used to be sold to buyers who would come to the border and purchase the goods to sell them in Nogales and other cities in Arizona.

Recommendations based on study findings

The position of female producers in San Ignacio could be strengthened by means of a variety of programmes and projects. The general and increasingly common problem of water scarcity could be addressed through a project similar to a current initiative in Mexico. A project in eight villages in the Mixteca region of Oaxaca, with a farmer organization called the Centre for Integral Small Farmer Development in the Mixteca (CEDICAM), provides a guide to what could be accomplished in the study area in Sonora. Farmers are planting native, drought-tolerant trees (raised in local nurseries) to help to prevent erosion, improve water filtration into the ground, provide carbon capture and green areas, add organic material to enrich the soil, and provide more sustainable, cleaner-burning wood for wood-burning stoves. CEDICAM is also working with farmers to construct contour ditches, retention walls, and terraces to capture rainfall to recharge groundwater and help to revive springs as well as to contribute to erosion control on the surrounding hillsides. Local production and use of organic fertilizers is also being undertaken; crop rotation is encouraged, as is the local selection of seeds. Women are taking part in the construction of these structures and may start reaping some of the benefits (Reider, 2006: 56). In San Ignacio, such measures would help to retain water from rainfall to recharge aquifers and might also increase flow in the springs. Such measures may also help to control erosion, which has been documented in other areas in Sonora due to heavy rainfall (Vásquez-León and Bracamonte, 2005) and is also a problem in San Ignacio. Planting trees on hillsides would also produce wood for cleaner fuel for the wood-burning stoves that are used to produce the quince jelly. This would make production cheaper in terms of labour time in collecting and/or purchasing wood. The use of organic fertilizers instead of chemical fertilizers, the production of less water-intensive crops, and a move away from monoculture (particularly evident in vegetable production in San Ignacio) would also help to reduce the negative impacts that chemical- and water-intensive agriculture have on climate (Shiva, 2005).

The scarcity of credit in general and the fact that women own little of the community's land which would serve as collateral creates a disadvantageous situation for women who wish to expand/improve their business. Credit programmes would help female producers to gain access to additional capital in order to produce more varied goods, helping to generate steady incomes to reduce vulnerability to environmental, economic, social, and demographic change. An initiative to aid female producers could include the formation of rotating savings and credit associations among groups of women and/or other forms of credit programmes such as government-subsidized credit programmes (Rogaly et al., 2004). Mainly male members of the association of spring-water irrigation users are members of a government-sponsored village banking association (*cajas solidarias*). Female canned-fruit and vegetable producers could become members of the *cajas solidarias*. Training designed for women would include courses in marketing and product design for new products and innovations in existing products. Research projects could be developed, using participatory research methods including research by the women themselves into existing marketing outlets and products on the market competing with their products; the project could also include the training of women by women in the community (Girón Hernández et al., 2004). Training courses could be varied and geared towards the needs of small groups of female producers, in order to reduce competition between producers and expand their customer base.

The establishment of a municipal market in Magdalena would attract customers from Magdalena and surrounding cities, towns, and villages. The fruit, vegetables, and processed products could be sold in the municipal market.

No additional product labelling would be necessary for the sale of these items in the market. The producers could also act as vendors of their products, or have a family member or employee sell the products from a stand in the market. This would minimize middle-man costs, as well as transaction costs such as transportation from San Ignacio to the market, because Magdalena is much closer to San Ignacio than the cities of Hermosillo or Nogales, where products are currently sold (mainly in their municipal markets). The existence of a municipal market would be likely to boost the local economy by providing greater inflows of people into the city of Magdalena. This would help to stimulate local businesses and services that would extend beyond the municipal market.

Conclusions

This study focused on gender differences in the benefits of agriculture-based activities in terms of labour supply, income, social ties, and food security. Labour is garnered from family members within and outside the community. Daughters-in-law constitute the most frequently unpaid family labourers, whereas male relatives tend to be remunerated for their work in quince jelly and jam production. Labour is also obtained by hiring labourers. For those tasks normally considered to be 'male', male labourers are hired; and for those culturally defined as female, women labourers are hired. The study also examined the gendered effects of water scarcity and climate change on these agricultural activities, which are an understudied area of inquiry and action. These have been studied mainly by male researchers, focusing on male actors, leaving out key information about the effects on women and their responses to these impacts. Water scarcity and climate change are threatening the sustainability of these agriculture-based activities. Higher temperatures are already affecting fruit trees and reducing supplies of water for irrigation, leading in turn to increased pest attacks. Orchards near homes have been more severely affected to date than those farther from residential areas, due to competition from domestic water use. Fruit trees are now frequently not replaced when they become too old to produce or are damaged by unfavourable conditions. Vegetable fields are increasingly left barren. Reduced local production limits the production of preserves.

Social networks that are currently fortified by means of extended family members working together to produce these goods, and by means of women's exchange or gifting of the products within social circles in the community, will be weakened. Women's vulnerability in particular is likely to increase in the absence of strong social networks. Women depend more than men on these social networks for a wide variety of needs, such as being able to fulfil their responsibilities towards their children, other household members, and family members living outside the community and even outside Mexico. In return, women gain the multi-dimensional support of these family members, which includes important inputs into the production process such as labour and containers. In an individualized system, women, and to a lesser extent men, would be responsible for earning an income and supporting themselves and household members with that income. However, the interviewees in this study reported that such income, derived from stable, well-remunerated employment, is impossible to obtain in this Mexico–US border area.

Women will also become more vulnerable without fruits and vegetables that they can retain for household food consumption. Canned products can currently be retained for times of the year when there is no harvest, thus increasing household food security as well as income flows. Women in the community of San Ignacio, like women worldwide, play a critical role in food provision and processing for their families, often achieving household food security despite low household income. It has also been well documented globally that women spend a far greater share of their income on the sustenance of their household members than men do (Elson, 1995; Dwyer and Bruce, 1988; Benería and Roldán, 1987; Eswaran and Kotwal, 2004). Thus, if women earn less from canned fruit and vegetable production and sale because these crops become less readily available in the area, their household members may be adversely affected because women will have less to spend on critical household needs.

In gender analysis of urban agriculture, it is important to ask how the urban and peri-urban context affects employment of different family members, their educational opportunities, social networks, marketing, processing, and product demand (Buechler and Devi, forthcoming, 2008). For this type of analysis, many different family members (of different ages, sexes, and places of residence) were interviewed in multiple locations (urban and peri-urban; home, back yard, and fields) at different times of day and season. This context has a strong influence on opportunities and constraints for women and men in peri-urban agricultural production. In this case study it was clear that the small city nearby had some marketing outlets. However, the volume of demand was limited by the size of the city and the services that it provided, such as markets, necessitating dependence on surrounding small and larger cities. The education of younger generations of women and men and the availability of alternative employment opportunities in the surrounding urban areas is causing a reduction in the numbers of people engaged in farming and in produce processing. However, these family members still depend on mainly female relatives engaged in the processing of fruits and vegetables for food, for additional income to supplement their low and often unstable urban wages, and for social support. The sustainability of this production, however, is at risk from water scarcity and climate change, as well as changes in labour markets and demand. Women and men will be affected differently by these physical, demographic, and social changes.

References

Benería, L. and Roldán, M. (1987) *The Crossroads of Class and Gender: Industrial Homework, Subcontracting, and Household Dynamics in Mexico City*, University of Chicago Press.

Bracamonte, Á. S., Valle Dessens, N., and Méndez Barrón, R. (2007) 'La nueva agricultura Sonorense: historia reciente de un viejo negocio', *Region y Sociedad* XIX, Número Especial, pp. 51–70.

Buechler, S. and G. Devi (forthcoming 2008) 'Gender, water and agriculture in the context of urban growth in central Mexico and South India', special issue on Gender and Water: *Gender, Place and Culture*.

Dwyer, D. and Bruce, J. (eds.) (1988) *A Home Divided – Women and Income in the Third World*, Stanford University Press, California.

Elson, D. (1995) 'Male bias in macro-economics: the case of structural adjustment', in Elson, D. (ed.) *Male Bias in the Development Process*, second edition, Manchester University Press, Manchester, England.

Eswaran, M. and Kotwal, A. (2004) 'A theory of gender differences in parental altruism', *Canadian Journal of Economics* 37(4): 918–950.

Hausmann, R., Tyson, L. D., and Zahidi, S. (2007) *The Global Gender Gap Report 2007*, World Economic Forum, Harvard University, and the University of California, Berkeley. Available from http://www.weforum.org/pdf/gendergap/report2007.pdf.

Hernández Girón, J. de la Paz, Dominguez Hernández, M. L., and Jiménez Castañeda, J. C. (2004) 'Participatory methodologies and the product development process: the experience of Mixtec craftswomen in Mexico', *Development in Practice* 14 (3): 396–406.

Instituto Nacional de Estadísticas, Geografía e Informática (INEGI) (2006a) *Conteo de Población y Vivienda 2005, Resultados Definitivos. Tabulados Básicos.* Available from http://www.inegi.gob.mx/est/contenidos/espanol/sistemas/conteo2005/Default.asp (accessed 20 February 2008).

Instituto Nacional de Estadísticas, Geografía e Informática (INEGI) (2006b) *Principales Resultados por Localidad 2005 (ITER).* Available from http://www.inegi.gob.mx/est/contenidos/espanol/sistemas/conteo2005/localidad/iter/default.asp?s=est&c=10395 (accessed 20 February 2008).

Magaña, V. O. and Conde, C. (2000) 'Climate variability and freshwater resources in northern Mexico. Sonora: a case study', *Environmental Monitoring and Assessment* 61 (1): 167–185.

Malkin, E. (2008) *Mexicans barely increased remittances in 2007*, New York Times (online), 26 February 2008. Available from http://select.nytimes.com/mem/tnt.html?emc=tnt&tntget=2008/02/26/business/worldbusiness/26mexico.html&tntemail0=y.

Preston, J. (2008) *Fewer Latino Immigrants sending money home*. New York Times (online). 1 May 2008. Available from http://select.nytimes.com/mem/tnt.html?emc=tnt&tntget=2008/05/01/us/01immigration.html&tntemail0=y.

Reider, R. (2006) 'Voices of the North and South: finding common ground', in Cohn, A., Cook, J., Fernández, M., Reider, R., and Steward, C. (eds.), *Agroecology and the Struggle for Food Sovereignty in the Americas*, International Institute for Environment and Development (IIED), the IUCN Commission on Environment, Economic and Social Policy (CEES), and the Yale School of Forestry and Environmental Studies, pp. 55–59. Available from http://www.iied.org/pubs/display.php?o=14506IIED.

Rogaly, B., Castillo, A., and Romero Serrano, M. (2004) 'Building assets to reduce vulnerability: microfinance provision by a rural working people's union in Mexico', *Development in Practice* 14 (3): 381–395.

Shiva, V. (2005) *Earth Democracy: Justice, Sustainability and Peace*, South End Press, USA.

US Census Bureau (2007) *Annual estimates of the population of metropolitan and micropolitan statistical areas: April 1, 2000 to July 1, 2007* (online), US Census Bureau, Population Division. Available from http://www.census.gov/popest/metro/CBSA-est2007-annual.html.

Vázquez-León, M. and Bracamonte, A. (2005) 'Indicadores Ambientales para la Agricultura Sustentable : Un Estudio del Noreste de Sonora', report for CONAHEC. El Colegio de Sonora. Sonora, Mexico, and the University of Arizona Bureau of Applied Research in Anthropology, Tucson.

Weaver, T. (2001) 'Time, space and articulation in the economic development of the US–Mexico border region from 1940 to 2000', *Human Organisation* 60 (2): 105–120.

About the author

Stephanie Buechler is Research Associate at the Bureau of Applied Research in Anthropology, University of Arizona, Tucson, USA.

CHAPTER 13

Urban agriculture and gender in Carapongo, Lima, Peru

Blanca Arce, Gordon Prain and Luis Maldonado

Abstract

The emergence in recent decades in Latin America of urban agriculture as an important strategy for both food security and income generation for poor urban households raises a number of questions about the roles of men and women in this phenomenon. This chapter explores these questions through a case study of Carapongo, a neighbourhood in the eastern shantytowns of Lima, Peru. The study used an integrated methodology, involving quantitative and qualitative characterizations of the agricultural production system and producer households through a baseline survey of 125 Carapongo producers from April to October 2004 and through use of participatory workshops and other tools, which were adapted to the urban gender characteristics of the producers of the study area. To facilitate equitable development of urban agriculture in Carapongo, co-operation is recommended between municipal policy makers, civil-society organizations, and the producers. Although a basic function of municipal leaders is to recognize and address the needs of different constituencies in their district, the needs of agricultural producers of either sex have only rarely been addressed in urban municipalities. It is essential that data generated through gender analyses are fed back into municipal decision making to improve the numbers and design of projects that address this sector.

Introduction

The emergence in recent decades in Latin America of urban agriculture as a strategy for both food security and income generation for poor urban households raises a number of questions about the roles of men and women in this phenomenon. Who has been most instrumental in gaining access to land for household food production? Who has invested more labour time in cultivation and animal production? Who determines crop and animal production choices, the sale or consumption of the produce, and the use of income earned? Most importantly, how do different gender roles in agriculture affect the livelihoods of the household as a whole, and the relation of agriculture to other livelihoods strategies? This chapter seeks to answer these questions through a case study of Carapongo, a neighbourhood in the eastern shantytowns of Lima, Peru.

Urban Harvest implemented the research project in collaboration with government and non-government research and development organizations and local municipal institutions, with financial support from the Spanish government, the City of Madrid, and the Province of Madrid through CESAL, a Spanish NGO.

The study area

Lima Metropolitan area has more than 7.5 million inhabitants, one third of Peru's total population. Among developing regions, Latin America has experienced the highest levels of urbanization over recent decades. The origins of this large urban growth are connected to import-substitution policies widely adopted after the Second World War, which led to rapid industrialization in urban centres (Lipman, 1977). A corresponding lack of investment in agriculture in rural areas resulted in high levels of rural–urban migration, as people sought access to the new industrial employment opportunities. This trend combined with relatively high overall population growth to create the large urban populations that we see today.

Not only has there been a major shift of total population from village to city over recent years, there has also been a migration of poverty, as cities have proved unable to provide full employment for newly arrived migrants and the natural growth of the urban population. Recent figures show that, whereas between 2001 and 2004 the level of poverty in Peru went down from 54.3 per cent to 51.6 per cent, in Lima it increased from 31.8 to 36.6 per cent over the same period (INEI, 2005). This may even underestimate the levels of poverty in the capital, since it is notoriously difficult to capture adequately the higher costs of the 'basic family basket' in metropolitan centres (Amis, 2002).

Carapongo, the neighbourhood involved in this case study, is located at about 200 m above sea level in the lower zone of the Rimac watershed. It covers approximately 464 ha, of which 46 per cent is cultivated land mainly under vegetable production, with limited areas devoted to large-scale livestock rearing or forestry (Table 13.1).

Table 13.1 Characteristics of the study area (2006)

Land type	Carapongo
Total area (hectares)	463.8
Population (000)	3,200
Land use (%):	
• Residential/roads	42
• Cultivated land	46
• Livestock	4
• Agro-forestry	2
• Uncultivated land	1
• Commercial/industrial,recreational	4

Source: Castro, 2007

Most of Carapongo's population is literate (95.6 per cent). Just over half of the population are migrants from the Andean region. The average Carapongo family has five members, and three quarters of the population are aged over 35 (76 per cent). Almost half of the male population is married, compared with 18 per cent of women. For both men and women, legal marriage is more common than co-residence (*convivientes*) (men 11 per cent, women 6 per cent).

Agriculture in Carapongo

Lima is located in the coastal desert of Peru, with one of the lowest rainfall regimes on earth. Carapongo receives water for agriculture mainly from the Rimac River, through a system of irrigation canals which permit farmers to produce three to four crops per year. Agriculture constitutes an important part of the urban population's income, in addition to other jobs, some in the public sector but most in the informal private sector.

Of the total population of Carapongo, 60 per cent of the population is involved in agriculture, mainly in the same neighbourhood. In these farming families, men were reported to be 'mainly responsible' for the farm in 70 per cent of cases, and women in 30 per cent. Just over half of the men and women involved in agriculture in Carapongo are immigrants from the rural areas of

Woman feeding piglets in Carapongo
By Urban Harvest Programme

the Andean region and were previously involved in agriculture. It is therefore probable that they continue to be strongly influenced by gender divisions of labour that prevail in rural areas. However, in Carapongo family labour is rarely supplemented by casual labour, which increases the responsibilities of the women in the household.

Forty-two per cent of the men and 34 per cent of the women indicate that farming is one of the few sources of regular employment available to them in the city. Almost 40 per cent identify a major benefit of farming as the ability to pay for education of their children, while 31 per cent regard it as their main source of food security.

Absentee landlords who rent to local farmers for agriculture hold a significant proportion of the land. Although 30 per cent of the agricultural land in Carapongo has been lost to urban sprawl over the past two years, farming persists, is still very important, and is mainly characterized by mixed cropping of vegetables. Farm plots are similar in size and planted with vegetables such as beet, lettuce, turnip, radish, basil, and other herbs (*huacatay*), mainly for commercial sale, while livestock is raised for family consumption and for sale. Poultry, guinea pig or *cuy*, and pigs are the most popular animals kept.

Carapongo people are likely to be part-time farmers (48 per cent), combining farming with other employment. Part of the reason for this is the limited access to land, with mean ownership of 1.9 plots or 0.81 ha per family. Many families cannot find sufficient farm employment in the locality. Among the working population, 45 per cent of men commute outside the Carapongo area to look for other work, mostly casual labour and especially in the informal and labour-intensive transport sector.

In Carapongo, agricultural production for subsistence/food security and production for income and employment generation are by no means mutually exclusive. They co-exist in a range of different combinations. In rearing animals, the women tend to emphasize the importance of production for subsistence, while men (husbands) emphasize it as a source of additional income. In crop production both emphasize the importance of the market, although they also consume a small part of the harvest. Three functional groups can be identified by their production systems: crop–livestock production systems (65 per cent); crop production systems (33 per cent); and livestock production systems (2 per cent).

Methodology

The study of the agricultural production system used an integrated methodology. At the farm level, data related to land-use practices, crop and livestock management practices, and gender division of labour (in agricultural work, non-farm work, and reproductive responsibilities) were collected. The data were then used to identify the main types of production system prevailing within the study area, and the gender division of labour in Carapongo households.

A baseline survey was conducted, involving 125 Carapongo producers. Of the households surveyed, the person mainly responsible for the farming activities was interviewed. In 70 per cent of households a man was identified as the main person responsible for the farm.

The random survey, implemented from April to October 2004, covered the following issues: (1) family composition and household characteristics, (2) migration, (3) production systems, (4) food security and health, (5) environmental attitudes and behaviour, (6) family planning and reproductive decisions, and (7) gender division of labour.

Also a wide range of qualitative participatory tools, adapted to the urban gender characteristics of the producers of the study area, was used both to facilitate and to complement quantitative data-collection methods (Table 13.2). These tools were used to identify the role played by gender in the division of labour, the access to and control over resources, and the decision-making processes in Carapongo households, and to identify gender-specific problems, constraints, and opportunities.

Work with key informants proved particularly important. We selected key informants from different categories of the farming population by differentiating the length of time that families had lived in the area, whether they had come directly from rural settings or had moved from other parts of the city, whether they were primarily crop producers or livestock keepers, etc.

Women proved to be the best informants in most situations, and methods were adjusted to adapt to their availability. Information or data from women were more specific and clearer than those obtained from men. This may be because women are more conscious of their different roles in the organization of the family, and their desire to improve the economic situation of themselves and of their families. In order to benefit from their contributions, given their limited availability, in some cases it was necessary to conduct interviews with groups of informants in workshops. However, this is also often a challenge in

Table 13.2 Specific tools applied

Specific tools	Main gender-related issue
Seasonal calendar Survey Daily activity clock	Division of (urban-agriculture related) labour, tasks, and responsibilities
Gender consultation	Decision-making power
Survey Transect walk Household resource-flow diagram	Access to and control over resources
Organizational linkages diagram (Venn diagram)	External factors
Problem drawing	Constraints, problems, and opportunities

the urban setting, where the time constraint seems much more severe than in rural contexts. Because of the difficulties described above, sensitization and motivation were important strategies preceding interviews with individual key informants or informant workshops. Also, intervention by staff of the municipality (local government) has been very important.

An important factor which influenced the study was the rapidly changing urban market. Frequently a decline in the value of agricultural commodities leads to one or other adult, mostly the man, leaving agricultural activities and seeking work in or near the city. Although in most cases the woman continues to farm, it may be in a different form.

Gender analysis of the local situation

Gender division of labour

In agricultural work

The division of tasks between men and women differs according to their agricultural production system, the cultural group to which they belong, the socio-economic status of the household, and the location of the household in the city. (The same factors are found to influence the decision-making power of women and men too.)

The division of labour in agricultural tasks between men and women is summarized in Table 13.3.

In only two types of task – land preparation and pest control – is there a clear assignment of responsibility to men. Only in special circumstances

Table 13.3 Primary responsibility for agricultural tasks in Carapongo

Activities		Men (%)	Women (%)	Shared (%)	No agricultural activity (%)	Total (%)
Crops	Land preparation	78	3	16	3	100
	Planting	36	6	55	3	100
	Fertilization, weeding, hilling, irrigation	30	6	62	2	100
	Pest control	87	3	6	4	100
	Harvesting	14	5	75	6	100
Livestock	Raising small animals	6	23	28	43	100
	Raising larger livestock	7	12	25	56	100
Purchase of inputs		46	22	30	2	100
Marketing products		23	41	36	0	100
Household		4	76	15	5	100
Day labourer (*jornalero*)		14	3	11	72	100
Off-farm activities		19	11	8	62	100

Source: Baseline survey project 'Agricultores en la Ciudad', Carapongo, Lima-Peru (2004); n=125.

are women responsible for these tasks or share responsibility for them: these women are either widows or are unmarried. Where land preparation is shared, women help to prepare vegetable beds, which cannot be done by ploughing alone. Many other tasks are more frequently shared than assigned specifically to men or women. Moreover, men and women do not always have the same perception of the division of tasks. For example, only a small proportion of the men recognize that women play an important role in the purchase of inputs and in the marketing of the products. Most men see this as a prime task of the men, while women on the other hand clearly see themselves as having the major responsibility.

Nevertheless, the data do suggest a general pattern in which men take on more of the crop-production activities (60 per cent, compared with 40 per cent) and women participate more in livestock activities (70 per cent, compared with 30 per cent). Men also play an important role in ensuring the availability of factors of production, purchasing inputs (46 per cent), and obtaining rental land for farming. In vegetable cultivation (beet, lettuce, turnip, for example), men take responsibility more frequently for land preparation, irrigation, and fertilizer application. Men are responsible for (a few) cash crops (beets 26 per cent and lettuces 15 per cent) to generate cash income for the family. In livestock management, women are actively involved in feeding, health care, and marketing. For some types of livestock such as poultry, women are principally responsible in almost two thirds of cases. On the other hand, men are more commonly involved with larger livestock like cows and goats.

Women's management of animals is often related to securing household food and nutrition. Poultry and guinea pigs in particular are household food assets, supplementing nutritional intake. Guinea pigs, sheep, and goats are also often considered as a 'savings bank': a ready source of cash for emergencies. These animals, as well as pigs and poultry, are also regular sources of monetary income, although, as will be discussed below, controls of that income varies according to circumstances. The production of pigs, as well as poultry, can also constitute a source of short-term food. Livestock provides both a means of risk aversion and a source of cash income through occasional sales. Also, animals provide manure, especially in the case of poultry, where the waste matter is an input for vegetable production.

The stronger role of men in vegetable production does not carry through to post-harvest and marketing activities, where women clearly play a bigger part (41 per cent of producers considered marketing the main responsibility of women, compared with 23 per cent who regarded this as the realm of men). In Lima, women are considered to be the better and tougher negotiators, although the survey results also show that a high percentage (36 per cent) consider marketing to be the responsibility of both sexes. Vegetable marketing generates economic earnings and employment, and its complex character helps to explain the importance of tough negotiation by women as well as the mixed responsibilities.

In marketing activities

Women are more commonly involved in the marketing of the vegetables and livestock than men, although men and women perceive the extent of women's control over this activity very differently. About one third of men interviewed thought that this is mainly a women's responsibility, 28 per cent said that men do the marketing, and 42 per cent that it was a joint task. Two thirds of women responded that they are the ones that do the marketing.

Beetroot, lettuce, turnip, basil, and animal products were identified as the most important local commodities. The major destination for the crop products is the Lima wholesale vegetable market (Mercado Mayorista) in the central district of the city, about 15 km away. However, the vegetables grown in Carapongo reach the wholesale market complex through a diverse array of producer marketing strategies. The small-scale producers themselves identify an important difference between 'conventional' practice, in which there is no stable business relationship between seller and buyer, and 'novel' practice, where the producer manages to sell directly to consumers who are often known to him or her, or else the producer sells to the same intermediary on a regular basis. Although this marketing strategy appears to demand more frequent harvesting, and hence greater crop planning and organization, it can result in a more steady income flow for producers and access to a more regular supply for buyers. In terms of gender implications, conventional practice places a high value on tough negotiation, since there is little trust involved in these ad hoc trading links, which results in the important role played by women, whereas the novel marketing strategy relies on building stable trading links, which may be done either by women or by men (Tesdell, 2007).

The prevalence of conventional, *ad hoc* marketing is an indication of a lack of social capital among producers in vegetable marketing. On the other hand, marketing of animal products, mostly by women, exhibited a remarkable aptitude for securing stable and local buyers for the meat. A wide range of services is provided by local traders, such as harvesting and washing of certain vegetables, collection and bulking up of produce, and transportation to different markets. Some services are especially offered to overcome post-harvest handling difficulties associated with particular vegetables. For example, the turnip is considered by producers to be a difficult and time-consuming crop to process for market. Turnip traders therefore specialize in buying the crop in the ground at a near-mature stage and then contract male and female labourers to harvest and wash the product in the early morning hours, before transporting it to the wholesale market by moto-tricycle (Tesdell, 2007).

In non-farm jobs

Over 60 per cent of the farming population engages in off-farm activities (work outside the farms) that provide a source of income with a guarantee of a minimum annual income. Of these, 80 per cent are men (working as drivers) and 20 per cent are women (working as market sellers and pedlars).

In reproductive tasks

In terms of reproductive activities associated with the household, women clearly play the key role. Women are caretakers of the family (100 per cent) and are responsible for ensuring that the household is able to reproduce itself over time. The daily activity profiles (see Figure 13.1) indicate that women have to combine a large number of activities during the day when they are at home, before and after going to the field. Women spend seven hours each day in agriculture activities and eight hours in household activities. Men spend nine hours working exclusively in agricultural activities. Women undertake the majority of environment-management tasks in urban households, including the purification of drinking water. In addition, most of the women have identified the need to learn how to better manage and recycle organic wastes to produce nutrient-rich fertilizer, including the treatment of black water from household sewage systems for use in vegetable production.

Thus, women carry out household care and maintenance regardless of the time that they devote to food production or other livelihood activities. This is particularly difficult for female heads of households, who bear the sole responsibility for both reproductive and productive tasks.

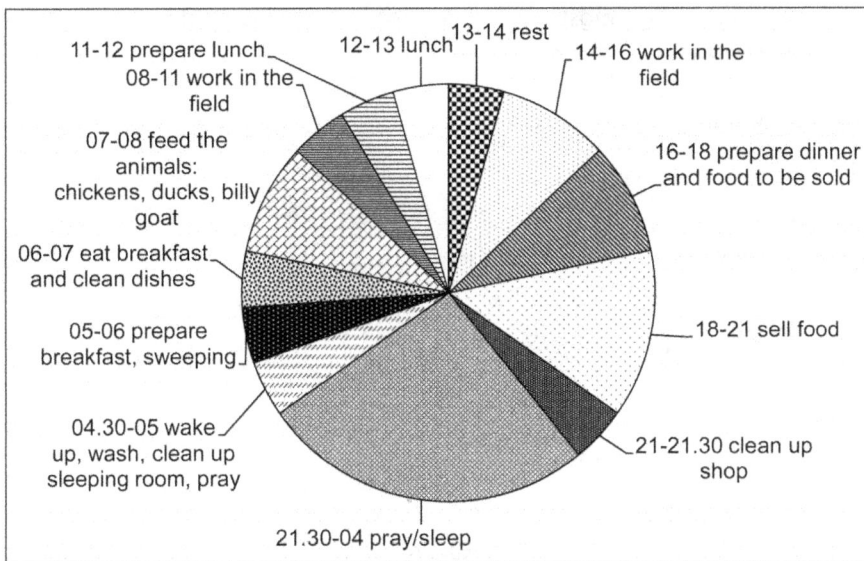

Figure 13.1 Daily timetable of a female producer in Carapongo
Source: Urban Harvest

Gender-based division of access to and control over resources and benefits

When considering access to and control over resources, three types of resource can be identified (see Table 13.4):

- productive resources used by the household, such as land, water, inputs, credit, technical and market information, training (capacity building);
- productive resources of organizations;
- benefits of production, such as cash income, food, and other products (for home consumption, sales or exchange).

Access to and control over land

In Carapongo, the key natural resources of significance to the urban producers are land and water. Land tenancy in the area is complicated by the history of land occupation and the current fluidity of land access. The old system of *hacienda* ownership was largely transformed into co-operative ownership during the agrarian reform process, starting in the early 1970s, although some households gained access to individual plots earlier than this. Subsequently the co-operatives were divided up as 'parcels' among individual households. Now there is a wide range of mechanisms to access land, based on ownership through direct purchase or inheritance, 'pre-inheritance' (*anticipo*), 'guardianship' (*guardianía*), and share-cropping (*al partir*). Some families also access land through informal squatting (*posecionarios*) of land located on the bank of the Rimac River. Although used for agriculture, these lands are in fact protected by law and are inalienable. Many producers also combine different types of land tenure, most commonly ownership and renting, resulting in higher average land holdings.

Table 13.4 Access to and control over resources in Carapongo, Lima-Peru

	Access		Control	
	Men	Women	Men	Women
Productive resources				
Household land	xxx	xx	xxx	x
Household access to water	xxx	xx	xxx	xx
Household agricultural inputs	xxx	xxx	xxx	xxx
Participation in organizations/access to credit	xxx	xx	xxx	xx
Capacity building, information sources	xx	x	xx	x
Benefits of production				
Income from sale of vegetables	xxx	x	xxx	x
Income from sale of animals	xx	xxx	x	xxx
Income from labour (from off-farm activities)	x	xxx	x	xxx

xxx Indicates complete access/control; xx Indicates partial access/control; x Indicates limited or no access/control

Source: Baseline survey and qualitative tools project, 'Agricultores en la Ciudad', Carapongo, Lima-Peru (2004)

Table 13.5 Land tenure by men and women

Tenancy	Men		Women		All	
	Area (ha)	%	Area (ha)	%	Area (ha)	%
Ownership	0.72 (± 0.42)	38	0.72 (± 0.67)	36	0.72 (± 0.49)	37
Rental	0.68 (± 0.39)	16	0.52 (± 0.40)	28	0.61 (± 0.39)	19
Posecionarios	0.82 (± 0.98)	9	0.84 (± 0.67)	6	0.87 (± 0.93)	7
Ownership + rental	1.0 (± 0.46)	17	0.34 (± 0.22)	5	0.96 (± 0.49)	17
Ownership + *posecionarios*	1.52 (± 0.80)	4	1.54 (± 1.37)	11		
Others*	0.85 (± 0.10)	16	1.06 (± 0.62)	14		
Total		100		100		

* guardianship, share-cropping, and combinations
Source: Baseline survey project, 'Agricultores en la Ciudad', Carapongo, Lima-Peru (2004); n=125.

Inherited resources are, theoretically, divided equally among children, both male and female, and owned land is the most important form of tenure for both men and women. Women can and do buy and rent land. The person accessing the land in one of these various ways is often the person who also administers it. As one man commented in a workshop on this issue: *'La herencia se respeta, sea hombre o mujer'* ('One respects inheritance, whether it is to a man or a woman'). On the other hand, there are cases in which one or other spouse is better capable of taking care of farming in the household.

Among the households where a man is seen as the main person responsible for the farming activities, 38 per cent live on their own land, but fewer than half of these have a formal title to it. Among the 30 per cent of households where women are mainly responsible for farming, the pattern is the same. Although 36 per cent have their own land, only 49 per cent of these have a formal title. For a little under half of male and female farmers in the two quarters of Campo Sol and Huancayo, the Landowners Association holds the title to the land, with members having usufruct rights. About 23 per cent of the members of these associations are women, but the leadership is primarily male. This appears to put women who are mainly responsible for farming in these neighbourhoods at a double disadvantage.

Responsibility for the farm is strongly influenced by whether the household is male- or female-headed. Of the 30 per cent of households in which women are mainly responsible for farming, 14 per cent are headed by women. In these households the woman holds absolute control over the household property. This is especially so for widows, who control land, house, and livestock. However, in this case they generally have a smaller average land area (0.31 ± 0.22 ha), and their plots tend to be of poorer quality and are consequently less productive. An important added challenge for households where women are mainly responsible for the farm is the constraint on their labour availability for farm activities, due to their heavy commitments to domestic chores.

With the city expanding rapidly, some agricultural lands are being converted into small residential plots, while in other places individual farm plots are being used wholly or partially for extraction of earth for brick-making or as construction material. So far there is no evidence that men are more or less willing to convert land in this way than women.

Access to and control over water

Access to water is a key component for productivity and success in urban agriculture. In Carapongo, the availability of water during the dry season in the mountains is declining. This results in low water levels in the River Rimac, from which irrigation water is drawn. However, only a quarter of male farmers and a fifth of female farmers identified water scarcity as a problem. But they did emphasize water quality. Increasing urban pollution and environmental contamination result in irrigation channels filled with garbage and pollutants.

Access to water at Carapongo is by household, and so tends to be a source of potential conflict between households, rather than between a couple within the household. Men mostly handle management of the irrigation systems, both at the watershed level and at the 'sub-sector' level of Carapongo. Only 8 per cent of women are members of the sub-sectoral irrigation commission of Carapongo.

Access to and control over inputs

Access to and control over the inputs for crop–livestock production depends on the purpose of the production. Both men and women invest significant inputs for commercial production (cash crops, animals for regular sale), whereas mostly women incur minimal expenditure for subsistence production (small-scale planting of root and tuber crops, beans, green maize, and herbs and rearing of small animals such as poultry and guinea pigs). Pesticides, organic and inorganic fertilizer, and hired labour are the major expenditures for cash crops, and commercial feed is purchased for some animals. Control over these inputs is a consequence of technical specialization or specific responsibilities assigned to men and women. The purchase and use of pesticides is largely a male responsibility, partly because of the physical exertion involved, but also because of the risk of contaminating children and food if the women handle pesticides.

Access to credit

In the study area, there are almost no formal sources of credit or loans, and 65 per cent and 73 per cent of households respectively claim not to have any access to either. Those that do access credit do so informally through small kiosks (bodegas) or, for agricultural inputs, through the agricultural supply shops and the manure traders. Families are the main source of loans, which are almost exclusively obtained for farming activities. In the few cases where loans are obtained from moneylenders, they attract very high rates of interest because the farmers do not have collateral to pledge. Male farmers have better access than women to agricultural credit because of their more frequent interactions with suppliers. In almost 90 per cent of cases, men are responsible for crop protection, both purchasing and applying chemicals.

Gender-differentiated participation in social networks

Participation in community organization and social networks that provide access to credit and loans or access to knowledge is variable between men and women. In Carapongo, women participate very little as representatives with decision-making power in public or community organizations. For example, although women use and manage water in farming (crop–livestock activities) and in the home more than men do, they hardly (only 10 per cent) participate in the irrigation commissions and committees. It is a tradition for men to hold the administrative positions on those committees, but women manage water for domestic use in Carapongo, where most households have access to a well.

There are seven associations of landlords, in which men participate more than women, since land ownership is the criterion for membership, which is male-dominated. However, women play an important part in local community organizations that relate to food security. These are community kitchens and the 'Glass of Milk' programme which are co-ordinated by the municipality through community-based committees of women.

A common perception among both men and women is that the local population is organizationally strong when making claims on official authorities, whether in relation to water management, land use, or social programmes, but social organization for those seeking improvements in agricultural production or marketing systems is weak.

Access to technical information and training

Regarding the gender aspects of human-capital formation, both men and women indicate a lack of access to training or information about crops and livestock-husbandry practices, although women are at a much greater disadvantage. Only 22 per cent of the farming population has received agricultural training, but of these 86 per cent are men and 14 per cent are women. Women are particularly interested in learning more about basic treatments pertaining to animal health, whereas men are more interested in information about sources of credit and government-sponsored training programmes.

Control over the benefits and risks of production

The predominance of women in the marketing of vegetables was confirmed in a mixed-sex workshop discussion which also highlighted the variability in decisions to sell and control of the proceeds of sales. Where a woman has control over land, through whatever means, she most often has the right to decide on the sale of the produce. On the other hand, since men are more commonly in control of land, they have more frequent authority in the decision to sell.

Women producers who are not land owners (almost 60 per cent) demand their share of revenue derived from production, because they are the ones who are responsible for the care of the family, principally children. However, when they are not successful in convincing their husbands to share the earnings, women retain part of the money from their sales of small animals and vegetable produce without the knowledge or consent of their husbands.

Given women's greater responsibility for small livestock, they tend to have more say and involvement in these sales. Even where a man has taken the initiative to become involved commercially in animal production, he may have a limited say. As one wife commented ironically: '*El decide, intelectualmente, a vender. Pero yo soy la que hago todo para la venta.*' ('He decides, in theory, to sell. But I am the one that does everything to do with the selling.') However, in the workshop discussion, women commented that they have to account to their husbands for money obtained from selling animals. Women make decisions on expenditure if the amount is a small sum. For all larger expenses, both men and women make the decision.

In the study area, 39 per cent of the men indicate that the principal constraint that limits their vegetable production is insect pests. Evidence from two thesis studies in Carapongo (Milla and Palomino, 2002; Maldonado 2006) indicates that there is wide use of toxic and highly toxic pesticides in this agricultural area, leading to health risks and negative environmental impact. Since men are mainly responsible for applying pesticides, the related health hazards are mainly affecting the male producers, although through handling and maintenance of pesticide-application equipment other household members can be affected (Yanggen et al., 2003). It appears that producers are constrained by a lack of knowledge of sustainable (integrated) pest-management practices, due to limited access to training courses offered by institutions or non-government organizations, or their limited exposure to sustainable commercial agricultural practices.

Bargaining power in decision making in Carapongo households

The way in which decisions are made within the family depends on how tasks are assigned within the farming system. The position of individuals within the household and the division of labour affect an individual's knowledge of the crop–livestock system which influences decision making. Commercial farmers make more decisions alone, and few in consultation with family members, while home-consumption farmers or non-land-owning households with only animals make fewer decisions alone and more in consultation with family members.

Table 13.6 Decision-making matrix in Carapongo households

	Men decide alone	Men and women decide jointly			Women decide alone	Comments
		Men dominate decision	Equal influence	Women dominate decision		
Inputs						
Who decides how the family labour will be used?		•				When men work off-farm, women spend more time in the field or hire labour for the farm.
What inputs to buy?			•			Who decides depends on the type of crop, type of of animal, and type use (food or cash)
To hire additional labour?		•				
Production						
Who decides which crops to grow?		•				A female land owner/renter may decide on crops.
When to harvest?		•				There is always some flexibility.
Number of animals to buy?				•		
Marketing						
Who decides what proportion of the vegetables is sold?				•		
When/which animals are to be sold?				•		In case of commercial production like pigs, men can influence the decision.
Investments						
Who decides to buy equipment and tools?	•					
To take a loan?		•				Depends on the purpose of the loan.
To buy or rent additional land?		•				
To buy more animals?				•		
Reproduction						
Who decides whether a child goes to school?			•			
To consult a doctor?				•		

Role of external factors on gender in urban agriculture

The policy environment has differential impacts on men and women involved in agricultural production. Although a detailed analysis is yet to be completed, there does seem to be a more positive environment in relation to women's situation compared with that of men; in the latter case there is quite a strong antagonism towards local authorities. The main contact between male producers and local government is through the irrigation committees, which tend to oppose certain policies adopted by the municipality. These primarily relate to the local taxation system and the desire of the Council to convert low 'rural' rates into much higher 'urban' rates. The higher urban rates are justified on the basis of the upkeep of urban amenities provided: parks, piped potable water, sewer systems, sanitation, refuse collection, etc. The Rimac Irrigation Committee is a focus of resistance to the conversion of agricultural land to urban status, for two reasons: the lack of such services in the agricultural areas, and the crippling financial burden that the urban rates would impose on families with only half a hectare.

The main points of contact between the Council and women are through the programmes dealing with Community Kitchens, the Glass of Milk programme, and the Mothers' Clubs. The Council also has a special programme addressing violence against women.

Recommended strategies to facilitate equitable development of urban agriculture in Carapongo

Municipal policy makers

A major function of municipal leaders is to recognize and address the needs of different population groups in their District, and there are often specific structures and functions addressing particular issues that affect women, youths, and children. However, it is still rare for the needs of agricultural producers of either sex to be addressed in urban municipalities. Thus a first and fundamental strategy is to recognize the existence and contribution of the agricultural sector to the local economy and society, and to develop policies that support safe and sustainable production, post-production, and marketing of crop and livestock products. With the growing city population, due to migration from rural areas and natural growth, the demand for employment is increasing; the demand for fresh products is also on the rise, especially for fresh vegetables and animal products. This provides a good opportunity for men and women farmers to increase their income. A positive effect of urban agriculture for women is that their important role in family food security and nutrition receives greater recognition. In addition, it helps them to become more independent, by generating some additional income from sales of surpluses (of guinea pig, for example, which is a novel opportunity market in Carapongo) and by saving cash on food expenditures which can be used for other purposes. However, gender-related characteristics of urban farming in Carapongo are an unequal division of labour, an absence of equitable access to and control over productive resources, unequal access to knowledge by both men and women, and inequality in power relations.

Recognizing the important role of urban agricultural production, and the existing gender inequalities, the municipality of Lurigancho-Chosica, of which Carapongo is part, could take the following measures to facilitate sustainable and equitable development of urban agriculture:

- Allocation of a budget for the recently established Sub-division of Urban Agriculture.
- Redesign of municipal land-use policies, encouraging co-operation between the Housing and Agricultural Associations, individual landowners and cultivators of the plots, and participation of male and female stakeholders in urban land-use planning, in order to increase security of land tenure for both male and female producers.
- Support for sustainable provision of irrigation water, through improved co-ordination and dialogue with the Irrigation Committees and SEDAPAL, the parastatal company responsible for water management.
- Support for social capacity building, through provision of training and development of producer networks and organizations, with emphasis on gender-sensitive approaches.
- Co-operation with the Ministry for Women and Social Development to replicate their gender-specific strategies which support the empowerment of women through a range of policies and institutional actions, especially its programme on poverty reduction (FONCODES), which addresses some aspects of gender in agriculture and has recently initiated some preliminary work in urban areas of Lima.
- Broadening the existing programmes of the Municipal Ombudsman for Women, Children and Adolescents (*Defensoría Municipal de la Mujer, Niño y Adolescente*) and the various Divisions and Sub-Divisions of the Council, such as Education and Culture, where empowerment of women is targeted, to include attention to women involved in agricultural production in the Municipality.
- Such gender-equity promotion programmes in urban agriculture should include active measures which could help women to gain greater access to credit services, training opportunities, and technical support. As part of collaborative R&D activities involving Urban Harvest, the Lurigancho-Chosica District Government, and the local irrigation users' committee, a micro-credit scheme has been launched which combines capacity building, development of micro-investment proposals, and the awarding of small, low-interest loans to local producers. There are currently 46 loan recipients, of whom 17 are women, 27 are men, and two are mixed-sex groups. Loans range from 800 to a maximum of 2,000 soles (approximately $280–$700).

- To improve the contribution of agricultural production to household food security and especially to improved nutrition of young children, some key policy issues would be:
  - enhancement of maternal skills in infant and child nutrition
  - inclusion of weaning foods in municipal food programmes
  - improvement of linkages between agricultural production and municipal food programmes
  - support for and recognition of small-scale animal raising by landed and landless households
  - targeted training, especially in animal production for women.
- To improve the economic efficiency of commercially oriented urban agriculture, some key policy issues are:
  - access to markets and market information
  - support for the productive use of recycled organic wastes and wastewater
  - zoning and permits for commercial pig raising.

Civil-society organizations

The inequities between men and women identified in the study in Carapongo cannot be dealt with at the policy level alone. Civil-society organizations can play an important role in the following ways:

- Mobilizing all parties involved in the development process, including academic institutions, non-government organizations, and women's groups, to improve the effectiveness of anti-poverty programmes directed towards the poorest and most disadvantaged groups, especially women –and, in the present context, urban women, female heads of household, young women and older women, and women with disabilities, recognizing that social development is primarily the responsibility of governments.
- Engaging in lobbying and establishing monitoring mechanisms, as appropriate, and other relevant activities to ensure gender mainstreaming and implementation of the recommendations on sustainable and equitable urban agriculture development as a strategy to eradicate poverty.
- Ensuring that data generated through gender analyses are fed back into decision making on the design of the projects to be undertaken. For example, findings from studies of urban agriculture and gender undertaken by Urban Harvest and local NGO and government partners, on which this case study drew, indicated the key role of women who engage in livestock raising mainly for subsistence or as security. The subsequent livestock interventions of the project have worked largely with women (74 per cent of those implementing livestock modules) to enhance the productivity of livestock, thus contributing to household

food and nutrition security and enhancing the income available to women. It also noted that a limited role for women in decision making about crop-production interventions in horticultural production began with a process of capacity building, using farmer field school methods, which tended to reflect the dominant role of men in this field (23 of 27 initial participants, or 85 per cent, were men). However, with the development of organic farming and the emphasis on alternative marketing strategies, increased efforts have been made to strengthen the participation of women. Female membership in the organic producers' groups has risen from 15 to 34 per cent.

- Developing gender-responsive working methodologies for use in research and the design of policies that recognize and value the full contribution of women to the economy, through their unremunerated and remunerated work, both on urban farms and off-farm.

References

CELADE (1993) 'Centro Latinoamericano de Demografía. Población, equidad y transformación productiva', CELADE-CEPALFNUAP, LC/G.1758 (Conf.83/3), LC/DEM/G.131.

Castro, M., Malena (2007) *Incorporación de la Agricultura en la Planificación Urbana del Distrito de Lurigancho-Chosica*, Final Report, Agropolis Program, IDRC, Canada.

INEI (2005) *Encuesta Nacional de Hogares (ENAHO): 2005*, Informe Técnico, Instituto Nacional de Estadística y Informática. Lima, Perú.

Lipman, M. (1977) *Why Poor People Stay Poor,* Harvard University Press, Cambridge, USA.

Maldonado, L. (2006) 'La agricultura urbana en Lima: Estrategia familiar y política de gestión municipal Caso: Localidad de Carapongo', Tesis de Maestría en Gerencia Social, Pontificia Universidad Católica del Perú, Peru.

Milla, O. and Palomino, W. (2002) 'Niveles de Colinesterasa sérica en agricultores de la localidad de Carapongo (Perú) y determinación de residuos de plaguicidas inhibidores de la acetilcolinesterasa en frutas y hortalizas cultivadas', Tesis para optar el grado de Químico Farmacéutico, Facultad de Farmacia y Bioquímica de la Universidad Nacional Mayor de San Marcos, Lima, Peru.

Tesdell, O. (2007) 'Marketing Approaches of Small-scale Urban Farmers: A Case Study in Lima, Peru', unpublished technical report, Urban Harvest, Lima, Peru.

Yanggen, D., Crissman, C., and Espinosa, P. (eds.) (2003) *Los Plaguicidas: Impactos en producción, salud y medio ambiente en Carchi, Ecuador,* CIP and INIAP, Lima, Peru.

About the authors

Blanca Arce is Project Leader at the Colombian Corporation of Agricultural Research (Corpoica), Tibaitata, Colombia.

Gordon Prain is Global Co-ordinator for Urban Harvest, International Potato Center (CIP), Lima, Peru.

Luis Maldonado is Research Assistant at the Impact Enhancement Division, CIP, Lima, Peru.

CHAPTER 14
Gender and urban agriculture in Pikine, Senegal

Gora Gaye and Mamadou Ndong Touré

Abstract

This case study is based on a review of exploratory research on urban agriculture in Pikine, conducted by IAGU in 2004, and additional data gathered in 2008 in the context of the small project 'Establishing a Proper Input and Equipment Supply System for Urban Agricultural Producers in Pikine-North', implemented by PROVANIA, a farmers' organization in Pikine, with the support of IAGU. A review of this pilot project shows that although gender aspects were taken into account during the design, planning, and implementation phases, the project nevertheless focused mainly on production aspects (dominated by the men) and pays little attention to the transport, processing, and marketing aspects (dominated by women). Priority constraints and interests that are specific to women and mainly related to processing and marketing were ignored.

Introduction

Background

This case study is based on the experiences gained in a pilot project entitled 'Supporting the Establishment of a Proper Input and Equipment Supply System for Urban Agricultural Producers in Pikine-North', which was formulated and implemented in 2007 and 2008 by the local farmers' organization PROVANIA, with co-funding from the international Resource Centres on Urban Agriculture and Food Security – Cities Farming for the Future programme (RUAF–CFF), regionally co-ordinated by the African Institute for Urban Management (IAGU) in Senegal.

This pilot project is a follow-up to a Multi-stakeholder Policy formulation and Action Planning (MPAP) process, involving several local institutions and urban agricultural producers' organizations in Pikine, supported by RUAF–CFF. This process started in 2005 by undertaking a (gender-sensitive) exploratory study of urban agriculture in the Pikine area, identifying the limited access of urban producers to agricultural inputs and equipment as one of the factors that limit the development and sustainability of urban agriculture in

Pikine, alongside land insecurity, poor access to credit, and other factors. The diagnostic phase was followed by joint development of a City Strategic Agenda on Urban Agriculture. This Strategic Agenda is being implemented now, co-ordinated by the participating organizations and institutions, including this pilot project by PROVANIA.

The area of Pikine

Pikine is located within a system of continental dunes, broken by depressions and valleys, named the Niayes Valley. The physical characteristics of the Niayes are fairly favourable for agricultural production. The annual rainfalls vary considerably, between 200 and 900 mm per year. The climate is tropical, characterized by alternation between a dry season – eight months, from early November to mid-July – and a rainy season from mid-July to the end of October. The natural vegetation of the Niayes is abundant (MUAT, 2004). These plots of land are sought after by building companies, although the groundwater is close to the surface, because buildings nowadays are constructed on good-quality embankments. However, the construction of infrastructure, such as water and electrical-supply facilities, is hampered by the physical characteristics of the area.

Pikine had around 800,000 inhabitants in 2002, 63.5 per cent of whom were aged under 25 (ANSD, 2004). Pikine was a dormitory town for a long time, but its economy has started growing in the past few years. However, apart from the free industrial zone, most of the economic activities still take place in the informal sector and, to a lesser extent, in horticulture and fish breeding in the Niayes (MUAT, 2003).

Urban agriculture in Pikine-Niayes

Urban agricultural activities in Pikine-Niayes include production, processing, and marketing activities. The main production systems are market-oriented horticulture (vegetables and fruits), small-scale (home) gardening, traditional fishing and fish breeding, tree nurseries and forestry, local cereal processing, and poultry farming.

Among the producers (over 1,500) in the Department of Pikine, horticulture is the main activity, due to the presence of water near the surface (*céanes*). In 2001, 501 horticulture farms were present in Pikine. The size of the plots varies between 100 m² and 5.9 ha. Women represent 21 per cent of the producers and are mainly involved in the marketing of products, although some of them also work on the farms. The main products are aubergine, lettuce, onion, chilli, parsley, leek, and strawberry.

The area of Pikine, especially the Niayes Valley, is state property, and therefore the producers in that area can be ejected at any time if the state decides to assert its rights. This land insecurity deters the producers from making investments.

Land is obtained through inheritance, by purchasing it (although it is strictly prohibited to sell plots of land in the area, some people sell/buy land, knowing that it has no legal value), by renting a plot, or by entering into a share-cropping arrangement (*métayage*).

Small-scale urban agriculture in Pikine area successfully contributes to better food security and nutritional status as well as generating income for the producing families. Nearly 95 per cent of the producers in the area claim that they are able to cover their family's daily cash needs with the income generated by urban agriculture (Dieng, 2004).

The main actors involved in urban agriculture in the Pikine are depicted in Figure 14.1. They include the following:

1. Direct stakeholders: the 1,500 producers and their organizations: PROVANIA and UPROVAN (larger associations of several local farmer groups), Mbou Gayif, Laaw Tann, Ndeck, Daan Doleé, and Feul Yeggo (smaller specific-purpose organizations), as well as the small merchants, larger land holders, agricultural workers, input suppliers, transporters, middle men, retailers, etc.
2. Institutional actors: ministries, local government, and decentralized state services (Ministry of Agriculture, City Of Pikine, Commune of Pikine Nord).
3. The Niayes and Green Zones Development and Protection programme (PASDUNES).
4. NGOs and other non-government supporting organizations (IAGU, Enda RUP, ENDA GRAF, ANCAR, ACDEV, CEREX Locustox, etc.) and CAT (the IAGU–PROVANIA co-ordination team for the pilot project)

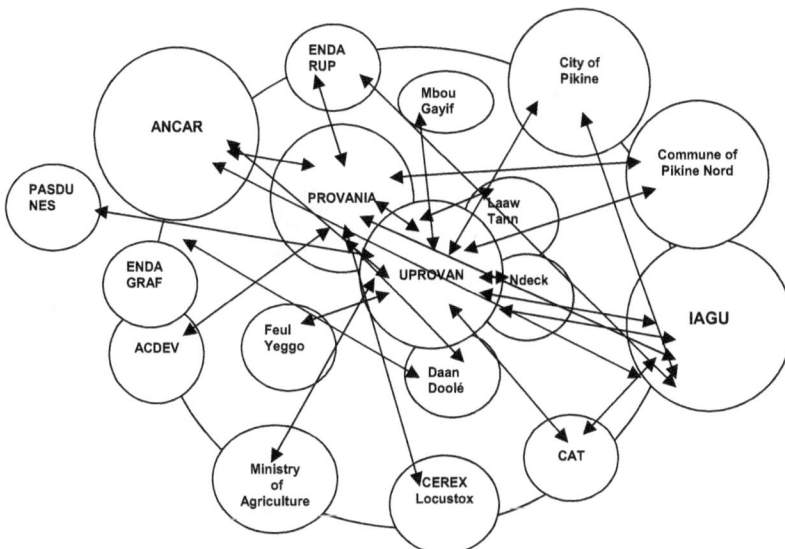

Figure 14.1 Actors involved in urban agriculture in Pikine

Figure 14.1 shows the various institutions that interact (to a certain extent) with the (second-level) farmer associations PROVANIA and UPROVAN, and the low level of interaction between these institutions, which sometimes creates duplication (as well as important gaps) in the implemented activities. The Multi-stakeholder Forum on Urban Agriculture that was established as a result of the MPAP process provides a platform for the farmer organizations to interact directly with these institutions and provides better institutional co-ordination.

The study area

The formulation and implementation of the pilot project took place in North Pikine on 60 ha mainly dedicated to vegetable gardening (and some floriculture). Some of the actors are engaged in fruit and vegetable processing. The majority of plot supervisors are men (72.2 per cent), while 21.8 per cent of them are women. Their age range is from 25 to 68 years. As regards land occupation, 35.8 per cent are tenants and 52.8 per cent are owners, but only 3.4 per cent have a title deed (Niang et al., 2005). The main farmers' organization in this area is PROVANIA, which developed the pilot project 'Establishment of a Proper Input and Equipment Supply System for Urban Agricultural Producers in Pikine-North' in co-ordination with IAGU and with the political support of the Pikine Municipality.

Field in Pikine-Niayes
By IAGU

Analysis of gender in urban agriculture in the area of North Pikine

Methodology applied

This case study is based on the following components.

1. **A review of available qualitative and quantitative data** relating to agriculture in the Pikine area, especially:
   - An early case study of gender aspects of urban agriculture in Pikine (IAGU, 2004).
   - A report on the Exploratory Study on Urban Agriculture (IAGU, 2006) carried out in the framework of the RUAF–Cities Farming for the Future Programme, presenting a diagnosis of the main urban farming systems and the actors involved, and the main constraints and opportunities for development of urban agriculture in Pikine area, as well as reviews of the actual legal and regulatory framework for urban agriculture in Pikine. The study includes a special chapter dealing with the gender aspects of urban agriculture.
   - A study on Access to Credit and Finance for male and female urban agricultural producers in the Niayes area (IAGU, 2007). Lack of means to finance their agricultural and marketing activities is one of the producers' major constraints, especially for the women.
2. **Focus-group meetings with male and female beneficiaries of the above-mentioned pilot project**, implemented by PROVANIA with the support of IAGU–RUAF. The aim of these meetings was to check the outcomes of the gender analysis contained in the available literature and to discuss to what degree women's (and men's) specific constraints and interests were taken into account in the design and implementation of the pilot project. Separate meetings were first held with six female producers and ten male producers, followed by a mixed forum (involving 24 women and nine men). The mixed meeting allowed men and women to compare their views on certain issues. During the focus-group meetings a number of participatory rapid appraisal tools regarding key issues in gender and urban agriculture were applied, including the following

   - A *Venn diagram* which highlights the way in which communities perceive (the importance of the services provided by) local associations and external institutions and helps to analyse the existing relationships between local organizations and the external institutions.
   - A *pyramid of constraints*, which is a tool for ranking the main constraints identified by the women involved in market-gardening activities.
   - A *daily calendar of activities* which shows the gender-differentiated distribution of tasks in the farm household.
   - A *profile of the access to and control over resources*, which allows researchers to analyse the productive resources to which men and women have access, and their degree of control over those resources.

- A *decision-making matrix* which makes it possible to understand better how men and women take part in the decision making, and in what aspects of urban agriculture men and women respectively are the main decision makers.
- A *priority-setting diagram,* used to identify the priorities of male and female producers respectively, and/or their views on the adequacy of identified solutions.

In the three focus-group meetings we first sought to establish a climate of trust in order to make the participants feel comfortable about the use of the collected data. Furthermore we applied a tactful approach when dealing with socially or culturally sensitive issues (for example, decision making in the household), in which we first raised a more general discussion on the issue and then slipped in the questions included in the tools one by one. During the meetings we had to deal with a lot of chattering and repeated digressions from the main subject under discussion, hence the need to be patient and refocus the debates on the main issues.

Results of the gender analysis

Distribution of labour

Men dominate production activities, and they normally manage most activities throughout the season. Women become especially involved when it comes to harvesting, transporting, processing, and especially marketing. Female heads of households (widows, unmarried, or divorced women) hire labourers (*sourgas*) for the production activities.

Men almost never go to the market to sell their produce. They often sell the produce on-farm to women who collect it directly from the field by hiring carriers to take the vegetable racks from the field to the main road, and horse carts to bring the production to the markets or processing centre. Transportation costs are relatively high, which affects the production costs and therefore limits profits. The women deplore the absence of access roads and insist on the need to have their own means of transportation, given the fact that horse-cart owners are often uncooperative.

Processing of the produce is also mainly done by women's groups, such as Dan Doolé (a Wolof term meaning 'Working to Earn One's Living'), which is processing cereals, vegetables, and fruits. There are several such groups in the area of Pikine which have organized themselves into a network of local organizations involved in the processing of agricultural products.

Another of the women's important roles is financing the purchase of inputs needed for the next production season, often using their own money earned in the marketing – which is also a way to ensure that the men are selling the harvest to them, rather than to outsiders.

Household duties are entirely performed by women, sometimes with the assistance of young girls. Some parents compel their daughters to interrupt

Table 14.1 Gender division of labour in farm-households in North Pikine

Activity	Women/Girls	Men/Boys
Obtaining a piece of land	x (consults men)	x
Guarding the piece of land		x
Preparing field (hoeing)	x	x
Obtaining/purchasing seeds		x
Preparing seeds	x (provides the funds)	x
Sowing	x	x
Transferring plants		x
Weeding		x
Localizing water source		x
Irrigation		x
Fertilization		x
Pest control		x
Harvesting	x	x
Threshing		x
Cleaning		x
Storage		x
Transporting products to the market	x (hires transport)	
Selling products at market	x	
Maintenance (e.g. irrigation system, warehouse)		x
Interacting with extension workers/local authorities		x
Exchanging information with other producers	x	x
Livestock feeding		x
Household chores	x	
Social/religious ceremonies	x	x
Community-political meetings/roles	x	x

or end their schooling in order to assist their mothers in the domestic chores.

Community activities are essentially divided between family ceremonies (christening, funerals, marriages, circumcision, etc.) and socio-cultural and political activities (associative structures, political activities, community activities). The entire family attends family ceremonies; or, if this is not possible, one of the spouses represents the other in equal proportions. In all the ceremonies, however, there are so-called 'female' components (for example, the festivities) and others that are more 'masculine' (for example, the religious activities).

Women are more inclined to participate in social and political activities than the men. However, they are rarely appointed as community representatives, and men dominate all the community authorities.

Access to and control over resources

Men control most of the resources; only a minority of women have their own farms, by virtue of an inheritance. However, lately an increasing number of women have access to land resources, by purchasing or renting plots of land.

Table 14.2 shows the predominance of men in the distribution of decision-making power. Single female landowners generally make decisions

Table 14.2 Access to and control over resources in farm-households in North Pikine

	Access		Control		
	Men	Women	Men	Women	Observations
Productive resources					
Land	75	25	75	25	Most of the land is used and controlled by men. Women have access to land either by inheriting it or by buying it, but do not easily have control over it, unless they are single.
Labour	75	50	75	25	Husbands often allow their wives to work outside the house with a view to the related economic gains. Widows and unmarried or divorced women have complete control over their activities and resources, and their male relatives have very little power over them. Such women are often involved in women's associations and hire labourers for the production activities if they can afford it.
Credit	25	75	75	25	Women have more access to credit than men (because several institutions especially target women engaged in small businesses), although both men and women judge it to be insufficient. The husband is often consulted before a woman takes a loan, and men often control its use.
Inputs	75	25	75	25	The men buy and control the use of the inputs, but it is the women who provide the funds to purchase the inputs
Equipment	75	75	75	25	Most equipment is collectively owned and controlled by the organization PROVANIA. Men dominate the use of equipment.
Training	25	50	75	25	Women have more training opportunities, especially courses in product processing and adult literacy, than men, but men often have a strong say in whether their wives may participate in a training course or not.
Benefits					
Profits	25	75	50	50	Men sell the produce to their wives, who market it, and so both get their share of the profit. However, husbands often dominate decisions on how the profits will be used, including profits managed by their wives.
Ownership of goods (e.g. house)			75	25	Normally the men are considered the owner of the house and other main goods. However, female heads of household can own a house (and land) through inheritance.
Community decision making	100	100	75	25	Women actively participate in social and political activities, but decision making is strongly dominated by men.

in consultation with the hired farm labourers, which clearly indicates the dominant position of men in the agricultural production.

Women's role in decision making

The decision-making power at the household level is strongly related to the control over means of production (land, capital, etc.), which are mainly in the hands of men, as has been shown above. The women have a much more important role in decision making when it comes to marketing activities, transport, and processing.

When women have the exclusive responsibility for a family (female-headed households), social norms allow them more decision-making power. Since the number of female-headed households is increasing, nowadays more women have gained decision-making power.

At the community level, women are active participants in social and political activities. However, the main decision-making authorities are male.

> *A female producer:* I have been a widow for 15 years and I have been working on urban agriculture for all that period. I am managing a household of 20 people. I manage to feed them and deal with all the needs of my kids and the rest of the family, which is made up of young women, married or single. To summarize, I can say that I manage to deal with all my needs thanks to the income I get from urban agriculture. Most of the female members of the association are in charge of families. I have a plot of land that I inherited from my father, who had a lease on it. This allows me to exploit the land.

It is important to note that women often are not sufficiently informed on important legal aspects related to agricultural activities, which weakens their position in a largely conservative society. The example of the woman quoted

Table 14.3 Decision-making matrix among farm-households in North Pikine (for households with husband and wife present)

Activities	Men decide alone	Discussions but the men's opinions prevail	Equal influence	Discussions but the women's opinions prevail	Women decide alone
Decisions on buying inputs or equipment	x				
Decisions on what crops to grow		x			
Decisions on taking a loan			x		
Decision about which part of the harvest will be consumed or sold			x		
Decisions on training activities		x			
Decisions on domestic activities					x
Family-planning decisions				x	
Decisions on socio-political activities		x			
Decisions on children's schooling		x			

above perfectly illustrates the case, as she was not aware of the fact that when her father passed away his death automatically cancelled the lease, and that she had to go through some formalities in order to get it renewed.

Main constraints encountered by male and female producers in North Pikine

During the exploratory survey in 2006 (IAGU–RUAF, 2006) two pyramids of constraints were elaborated, one by the women and another by the men, in order to identify the main constraints encountered by women and men respectively in urban agriculture in North Pikine. The most important constraints appear at the base of the pyramid, and the less pressing ones at the top.

We observe that a number of problems are seen as priority constraints by male and female producers alike, notably insecurity of tenure, shortage of irrigation water, and limited access to credit. These are serious obstacles to sustainable urban agriculture in North Pikine. According to the agricultural producers, agriculture might disappear altogether, under pressure from rapid and uncontrolled urbanization and the subsequent building of housings and collective facilities.

We also see some important differences between the main concerns of men and women, with the women being more preoccupied with marketing-related issues (poor access roads, limited market places, problematic consumer relations, etc.) or issues related to their situation as female heads of households (paying hired labour); whereas the male producers are more preoccupied by production-related issues (lack of good seeds, crop diseases, crop and soil

Figure 14.2 Pyramid of constraints encountered by female producers in North Pikine

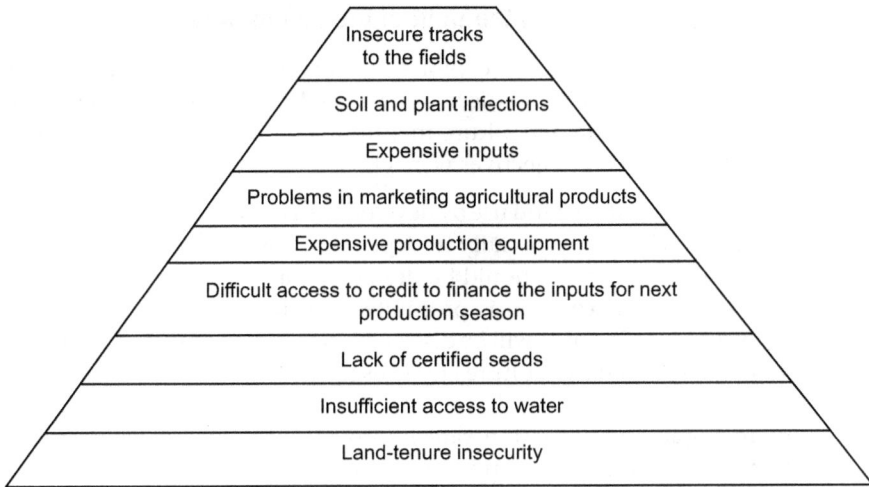

Figure 14.3 Pyramid of constraints encountered by male producers in North Pikine

contamination, and the high cost of equipment and inputs). Women also raise issues related to the employment of paid labour. Problems often occur during the production period, when they are faced with difficulties in selling their products.

Priorities of female producers

The female producers from North Pikine involved in the focus-group meetings (and the pilot project) suggested a number of measures that could improve their activities:

- Organize the women involved in marketing in more formal groupings, in order to put pressure on the public authorities to create more local vegetable markets and to lease these out to them formally.
- Set up a savings and credit system (*mutuelles*), shared by the women's groups, to provide loans to the producers.
- Provide more capacity-building programmes, to enable women to improve the management of their activities, at organizational, financial, and technical levels.
- Enhance the security of the agricultural lands.
- Provide motor-pumps to facilitate access to water resources and increase the value of non-cultivated plots of land.
- Assist the female land owners to shift to a share-cropping system so that they can avoid the problems related to the use of paid labour.
- Improve the transport of the products from the fields by creating/ improving access tracks and re-organizing the plots in such a way that all plots can be reached.

Review of the pilot project in the light of the gender analysis

Drawing on the results of the exploratory survey implemented in 2006 (IAGU–RUAF, 2006) the pilot project was focused on the main obstacles to the development of urban agriculture mentioned by both men and women. Accordingly the following objectives were chosen.

1. Enhancing security of land use by developing a plan for the regularization of agricultural lands, especially in depression zones and areas with high ecological value. These should be designated as permanent agricultural zones (where construction is prohibited) in the city development and land-use plans. The plan will be discussed in the Pikine Multi-stakeholder Platform on Urban Agriculture and subsequently presented to the Pikine City Council.
2. Enhancing access to irrigation water by providing the farmers' organization PROVANIA with a pump and irrigation equipment and establishing a savings scheme to pay for its maintenance and replacement. (A planned scheme for safe reuse of wastewater had to be postponed, due to some problems with a key institution involved in its implementation.)
3. Enhancing access to good-quality inputs by setting up a system managed by PROVANIA for collective purchase of seeds and tools/equipment and their provision to the producers on credit (inputs) or on hire (tools/equipment).

So, with the results of the gender analysis now in hand, what can be said about the gender responsiveness of this pilot project? During the preparation of the project, an attempt was clearly made to understand the roles and contributions of men and women in urban agricultural production, processing, and marketing, before designing the project. The project also involved male and female producers in making decisions about the objectives of the project, resulting in a focus on issues that were identified by male as well as female producers as key constraints limiting the development of urban agriculture. During implementation, equal access of male and female producers to inputs and equipment supplied by the project through PROVANIA was ensured. By doing so, the role of women in the production activities, their access to resources, and their participation in PROVANIA have improved.

But we also observe that the pilot project will benefit mainly women who control some land themselves (since they are more involved in production, can become members of PROVANIA, and have a share in the distribution of inputs and equipment). But there are only a few women in this category (mainly single women who have inherited some land). An important constraint indicated by these women, their problems with hiring/paying male farm labour, has not received attention. Another problem is that the project design neglects the fact that the women's main roles in urban agriculture are in the organization of transport, processing, and marketing of the products, and only to a minor extent – and in a secondary role – in the production. The project focuses more on production aspects (dominated by the men) and gives little attention to the transport, processing, and marketing aspects (which are of equal importance to the men, since they sell their produce through the women and receive from them the financial means required to buy the required inputs). Priority constraints and interests that are specific to women and mainly related to processing and marketing (better access roads, more market places, better organization of trading women and their access to sources of financing, more capacity building related to marketing and processing, etc.) were left unaddressed.

The way forward

PROVANIA is now taking up the above-mentioned issues in the context of the Multi-stakeholder Platform on Urban Agriculture, identifying the institutional actors who can contribute to solving these issues. For example:

- Efforts are made by the Pikine municipality to provide specific places in the city markets for women to sell products from urban agriculture. The municipality also provides better facilities (especially credit) for female vegetable traders through a municipal funding project.
- The credit institution PAMECAS is experimenting with a new financial tool for urban agriculture, specifically women involved in the processing and marketing of vegetables and other products.
- The Strategic Action Plan on Urban Agriculture developed by the Multi-stakeholder Forum gives explicit attention to gender issues; and several institutions, when planning actions, are now taking into consideration the specific constraints, opportunities, and benefits for men and women involved in urban agriculture.

In the design of a follow-up project by PROVANIA, the specific interests of women involved in processing and marketing of agricultural produce will be given more attention. A gender expert working with National Agency for Agricultural Development and Rural is providing advice to PROVANIA.

Main lessons learned

The main lessons from the process of integrating gender in this urban agriculture project are as follows:

- It is essential to demonstrate to all parties the crucial value of women's contribution to the agricultural production and marketing process and the income that it generates. The participatory diagnosis clearly showed the important role that women play in pre-financing the new season, in the production process during peak labour periods, and especially in the (organization of) transport, processing, and marketing of the products.

- It is important to resist the tendency to focus diagnosis and project formulation on production-related aspects. Processing and marketing aspects should be given equal attention.
- It is important that, during the formulation of a local project, male and female producers can independently formulate their priority interests and preferred actions, and that they jointly conclude which actions will be included in the project. Although initially the men still may dominate the final prioritization (as was the case in the PROVANIA project), such procedure establishes the mechanism that women's specific interests are explicitly taken into account and that men accept and support the implementation of related actions.
- A gender-sensitive situation analysis does not automatically ensure that a project will be designed to serve the main interests of women. Gender-responsive planning and monitoring requires more attention. Gender mainstreaming is to be understood as a repetitive process which results in small steps forward during each phase of the process.
- To tackle priority issues regarding gender in urban agriculture requires complementary actions by various institutions. The Multi-stakeholder Platforms on urban agriculture and food security can play an important intermediary and co-ordinating role.
- Tools used in the diagnosis stage (like the access to/control of resources tool, the decision-making matrix, and the distribution of benefits map) can be effectively used in the implementation stage to monitor gender-differentiated distribution of participation in decision making and distribution of benefits of a project.

References

ANSD (2004) *Situation Économique et Sociale de Dakar*, Agence Nationale de la Statistique et de la Démographie, ANSD, Dakar.

Dieng, M (2004) *La viabilité financière des exploitations agricoles dans la zone des Niayes de Pikine* (mémoire de fin d'étude pour l'obtention du Diplôme d'Ingénieur des travaux en Planification Économique et en Conseiller en Gestion des Organisations), École Nationale d'Économie Appliquée (ENEA), Dakar.

IAGU (2004) *Genre et agriculture urbaine dans la vallée des Niayes de Pikine (Sénégal)*, Étude de cas Exploratoire, IAGU-RUAF, Dakar.

IAGU (2006) *Rapport de l'étude exploratoire sur l'agriculture urbaine dans la ville de Pikine (Sénégal)*, IAGU–RUAF, Dakar.

IAGU (2007) *Le financement des agriculteurs et agricultrices urbains de la zone des Niayes*. Étude de Cas, IAGU, Dakar.

IAGU–RUAF (2006) 'Synthese du rapport de l'etude exploratoire sur l'agriculture urbaine dans la ville de Pikine (Senegal)', IAGU, Dakar.

MUAT (2003) *Plan directeur d'urbanisme de Dakar horizon 2005*, Direction de l'urbanisme et de l'architecture, AUS-BCEOM, Dakar.

MUAT (2004) *Élaboration du plan directeur d'aménagement et de sauvegarde des niayes et zones vertes de Dakar. Programme d'actions pour la sauvegarde et le développement urbain des Niayes, Rapport sur les études diagnostiques,* Direction des Espaces Verts Urbains/DDH Ltée/Cabinet Prestige/GEOI, Dakar.

Niang, S. et al. (2005) *Treatment of Health Risks Linked to the Use of Waste Water in Urban Agriculture, (Dakar –Senegal),* IFAN–ENDA RUP, Dakar.

About the authors

Gora Gaye is Project Manager at the African Institute for Urban Management (IAGU), Dakar, Senegal.

Mamadou Ndong Touré is Geographer for the Cabinet d'Architecture et d'Urbanisme du Sénégal (CAUS), Dakar, Senegal.

PART II
Guidelines for Gender Mainstreaming in Urban Agriculture Research and Development Projects

CHAPTER 15
Incorporating gender in urban agriculture projects

Abstract

This chapter suggests how to include gender systematically in various phases of the project cycle: from diagnosis to design, action planning, implementation, monitoring and evaluation, and going to scale. For each project phase important steps are highlighted. At the end of each section some specific tools are suggested which may be used in this particular project phase (these tools are discussed in detail in Chapter 17), in addition to questions which can be used as a checklist during that phase.

The urban agriculture project cycle

As discussed in previous chapters, the aim of gender-sensitive management of research and action projects is to enable men, women, and youth to participate in development processes on equal terms, both as agents and as beneficiaries. The case studies featured in Part I reflect this aim, and the results and experiences stemming from the cases form the foundation of this chapter. Specifically, lessons learned from original testing of guidelines have been used to adjust and improve those presented below; examples from case studies presented in Chapters 2 to 14 provide illustration of key points about gender within each specific phase of the project cycle.

Gender sensitivity and mainstreaming begins with the following:

- Acceptance of the principle of '*equal human rights for all*'.
- Acknowledgement of the real *value of women's contribution* to development: production, food security, income, etc.
- Recognition of *women as independent actors in and beneficiaries of* public policies and projects.
- Recognition that the *needs of men and women are different,* and that women's access to and control over resources and their participation in decision making are restricted by socio-cultural and institutional traditions.

- Recognition that public policies and projects, as well as economic and technological trends, can have *differential effects on men and women.*
- Recognition that *affirmative actions are needed* to ensure that women (and men) can reap equal benefits from public policies and projects.
- Recognition that advancing a gender-sensitive approach requires *cultural tact and diplomacy* if embedded constraints (such as traditional cultural norms and institutional sexism) are to be overcome and resistance minimized.

An effective framework for gender-sensitive project management, and the one chosen for the case studies and this publication, is the project cycle. A project 'encompasses a specific range of resources and activities which are brought together to generate clearly defined outputs within a given budget and a specified period of time. Compared to a programme, a project is more specific and has more defined targets and time frames' (GWA, 2003). The project cycle consists of a set of steps that provide the basis for adequate preparation and implementation of a project, as well as opportunities to learn from its results.

The outputs for urban agriculture projects will be somewhat different and the project cycle will vary, depending whether the project is more research-oriented, such as the case studies of Ghana (Chapter 4) and Carapongo (Chapter 13), or development-oriented. Even in strongly development-oriented projects, such as the case study of Villa María del Triunfo (Chapter 8), which features assessment of a multi-stakeholder action-planning process, research almost always forms at least some part of the cycle, for example in the diagnosis phase and in contributing to the design, planning, and implementation processes. In projects that are primarily research-oriented, not all phases of the project cycle will always be included; or there may be shifts in the sequence of phases. For example, projects may involve a research-design phase (hypotheses, definition of research objectives, design of research methodology), which precedes diagnosis and implementation.

For clarity, we present the broad phases of the project cycle as follows, highlighting where necessary special considerations related to research or development:

1. **Diagnostic research.** In this initial stage, local needs, problems, and opportunities are determined by reviewing the results of earlier research and undertaking additional studies (especially participatory analysis activities, applying rapid appraisal methods). Diagnosis focuses on the analysis of existing urban farming systems and broader livelihood strategies of farming households, as well as local conditions in the target areas, in order to define important problems, needs, perspectives, and opportunities.

2. **Design.** During this phase the project's goals and strategies are identified through a joint process of 'visioning' the desired development and project outcomes, and evaluating alternative strategies that might be applied to

realize those outcomes, by looking into their viability and effectiveness to produce the required outcomes. In research-oriented projects, some of the strategies identified may include further (participatory) research into, for example, technical options.

3. **Activity planning.** During this phase the goals and strategies are operationalized by identifying activities needed to implement the strategies, developing methods and tools to be applied, dividing responsibilities and tasks among the participating stakeholders, defining co-ordination and monitoring mechanisms, and developing a budget and timeframe. Also all necessary practical arrangements for the actions are made. Action planning may include action research to test alternative solutions to a given problem.

4. **Implementation.** During this phase the planned activities are implemented and, if need be, adapted to fit the specific context better.

5. **Monitoring and evaluation.** During implementation the project is monitored and periodically (auto-)evaluated in order to:
   - assess whether the project is on track with regard to the realization of planned goals, strategies and activities as set out previously;
   - solve problems that emerge during implementation;
   - enhance learning-from-actions by periodic reflection on experiences gained from implementation processes, as well as results obtained (for example, how to get more/more relevant/better-quality results; how to enhance cost-effectiveness; how to enhance participation, local ownership, and sustainability and out/up-scaling of the results, etc.).

   In larger projects, external reviewers/evaluators may be involved to realize a mid-term and/or end- of-project evaluation, to systematize results obtained, and draw lessons for uptake in policies and future action projects.

6. **Going to scale.** During this phase the focus is on planning follow-up actions, dissemination of results/information to relevant stakeholders, and influencing policy making in the same city or similar areas (out-scaling) or at the national level (up-scaling). Learning from the project is used to plan new research or development projects and to develop new policies (or revise existing policies) on urban agriculture.

It is important to note that although this chapter presents key phases of the project cycle as a linear process, the cycle is more reflective of a spiral (Figure 15.1), with the project passing several times through the phases of planning, action, observing or monitoring results, reflecting on lessons, and re-planning. This iterative process also underlines the fact that one or several phases of the cycle may appear more than once within a project cycle, or may have strong links and cross-over with several other phases. In particular, monitoring and evaluation frameworks, specifically the indicators or questions used to assess realization of goals, strategies, and activities, often become an

OBSERVE OBSERVE OBSERVE OBSERVE

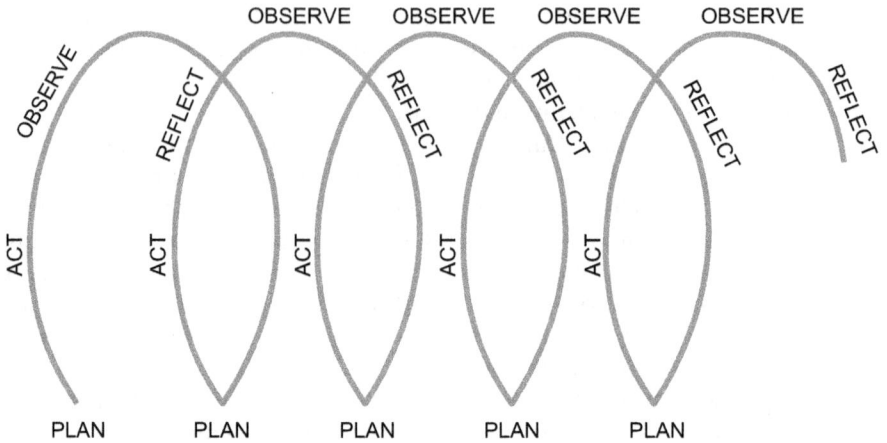

PLAN PLAN PLAN PLAN PLAN

Figure 15.1 The project cycle as a spiral process
Source: Gender and Water Alliance, 2003

integral and central focus of design, planning, and implementation phases of the project cycle. As such it is important to recognize and operationalize the fact that monitoring and evaluation is not necessarily the 'fifth phase' of the project cycle but rather underlies, and becomes a foundation for, the entire project cycle. The uptake of project results into local or national policies or programmes can most effectively be managed if there is frequent communication during planning and implementation phases with relevant stakeholders to facilitate up- or out-scaling. Or further, results of research and diagnosis might be fed directly to policy makers or other planning projects; monitoring and evaluation during implementation can likewise identify new research questions and be shared with available research organizations.

Above all, the project cycle in its entirety and throughout its duration must reflect a commitment to gender sensitivity and mainstreaming. This requires, first of all, a commitment in the spirit and content of project-cycle components such that the focal point at all phases is illuminating the differences and potential inequalities between men and women in urban agriculture, and developing and operationalizing actions that may lead to more gender-sensitive planning and policy-making. Second, this requires a commitment to training and capacity building of personnel in the use of gender-sensitive tools and guidelines. Those persons adopting the tools presented in this book, for example, must learn how to lead and facilitate, for example, a focus group or one-on-one interview in a way that will build trust among project participants and encourage people to share details about the roles, expectations, and experiences of men and women within a particular context. This also means that participants must be clear on and committed to the importance of highlighting gender in urban agriculture activities: the benefits of participating in a gender-sensitive project cycle must be clearly stated up-front. Commitment to gender training and facilitation necessarily means that project or even programme budgets must be allocated specifically to such activities.

In the remainder of this chapter we will describe the steps to be taken in each phase of the project cycle. The description of each phase ends with an overview of the tools that can be particularly useful in that specific phase. A full description of these tools and how they can be applied is provided in the Tool box in Chapter 17. While certain tools have been suggested for use in a certain phase, it is possible that they may also be relevant to other phases of the project cycle. It is also possible that guidelines or tools may need to be adapted to local contexts to address specific needs and interests. We encourage practitioners to approach both the guidelines and tools with flexibility and creativity, to ensure that they are as relevant and insightful as possible. This spirit was evident during the tools-testing phase, as described in the Preface; project partners took a mix-and-match approach to the tool box, adapted tools as needed, and even created new tools by combining elements of those presented in the text. The result was positive, with important gender dynamics highlighted at different phases of the project cycle.

Phase 1: diagnostic research

Although there are significant differences between the tool boxes used in mainly research-oriented projects and those primarily oriented towards development goals (to be further discussed below), both depend on some form of diagnostic research to make an assessment of the social, economic, political, and environmental circumstances in which people live and work, to analyse their farming systems and livelihood strategies, and to define the main problems, needs, resources, and opportunities.

The framework of gender issues presented in Chapter 1 provides a guide for a *gendered* analysis of the local situation. Gender analysis provides an examination of the respective interests, problems, and needs of men and women in the community, and their implications for urban agriculture. Starting with a gendered analysis of the local situation offers a snapshot of the project's point of departure. It recognizes that each project situation is unique: there is no other situation with the exact same combination of gender roles, resources, political circumstances, etc. Important questions to be answered in this phase focus on a community's problems and challenges, but also on its available resources, capacities, and opportunities for development. The findings of this phase provide the project team with valuable information on what kinds of intervention would be useful for whom, and on how to conduct these interventions. As the findings of this phase lay out the map for the next phases, it is very important that diagnostic research be carried out with great care, fully involving all community members, including men and women equally.

Some of the important steps to follow in this phase are (1) the identification of key themes, methods of data collection, and participants, and (2) the collection of data itself. These are detailed below.

Identification of key themes, methods of data collection, and participants

The process normally starts with a review and synthesis of available data (see Tool box on review of secondary data) on urban agriculture in the city or target location. These data may include land-use data or maps (e.g. zoning, actual use, planned use), socio-economic data (e.g. the number and location of households below poverty line or with food-security problems), data on food habits and nutrition status (e.g. presence and location of undernutrition or malnutrition in the area or within specific households, detailed purchasing and consumption patterns, farming-activity typologies, etc.).

The data review will help in the identification of important *key themes* for the participatory diagnosis of an action project, or key questions and working hypothesis for a research project. It will help to identify key informants and the different focus groups with which one will be working during participatory diagnosis, or the main categories of respondent for research activities. For example, in Rosario (Chapter 10), documenting and assessing the Urban Agriculture Programme in terms of its objectives and outcomes formed the basis of subsequent gender-data collection and analysis in the city. In Nakuru (Chapter 11), the review of secondary data and discussions with community-based organizations helped to focus attention on land, income, and food security, which became central elements of individual interviews with men and women involved with urban agriculture activities.

Also during this step the *methods* to be used to collect the required information will be selected. Different methods are available, with different strengths and weaknesses. For example, household surveys using questionnaires have the advantage of quantifying findings, but also the weakness of expense, duration, and superficiality. The disadvantages of formal surveys have been commented on at length (e.g. Pretty, 1995), but if there is a need for a baseline for measuring change, then they are very important (see Tool box section on questionnaires). In another example, more rapid and participatory appraisal techniques (e.g. seasonal calendars, decision-making matrices, etc.) have the advantage of speed, low cost, and better understanding of processes, but also the disadvantage of limited quantification and ability to generalize results.

Using both quantitative and qualitative tools can be an effective way of ensuring that data collection is comprehensive and systematic. While statistical figures are attractive to many key stakeholders, they too often lose their significance in terms of what they mean for people on the ground. Qualitative or textual accounts of people's experiences can go a long way towards establishing a compelling argument for a particular project strategy or development intervention. For example, the Pikine case study (Chapter 14) presents findings about men's and women's access to and control over resources first in terms of percentages, and second in terms of brief explanations of what these percentages mean (see Table 14.2). While the statistical figures illuminate gender imbalance in land access (with men having access to and control over 75 per cent of this resource), qualitative details provide insight into the fact that single women are in a better position than married women in this regard; as a result, gender differences between women have been identified in this particular case study which may be important for understanding the local context and ultimately for development planning.

How to measure particular aspects of men's and women's circumstances, experiences, and perceptions, and how to communicate them to an audience through information dissemination, requires some attention to the rating or ranking system that is used in conjunction with the methods available in the Tool box and elsewhere. Researchers and project staff should pay attention during the diagnosis phase to ensure that research findings reflect the data collected. For example, by using a questionnaire survey, it may be revealed that farmers express varying degrees of satisfaction with urban agriculture extension workers from a government department. In order to measure 'degrees of satisfaction', a rating system based on five stars is developed whereby one star indicates complete dissatisfaction and five stars complete satisfaction. Standardizing such rating systems throughout the diagnosis phase will ensure that people's perceptions on the ground are adequately captured through data collection. The rating system itself must be explained in any documents or scenarios where data dissemination takes place, to ensure clarity.

In the diagnosis phase of urban agriculture projects Participatory Appraisals (PA) rather than formal surveys will normally be applied, with additional advantages of flexibility and innovation. Teams using PA should determine the best combination and sequence of techniques to be used and adapt existing techniques or invent new techniques when needed. The Tool box in Chapter 17 provides a selection of gender-sensitive PA tools. More examples of gender-sensitive PA tools and discussion of their application in the diagnosis phase of urban agriculture projects can be found in De Zeeuw and Wilbers (2004) and Martin, Oudewater, and Gündel (2002). Two guides that give attention to making PA tools more gender-sensitive are Pretty (1994) and FAO (2001).

It is preferable to select techniques that complement each other in the type of information they supply and/or the effect they have on the participatory process. Some techniques are applied in work with individual households (e.g. farm diagram, household decision-making matrix, etc.), while others are applied with a small group of informants (e.g. transect walk, resources mapping, organizational linkages diagram) or a specific category of the population, to dig deeper in the analysis of certain problems/causes (e.g. focus-group discussions) and the perspectives of informants, and to prioritise certain problems or opportunities (e.g. ranking exercises). The case studies in Part I of this publication draw on a range of techniques used in diagnosis, with many individual cases drawing on a combination of approaches to generate holistic diagnostic results that inform subsequent stages of the project cycle.

Although most experience with PA is in a rural setting, the techniques can also be applied in an urban situation. Urban Harvest partners have developed the methodology for and carried out a Participatory Urban Appraisal (PUA) in Kampala and other towns (Prain et al., forthcoming). This and other experiences underline the need to take care to adapt the PA techniques to the specific urban conditions.

Some special considerations in this respect, taken from Prain and De Zeeuw (2007), are listed below.

- Whereas in the rural context PA is normally implemented in villages where people know each other and the local social and economic fabric reasonably well, this might not be the case in the urban situation, where communities are more fragmented, more complex, and subject to more rapid changes.
- Whereas in the rural situation farming families and their resources are often easy to identify (although there are exceptions), this frequently turns out to be a fairly complex issue in the urban context, where the house of a family and the plots farmed may be far apart, plots may be illegal and so not mentioned, and number and size of plots or animals raised may vary considerably over time.
- Whereas in rural farming households most members who are present consider farming or directly related activities their major occupation (and, for many, their only occupation), in urban households livelihood strategies often are more varied, with farming often practised by one or two members only and as secondary to non-agricultural activities. Many poor urban households are broken or scattered.

Since PA techniques enable local people to express their needs, views, problems, perspectives, and priorities, they are also very useful for the analysis of gender differences in a community if the techniques are applied in a gender-sensitive way. This often implies that the PA exercises are implemented with men and women separately (often followed by discussion of results in a mixed group), as will be discussed in the next section.

It is also necessary to select the persons who will do the fieldwork and, if needed, provide them with PA and gender training (including training on key gender issues in urban agriculture, adequate communication with men and women, the adequate use of the selected data-collection techniques, 'what to do if ...' situations, etc.). Authors of the Manila (Chapter 3) and Kampala (Chapter 5) case studies noted in their comments on guidelines testing that training of staff to conduct gender diagnoses is a particularly important element of mainstreaming in order to generate meaningful gender-disaggregated and gender-analysed data, and to ensure update of gender insights into subsequent stages of the project cycle. This training can be part of the participatory planning of fieldwork. Women researchers and local staff may have better access to and rapport with local women than do men. It is possible that issues related to gender dynamics may be particularly sensitive and that people may not be willing to share details or information; this is often dependent on who is asking the questions, and whether or not they are perceived as an 'insider' or 'outsider' by the community at hand. In these instances it is particularly important that researchers and staff are properly trained to unearth local gender dynamics in a way that is seen as helpful rather than harmful to participants.

Finally, it is important for those responsible for diagnostic research to bring back their findings to the community and to engage with ongoing processes of change in order to increase the impact of research findings – as, for example, was done in the case of the Urban Harvest's research work in Kampala and Nakuru (Prain et al., forthcoming). The Nairobi case study (Chapter 9) also features dissemination of research results as a methodological strategy; feedback workshops provided an opportunity for community members to discuss results and for further information to be collected on gender circumstances and experiences.

Data collection and analysis

Men and women have different views of reality, encounter different problems, have different interests and needs, and use other criteria to judge 'solutions' or 'innovations', all based on the gender differentiation of roles, tasks, and responsibilities. Hence, in order to arrive at a thorough understanding of the local situation, and a gender-balanced identification of development problems and opportunities in urban agriculture, women and men should be equally involved in the diagnosis or research process.

Only when gender-differentiated information is available (for example on the tasks, role in decision making, access to resources, problems encountered, needs, interests, perspectives of both men and women in urban agriculture) can project planning be made more gender-responsive, as it allows one to choose the right strategies for/with the right sub-sets of the urban target group. It further allows for monitoring and evaluation of the impacts of the project on men, women, children, and other groups.

Hence, gender-disaggregation of data is an absolute requirement of getting good information for urban agriculture projects. The most important thing about collecting gender-differentiated data is asking gender-related questions in the first place, and this is something that has not always happened in the past. There are numerous examples of gender-disaggregated data within the Part I case studies, including detailed information on division of labour, access to and control over resources or benefits of resources, and gender-differentiated opportunities and constraints.

Furthermore, it is important to realize that information about what men and women *do* in urban agriculture (i.e. gender division of labour) can be gathered from anyone around who knows the situation. However, what men and women believe, prefer, and prioritize, and their perspective on the desired developments, can be stated only by themselves. This means that when looking for more general factual information (e.g. Is there a well here? Is the group leader a man or a woman?), we may either interview a man or a woman or find the answer by making observations. However, for types of information where the answers depend on personal knowledge or preferences, we will have to interview the specific persons (men or women) who hold that personal information or those personal views.

PA techniques are the best ones to apply in many situations where information is needed about the different practices, preferences, and priorities of men and women involved in urban agriculture in a particular place. It is advisable to consult separate groups of women and men, since women often feel constrained when speaking out in public with men present. PA case studies often involve certain levels of quantification about a specific location (for example, the participants in the focus-group meeting said that about 75 per cent of the households keep chickens and that it is normally a woman who looks after them) but without making it possible to generalize to the whole population or city.

Where numbers or percentages are needed, household surveys among a representative sample of the population are usually carried out, using written questionnaires and structured interviews. These provide numerical information about those households' levels of wealth and health, and what they do as urban producers. To achieve gender-disaggregation, the results of the survey are usually divided into findings from men-headed and from women-headed households. While providing much useful data, disaggregating by the sex of the household head will not tell you everything you need to know about differences in gender behaviours. Logical steps must be applied to get really useful results. For example, a household survey in Manila, Philippines (see Table 3.2) tells us that according to both men and women, women have 90 per cent access to and control over the income from sales. Further questions reveal that 10 per cent of income is specifically allocated to men's social activities, over which the women have no say whatsoever, while women are not offered an equivalent benefit. This type of insight is important to investigate in order to get a clear picture that goes beyond numbers into actual dynamics between family or community members.

In many situations, equal involvement of men and women in the diagnosis is not easy to obtain, owing to the low participation of women in public discussions and decision making, their low level of literacy and education, cultural restrictions or isolation, etc. It is sometimes the case that elders are reluctant to name women as key informants, if tradition casts men in the role of community spokespersons; or men do not value the contribution of women, while women themselves are not always convinced about the usefulness of expressing their views and ideas. In some instances, those implementing the diagnosis phase themselves may unintentionally bias participation towards men at the expense of women. For example, if the project highlights those farmers involved in commercial urban agriculture (in many contexts dominated by men), then researchers may inadvertently miss out on women's involvement in urban agriculture (which in many contexts is subsistence-oriented). Or if the target population to participate is identified as those who 'own' land, then more men than women (in many instances) will be selected to participate, given their greater access to landholdings. Women are often excluded from data-collection efforts when the category of 'head-of-household' is used to identify a participant; men are usually identified as household heads even if women are those predominantly involved in urban agriculture activities (and thus their insights and perspectives are neglected).

Therefore, special attention needs to be paid to involving women, by taking the following measures:

- Choose a time and place convenient for both men and women.
- Use adequate techniques that appeal to women and encourage their participation.
- Use male and female staff, since the latter establish easier contacts with urban women.
- Ensure that the agenda of interviews and focus-group discussions includes items which are of primary interest of women.
- Consider the language used, given that women often do not speak the official language. (In such a case, use translators and/or use visualization techniques)
- Combine a variety of techniques so as to get insight in each of the main gender aspects. Table 15.1 presents an overview of the main gender issues in urban agriculture (see also the framework presented in Chapter 1) with a suggested PA tool to study each issue. The tools have been suggested for their cost-effectiveness and their ease of application.
- Pay attention also to the historical perspective and trends to get more insight.
- Make sure that all data collected differentiate between men and women (but often also for age group, socio-economic status, etc., since the

Table 15.1 Overview of main gender issues in urban agriculture and recommended PA tools to be used

Main gender issues in urban agriculture	Recommended tools (see Tool box, Chapter 17)	Eventually combined with (see Tool box, Chapter 17)
Division of agriculture-related labour, tasks, and responsibilities	Gender-activity analysis chart Seasonal calendars	Direct observation Semi-structured interviews of male and female members of selected households Group interviews with women and men
Decision-making power and distribution of benefits	Decision-making matrix Benefits analysis	
Access to and control overresources	Resources analysis (chart and mapping)	
Constraints, problems, opportunities	Problem and opportunity ranking and analysis	
External factors	Analysis of institutional linkages and timeline variations	

conditions and interests of men and women are not always the same). In that light it will be required that data are collected about (or from) men and women separately: interviewing men and women in separate (individual or group) interviews (eventually followed by a discussion in a mixed group).

Analysis of the information generated in the diagnosis phase of a project will focus on the following:

- Characterization of the urban farming system(s) concerned (crops grown and animals, type and level of technologies used, resources used and their origin, main cultivation practices, degree and channels of marketing, etc.).
- Characterization of household livelihoods (resources, activities, vulnerabilities, internal and external relationships, etc.).
- Characterization of the institutional context of urban agriculture.
- The identification of key problems in the functioning of these urban farming systems (and their causes and consequences) as well as existing potentials and opportunities for their development.

It is important at this stage, or ideally earlier, to reflect on those participants featured in the diagnosis phase of the project cycle. Which social groups have been included or excluded? For example, does the diagnosis account for the circumstances, experiences, and opinions of landholders but neglect to account for those who cannot gain access to land? Or what household circumstances are left unexplained? Are only households with a male–female couple included, to the exclusion of female-only households? These sampling choices within the diagnosis phase will have a major impact on information gathered at this stage, in terms of what one can say about gender and urban agriculture dynamics, which will ultimately be used to design, plan, and implement project activities in subsequent phases. For example, in the Mexico case study (Chapter 12), women and men were interviewed separately to ensure that each social group had the freedom to speak without being interrupted or overshadowed by particular individuals; this sampling strategy aimed to account for both men's and women's opinions equally. Or in the Nairobi case study (Chapter 9), particular attention was paid to youth activities in urban agriculture as a way of diagnosing the circumstances and experiences of boys and girls, in relation to those of the community in general.

In PA-based diagnosis, the analysis of the collected information will already start during the fieldwork. In team meetings, early results of the fieldwork are reviewed in sequence, which may reveal gaps and contradictions in the collected data or questions about how to interpret certain data correctly. This will lead to complementary information collection and 'digging deeper' during farm visits, interviews, and focus-group discussions, in order to achieve a better understanding of existing problems and opportunities, causes and consequences, and related gender differences.

The analysis should be made from a gender-sensitive point of view, identifying the following factors, among others:

- The roles and responsibilities of men, women, and children.
- Who has access to and control over available productive resources.
- How men and women participate in decision making at farm, household, and community levels.
- How knowledge/skills, problems, and opportunities differ between the men and women involved in these urban farming activities.
- What the strategic and practical needs and priorities of these women and men are.
- Social, cultural, political, legal, and economic factors related to gender-based distinctions found in this local situation. Which of these factors can be expected to change in the coming period?
- What the project could/should do to incorporate gender considerations in the project design.

To achieve analytical depth, the key question to be raised continually is 'Why?' It is not enough to document 'what' the differential roles and responsibilities of men and women in a particular context are, or 'how' those affect urban agriculture activities. It is important to understand 'why' this is so. For example, limiting diagnosis to a listing of who does what in terms of urban agriculture tasks can help to set the scene so that resources from development interventions can assist those in particular tasks (e.g. if women are involved in water collection, then an appropriate intervention may target improved access to water pumps within a neighbourhood). However, as revealed in case studies from Kampala (Chapter 5) and Kisumu (Chapter 7), if one wishes to understand why it is that only women collect water then asking 'why' may reveal deeply rooted assumptions about 'women's work' and 'men's work'. In this instance, 'women's work' may be seen as that which is vital to household well-being yet is not directly paid; 'men's work' may be seen as that which is rewarded with income, so husbands might be preoccupied with cash-crop production. A development intervention aimed at increasing women's access to water may actually have the effect of providing men with closer sources of water for their production efforts, and thus the water source becomes controlled by men.

In-depth analysis of gender dynamics in urban agriculture also must involve an investigation of intra-household relations so that the 'behind the scenes' divisions of labour, decision making, resources, etc. and their impacts on men's and women's lives can be better understood. It is important in gender analysis to make distinctions between social ideals and the realities of men's and women's lives. In some instances, when asked about gender roles and responsibilities, both men and women may speak to the broader social context (in other words, what men or women *should* be doing) rather than detailing what they *actually* do. Going into analytical depth allows researchers and staff to investigate differences between perceptions and reality.

Beyond collecting in-depth gender-sensitive information regarding urban agriculture, it is important to recognize the broader context in which such dynamics operate. For example, urban agriculture is only one livelihood strategy that urban residents engage in to generate income and provide for their families. When investigating this phenomenon specifically, one can also consider the other types of livelihood activity that an individual or a household may be engaged in, particularly to consider how these activities may be traded off or used in combination to achieve household food security or general welfare. In other words, what other activities beyond urban agriculture do urban residents participate in, and how do these activities affect – or how are they affected by – urban agriculture?

Similarly, when investigating gender specifically, it is possible that other key identifiers such as age, class, ethnicity, etc. also influence urban agriculture roles, responsibilities, and activities. In some contexts, then, it will be equally important to understand age *and* gender and to ask how these two identifying characteristics shape different roles and responsibilities for women/girls and men/boys. For example, the case study from Sonora, Mexico disaggregates the division of labour by both gender and age (Table 12.1), revealing the significant extent to which young boys and girls, as well as elderly persons, are involved in urban agriculture activities.

In other situations it may be important to consider not only a contemporary snapshot of gender dynamics but also a historical overview of how urban agriculture has developed and how gender differences have evolved; gendered histories reveal that men's and women's roles and responsibilities are deeply entrenched in socio-cultural traditions (e.g. position within the household or community) and political-economic structures (e.g. land, legal systems). For example, the rich institutional history of the Musikavanhu Project in Harare (Chapter 6) provides insights into the key stakeholders and programme structure involved, and identifies opportunities for gender mainstreaming based on these elements. The case of Mexico (Chapter 12) notes the ways in which men and women involved in urban agriculture make use of cross-border social networks and economic scenarios to purchase inputs for production or sell goods, and how that has changed over time. Thus while researchers and staff should remain focused on gender and urban agriculture specifically, it is vital to do so within the larger context of various livelihoods, identifying the characteristics, histories, and political-economic structures that shape and are shaped themselves by gender.

Preliminary analysis normally will be 'validated' by organizing feedback meetings with the target population in the locations studied, to present main findings and give informants the chance to correct the team's observations and interpretations, as well as to jointly 'dig deeper' in the analysis of the main problems and potentials identified. In this way the stakeholders themselves analyse their situation and deepen understanding of these issues. Such feedback workshops often mark the transfer to the design stage, since in the same workshop the priority of the identified problems and opportunities may be defined, as well as the acceptability of and preference for certain solutions and intervention strategies. Note that the diagnosis should focus not only on the analysis of problems (and identification of potential solutions for such problems) but also on the identification and analysis of opportunities: local innovations that can be further developed, under-utilized resources that might be used, new market opportunities, etc.

It is important to note that all sources of data can be valid for gender analysis, even if they are not obviously gendered or gender-sensitive. For example, documenting men's circumstances, experiences, and perceptions is as valid as documenting those of women, given that gender analysis requires attention to both men and women. Or interviews with key informants who are *not* gender-sensitive can be as enlightening as interviews with gender experts, largely because of what is not said, or in order to get a sense of how gender-insensitivity operates. Similarly, secondary data (e.g. census, documents, reports, etc.) that are gender-insensitive require a 'reading between the lines' in terms of what is not said about gender and making its invisibility visible. Finally, one cannot assume that women know it all in terms of illuminating gender dynamics in a particular context, as they may not be aware of these themselves.

Box 15.1 Recommended tools for diagnostic research

- Activity-analysis chart
- Seasonal calendar
- Resources-analysis chart
- Resources mapping
- Decision-making matrix
- Benefits chart
- Problems and opportunities identification and ranking

Box 15.2 Gender-mainstreaming questions for diagnostic research

- Did men and women actively participate in identifying and analysing the local situation?
- What methods were used in the identification of the problem or situation? Did the analysis yield gender-disaggregated data on all issues investigated?
- Was the situation or problem analysed from a gender perspective? Did the analysis identify:
  o Roles and responsibilities of men, women, and children?
  o Who has access to and control over available productive resources?
  o What the strategic and practical needs and priorities of women and men are, and whether these differ?
  o Social, cultural, political, legal, and economic factors related to roles and responsibilities, and access to and control over resources?
  o What the project could/should do to incorporate gender considerations?
  o What changes in gender-based distinctions found in this local situation can be expected to change?
- Were women involved in decision making on the priority issues to be attended?
- Who are most affected by the problems selected as key priorities?
- Who will benefit most from the opportunities selected as key priorities?
- What is the level of preparedness of men and women to get involved in the project?

Phase 2: project design

In the design phase, the project goals are defined, as well as the strategies through which these goals will be attained, taking as a starting point the needs, problems and opportunities of the local producers (men and women) that were identified during the diagnosis phase.

It is important that female urban producers and male urban producers are consulted equally during the design of the project, so that the needs and interests of varying social groups within a particular context are identified, strategized around, and eventually addressed. This may require special efforts and creativity on the part of the institutions involved in order to ensure the required conditions; for example through the use of female staff, working in separate male and female groups, selecting appropriate times and venues for meetings, adapting particular languages, using visual aids, etc. The Rosario case study (Chapter 10) highlights steps taken to gender-mainstream project design through 'encounters' with female producers; these meetings helped to visualize the role of women in the project-design process and provided skills training to enhance leadership capacity and self-confidence.

The issue of gender training and capacity building of staff involved in the project team is important in the design phase, particularly in terms of facilitating participatory processes that encourage active collaboration during the design phase and give voice to those community members, in many instances women, who may not have had opportunities in the past to identify their concerns or state their opinions. The inclusion of gender balance in the project team is significant in this participatory process, given that women may share information more easily with other women, while men may be more likely to share with other men. Also prominent at this stage of the project cycle is the issue of group dynamics and cohesion. Facilitators well trained in recognizing and mediating gender dynamics will also be aware of how power relations among individuals can shape project design and resulting strategies. Groups that can come together in a concerted vision, acknowledging and respecting views and circumstances of both men and women, will generally be more successful in their design efforts than those who cannot diffuse internal strife and conflict.

Throughout the design phase, it is useful to take a multi-stakeholder approach to identifying problems and opportunities within a local community or context, as well as formulating objectives, results indicators, and project strategies to address issues raised. Engaging with persons, organizations, or structures beyond the locality is a step towards ensuring sustainability of the project as a whole and up-scaling particular strategies. A multi-stakeholder consultation process during the design phase, as was conducted for example in the Nakuru case study (Chapter 11), may involve producers, women's groups, food-security coalitions, city counsellors, government ministries, NGOs, etc., allowing each actor to assess the opportunities and constraints emerging from design discussions among local male and female producers. Collaboration early on in the project cycle can strengthen the possibility that more informed, appropriate, effective, and efficient design of strategies occurs. An additional consideration here is the fact that urban agriculture activities necessarily take place within a larger urban context. Those tasked with project design should keep an eye on broader issues of, for example, land and water allocation/access, regional politics, economic trends, city planning, etc. In essence, the city influences the opportunities and constraints facing both male and female urban farmers.

Important elements in the design phase are: (1) problem and opportunity analysis and prioritization, (2) definition of project objectives and results indicators, and (3) strategy development. These are detailed below.

Problem and opportunity analysis and prioritization

In a workshop with representatives of local producers (men and women), the inventories of problems and opportunities are drawn up and analysed, using the results of the diagnosis as well as inputs from participants. Both problems and opportunities in the inventories are then prioritized.

The ranking will be conducted separately for women and men and then compared and discussed in order to agree on priorities that reflect the interests and needs of both equally. Important criteria for prioritizing may be, for example, urgency, number of people affected/who will benefit, availability of practical solutions or market demand, availability of local resources, etc.

Priority problems will be analysed for key causes and consequences in order to identify alternative strategies to solve the problem, and to discuss their acceptability and preference for certain solutions and intervention strategies. Various options in terms of what could or needs to be done to develop such potentials and use such opportunities will be explored.

Definition of project objectives and results indicators

Based on the outcomes of the above step, it will be easier to define the expected project results and to formulate the project objectives that must be realized to achieve the said results. To do so, the objectives should be formulated in a SMART way such that they are: specific, measurable, appropriate, realistic, and time-bound (see also the Tool box in Chapter 17). This formulation process is also conducive to developing monitoring indicators and tools that coincide with both objectives and results within the design stage (see phase 5 for more details on monitoring and evaluation).

Once the objectives have been formulated, one should check whether the problems and opportunities prioritized by the women producers are still well represented in the project objectives. Experience shows that the needs and interests of women tend to disappear as soon as information is integrated and we move closer to concrete action planning. As an extra assurance that gender will be given sufficient attention during project implementation, an objective can be included that is specifically focused on enhanced gender equality and for which a special allocation of resources is made.

Strategy development

In this step the courses of action and related working methodologies that are expected to realize the objectives, based on the results of the two steps above, are identified.

The proposed strategies should be carefully checked for the degree to which:

- The strategy will solve the key problem or develop the identified potential/opportunity.
- Economic resources will be needed to implement the solution. Attention here should be on the expected benefits for men relative to women beneficiaries (level of benefits, number of people involved), as well as negative trade-offs and costs for men relative to women (level of benefits and negative trade-offs/individual costs, number of people involved).
- Implementing organizations have the available expertise and capacity to implement this solution adequately.
- Producers (men compared with women) have the knowledge, skills, land, water, labour, cash, and other requirements of this strategy; and the degree to which the intended beneficiaries have these available to them.
- The proposed strategy contributes to enhancing gender equality at farm-household and community levels.
- This solution makes use of local knowledge and innovation capacity (of both men and women) and the available under-utilized natural resources.
- The proposed solution meets existing/expected market demands.

The selection of strategies will depend on the type of project, but it should involve a dialogue between the project team, male and female producers, and other local stakeholders. Both women and men, including the young and the old, should be involved in the selection of strategies, since their conditions are different (e.g. responsibilities, access to resources, knowledge, etc.) and special constraints for women have to be taken into account (e.g. restricted mobility, high degree of illiteracy, inexperience with speaking in public or management functions).

In projects that are more research-oriented, strategies may include participatory evaluation of alternative technologies or practices. This requires careful assessment of whether women and/or men are the main users and main sources of innovation associated with the technologies.

Where men and women have different interests and activities, distinct men's and women's strategies may be identified and carried through to implementation. Furthermore, it should be verified whether gender-affirmative strategies are necessary to overcome existing inequalities and barriers for female participation in order to be able to realize the project objectives. During the diagnosis phase, special constraints encountered by women or other disadvantaged groups might have emerged, and factors that limit their participation in the project or reaping the project's benefits. Affirmative actions will have to be designed to overcome such constraints.

For instance, in order to enhance participation of women in urban producers' organizations and project management, most projects create special opportunities for women to build up their self-esteem and participatory competencies, such as the establishment of women's groups to discuss their interests, and training of women leaders to enhance their capacities in management and participation in public discussions and decision making. Other examples of such affirmative actions include guaranteeing a percentage of available credit for female producers, applying differential minimum requirements for male and female project staff to ensure gender balance in the team, and/or creating a 'fast track' for further capacity building and promotion to higher levels of responsibility for women with good potential.

As was emphasized above, although these guidelines give the impression of a linear process, project design is an *iterative* process in which one often goes through the various elements more than once and not necessarily in the same order. It is useful, once the activity or project has been designed, to subject it to an evaluation to assess relevance to the critical gender issues, problems, and opportunities defined during situation analysis and for differential impacts on men and women. This could be done in a session headed by the gender specialist.

It is important to check whether the identified strategies logically lead to the desired results or whether alternative or complementary activities are needed. The preparation of an 'intervention logic map' or 'logical framework' might be very helpful at this time.

Box 15.3 Recommended tools for project design

- Problem- and opportunity-analysis chart
- Problem tree; opportunity tree
- Formulation of results-based objectives
- Group definition
- Institutional linkages analysis

Box 15.4 Gender-mainstreaming questions for project design

- Did men and women actively participate in designing the project?
- Were men and women equally involved in decision making?
- Do project objectives reflect priority problems and opportunities identified by both men and women?
- Will the project contribute to achieving strategic gender needs and greater gender equality? How so?
- Are project strategies well adapted to the practical conditions, knowledge/skills, access to resources, etc. of both men and women?
- Have distinct men's and women's strategies been identified in cases where men and women have clearly different interests?
- Have gender-affirmative actions been planned to overcome existing inequalities and barriers against female participation in the project and its benefits?

Phase 3: activity planning

In this phase of the project cycle, the strategies defined in the design phase are worked out in detailed action plans, responsibilities are divided among project partners, time schedules are defined with critical dates for the delivery of certain products, and the budget is prepared, together with identification of financial sources and other means required. Attention to gender issues already identified is required throughout this planning phase. The Villa María del Triunfo case study (Chapter 8) provides an overview of a multi-stakeholder planning process and incorporation of gender into a platform and strategic agenda focused on enhancing urban agriculture activities. In particular, it emphasizes the need to ensure gender-balanced participation in a forum organization, encourage collective efforts among producers, and build cohesive vision among various groups.

Important steps in this phase are: (1) detailed activity planning and division of labour, (2) defining the project management and co-ordination structure, (3) developing the monitoring and evaluation plan, (4) budgeting, and (5) thinking through and the final revision. These are detailed below.

Detailed activity planning and division of labour

Here the strategies formulated and selected in the design phase are worked out in detailed action plans, in terms of who (specified in terms of men and/or women) will do what when, where, how, with what means, and resulting in what outputs. Sometimes this is done in detail for the first project year and more broadly for the years thereafter.

Defining the project management and co-ordination structure

This is done by:

- Defining how the project will be internally structured and how decisions will be made and activities be co-ordinated to ensure successful, participatory, gender-sensitive, and timely implementation. There may be equal gender representation on a joint committee and/or separate men's and women's management (sub-) committees.
- Defining the roles and expected contributions of each of the project partners, including milestones (moments in the project when certain steps are concluded and new steps are prepared), deliverables (concrete outputs/products to be delivered), and deadlines. Ensure that the project's beneficiaries (men and women urban producers and disadvantaged groups/consumers) have also been assigned a role and contributions in the project. Ensure that in the description of the role and expected contributions proper attention is given to the gender aspects.

- Defining who will report when to whom, and what the contents of the reports should be. Include gender as an aspect of all main items that must be reported on (e.g. progress and results per objective) as well as a separate reporting item (e.g. progress and results regarding the specific gender-equality objective and gender mainstreaming in general).

Developing the monitoring and evaluation plan

Together with the beneficiaries, gender-specified indicators for project success will be defined for each of the project objectives (how can we measure whether the expected results have been realized?). This will be relatively easy if results-based objectives were formulated and in a SMART way. How will data regarding the indicators be gathered: by whom, with help of which methods, with what frequency? Such information can be summarized in the logical framework.

Gender-specified indicators allow differentiation between the impacts of the project on men relative to women (benefits as well as negative trade-offs/costs) for each of the project objectives (indicating the expected results of the project). Also one or more specific indicators might be included to monitor the degree to which the combined project results contribute to enhancing gender equality and strategic gender issues.

The indicators should preferably be developed in a participatory way, including both men and women. The number of indicators should be kept to a minimum and should cover both qualitative and quantitative aspects of the objectives. Both male and female urban producers should be included in the monitoring methodology, both for the collection of monitoring data as well as in the periodic reflection on progress made/results obtained.

As indicated earlier, it is very important that the reporting guidelines oblige all partners to report on 'results obtained', disaggregated by sex, and that the reporting format includes a paragraph on gender issues and the identification of gender-related project failures and successes. It should be noted that a solid monitoring and evaluation plan (see phase 5 below for more details) requires sufficient resources in order to be implemented well.

Budgeting

Identify the means needed to implement each activity, such as people, travel, equipment, materials, and office costs. Ensure that sufficient funds are included for gender-affirmative actions and specific women's project strategies (based on their specific interests and activities). Ensure that undifferentiated project resources will be available for both men and women. Gender-specified targets and allocations may be defined (e.g. the minimum percentage (or absolute number) of the land or credit that will be allocated to women, and a minimum number of female producers involved in training activities).

Budgeting is a particularly crucial element of effective and efficient gender mainstreaming. Often, in order to be taken seriously, specific budget lines must be allocated for gender-focused activities or resources so that they are not overlooked. Monies for training manuals, capacity-building sessions, consultative meetings, data collection, facilitation, etc. must be earmarked and visible within project budgets, and even labelled as such (e.g. *gender training* rather than simply *training*). Minimum spending levels for women's activities may go a long way to ensuring commitment to gender equality in the entire project cycle. Further, it is important that gender-sensitive budgetary elements are flagged for all partners, in particular within multi-stakeholder contexts where organizations involved may not be as committed to gender mainstreaming as the project team itself.

Thinking through/final revision

In this step, the project team should reflect on the following issues:

- **Assumptions/risks.** These are external factors beyond the control of the project implementers. They are related to each of the intervention strategies and identify 'what if' strategies (contingency plans). Include the results of this analysis in the logical framework. Adapt the project design if certain risks are too big or certain assumptions are not realistic.
- **Continuity/sustainability.** Analyse whether/how the results of the project can be sustained after it finishes, and how the project will work towards this goal. Will both women and men be satisfied and involved? The way in which projects are financed often has implications for their sustainability and for the feasibility of increasing scale and replication later on. Adapt the project design where needed in order to make it more sustainable.
- **Dissemination/replicability.** Analyse what relevance the project results will have for wider application (beyond the direct beneficiaries of the project) and the conditions for successful replication by other organizations and in other areas. Also determine how the project will work towards dissemination of the project results and their application to potential users. Include such dissemination activities in the project design and budget.

Box 15.5 Recommended tools for the activity-planning phase

- Activity-planning matrix
- Participatory budgeting
- Scheduling the work

Box 15.6 Gender-mainstreaming questions for the activity-planning phase

- **Planned activities**
  - o Will the activities be undertaken in such a way that the special constraints that often restrict women's participation in decision making and implementation are diminished (e.g. appropriate timing and location of meetings, language used, allowing women to develop their viewpoints in homogeneous subgroups before meeting at organization or project level, use of female facilitators, etc.)?
  - o Have both men's and women's interests been addressed and gender-affirmative actions planned?
- **Management structure**
  - o Have male and female farmer representatives been included in the project management structure?
  - o Have the project implementing partners agreed on a number of measures to ensure gender-responsive implementation of the project by their organization and staff (gender policy, staff training on gender and UA, gendered Terms of Reference for staff, gender balance of project staff, gender on the agenda of project meetings, etc.)?
- **Monitoring**
  - o Are the indicators gender-specific, so that the differential impacts of the project on men and women can be determined?
  - o Have indicators for the combined effects of the project on gender equality and strategic gender issues been defined?
  - o Is the direct participation of women (and men) in the monitoring and periodic evaluation foreseen?
- **Budget**
  - o Are the relevant budget items gender-specified?
  - o Are both men's and women's priority activities funded?

Phase 4: implementation

To ensure the active participation of women and men in the implementation phase, it is helpful to consider gender mainstreaming at two levels. Through the implementation *process*, tools and approaches will be generic or common to all kinds of situation. The process involves issues such as working with existing structures, group dynamics, mutual respect, empowerment, and capacity strengthening (especially of the marginalized), and the 'rules of the game' for implementation through written agreements or a constitution (see below). At the second level, particular types of implementation are engendered through attention to *content*, for example participatory technology development for crop-production innovations, addressing marketing policy, addressing gender issues in livestock production, in agro-processing, in women's land rights, etc. These levels and elements are described in more detail below.

Examples of gender-sensitive implementation phases can be found in the Nairobi case study (Chapter 9), which highlights both the dynamics of men and women participating in day-to-day planning and budgeting of project activities (process level) and efforts to engage both in testing of energy briquettes (content level). Similarly, the Nakuru case study (Chapter 11) details how social-cultural diversity among the beneficiaries in agricultural interventions affected participation in terms of labour distribution (process level), and how a Community Organizational Development and Institutional Strengthening (CODIS) course was used to address these challenges (content level).

Process: adapting/'deconstructing' existing structures

For community-level interventions, project members need to work with existing structures or organizations and to insert gender more strongly into them. This may involve working with urban agriculture programmes or forums, or tapping into producer organizations and informal farmer networks. An important opportunity exists to empower women in agricultural contexts through linking women's organizations to government structures or linking agricultural functions to women's organizations which are already connected to government (e.g. government-linked community kitchens run by women's groups are being supported to add individual or communal agriculture activities).

In order for existing organizations to transform themselves into engendered groups for implementation activities, there may need to be a 'deconstruction' of the group in terms of its life history, its political and economic structures, and its decision-making processes. Diagnostic tools such as timelines, calendars, bio-sketches, etc. can be used to help in this deconstruction process, in order to detail the gender dynamics.

Process: group dynamics and group governance

Whether the group already exists or is formed for the intervention, the success of action research projects is very much influenced by the stability and performance of the organizations involved. Group composition during the implementation phase should have an even mix of men and women, or there should be an effort to combine men and women from same households. The latter option can reduce the drop-out rate, which is likely to be problematic in implementation activities.

Group-dynamic issues need to be identified and addressed in preparations for and during implementation activities. Issues such as leadership, roles and benefit sharing, and participation in group meetings, project management, and financial management have been noted to cause problems among group members of mixed groups, resulting in deterioration in performance or member drop-out.

One tool for addressing such group-dynamic challenges is the clear establishment of the 'rules of the game' in the form of a group constitution to govern the behaviour of members. The constitution or by-laws should stipulate the following elements:

- self-definition of the organization: who are we?
- gendered objectives
- gender-responsive activities
- gender equity in leadership
- gender equity in participation in activities and benefit sharing.

To mainstream gender and age in leadership and decision making, for instance, the constitution may state that when the chairperson is a man, the vice chairperson should be a woman, and vice versa in mixed-sex groups, while in same-sex groups, when an adult person is the chairperson the vice chairperson should be a youth, and vice versa. This rule is applied to all executive posts, including treasurer and secretary.

Gender equity in benefit sharing based on members' participation in activities is an area that requires development of rules and norms to avoid the possibility that some members take advantage of those highly committed to the group's objectives and activities. A constitution may state, for example, how much money a participating member is making from sales of the fuel briquettes, in order to motivate other members to contribute in terms of labour.

Process: capacity development

To facilitate positive group dynamics, the implementation phase needs to consider capacity development of group members in leadership roles and benefit sharing, business and activity planning, project and financial management, and gender-responsiveness skills among others. These can be done through designing courses suitable for local organizations. Research organizations, for example, may implement these community-based organizational development and institutional strengthening courses through partnering with non-government organizations involved in development work.

Content: technology innovation, capacity development, and empowerment

Capacity building and training of women leads to empowerment and changes in gender dynamics, and to tangible benefits for women through technological innovation. Specific interventions have multiple components with gender implications, and diagnostic tools can be generally adapted for the implementation phase, with the same gender considerations. For example, a diagnostic workshop can be held to prepare a gendered curriculum addressing crop-management issues of importance to both men and women. Checklists tailored to the specific intervention can be used for a similar purpose. The development of gender indicators is similar to the elaboration of a checklist, but with greater precision and quantification, and with a time dimension.

During the implementation phase, regular monitoring according to the previously developed monitoring and evaluation plan allows one to see whether the project is on track with regard to the realization of planned activities and the realization of achievements as set out previously, to solve problems that emerge during implementation, to keep track of the results achieved to date, to reflect on the effectiveness and efficiency of the strategies and working methods applied, and to make adaptations where needed. Such learning-from-actions is of special importance to enhance the relevance of research and development projects for women producers.

Box 15.7 Recommended tools for the implementation phase

- Implementation checklist
- Urban Producers Field Schools
- Participatory innovation development

Box 15.8 Gender-mainstreaming questions for the implementation phase

- Are representatives of men and women producers actively involved in periodic project planning and evaluation meetings and decision making on the course of the project?
- Is there active participation of both men and women of the target group in the implementation and monitoring of the project activities?
- Is there equal access for men and women to resources provided (training, credit, tools, land, seed, irrigation water, etc.) during project implementation?
- Are affirmative actions taken to counterbalance gender inequalities?
- Are activities being implemented to promote gender mainstreaming in the participating producer organizations?

Phase 5: monitoring and evaluation

As indicated earlier, the monitoring and evaluation activities begin during the design phase, during which – in relation to the project objectives – adequate monitoring and evaluation indicators are defined, as well as appropriate measurement methods selected.

When planning the project activities, it will be defined who will collect what monitoring information, when and how it will be collected, how these data will be stored and processed, at what moments the data will be used to reflect on the achievements of the project to date, and who will be involved in such meetings (monitoring and evaluation plan).

During the implementation phase, regular monitoring according to the previously developed monitoring and evaluation plan shows whether the project is on track with regard to the realization of planned activities and realization of achievements as set out previously. It helps to solve problems that emerge during implementation, keeps track of the results achieved to date, and reflects on the effectiveness and efficiency of the strategies and working methods applied, so as to make adaptations where needed.

Monitoring and evaluation should be understood and organized as 'learning-from-actions by periodic reflection on experiences gained'. Reflection on the monitoring data allows us to identify how we can get more results and of better quality, how to enhance cost-effectiveness, how to enhance participation and local ownership, and how to enhance sustainability and up-scaling of the results, etc.

Such learning-from-actions is of special importance for enhancing the relevance of research and development projects for women producers and their empowerment. It should be seen as a specific tool of both gender analysis and women's empowerment, in that it provides a forum where community members, both men and women, direct attention to and discuss these issues.

When *selecting indicators,* it is important to ensure the following:

a. All indicators are *gender-specific.* This should be reflected already in the objectives in terms of the expected results for men and women; for example, if the objective is to enhance access to credit, how many men and women have obtained a loan (and maybe what is the size of those loans) and not just the number of households that received a loan. The Nairobi case study (Chapter 9) uses the following gender-sensitive indicators to monitor project impacts: number of men and women participating in meetings and training sessions, number of issues raised by men and women during such meetings, number of men and women occupying leadership roles, involvement of men and women in production and selling activities, as well as changes in behaviour of men and women observed qualitatively (e.g. men's views on women's involvement in leadership).

b. All indicators include sufficient indicators to *monitor the effects of the project on local gender relations and the empowerment of women.* Even when all indicators are gender-specified, these do not necessarily reflect how the project influences the position of women. Therefore it is important to check this element through, for example the following indicators (the numbers refer to suggested rating scales explained below):
   - Plots allocated to women (1 or 3).
   - Women's control over the benefits (products, income) raised in urban agriculture activities (decisions on the use of the produce and income raised) (2 or 3).
   - Women in leadership positions in producers' organizations, urban agriculture committees, and/or institutions established (1 or 3).
   - Women in leadership positions in community organizations and activities (1 or 3).
   - Degree to which women experience constraints in urban agriculture not applying to men (2).
   - Degree to which conventional constraints on women's participation have been removed or altered for the better after the project (3).
   - Women's income from crop and/or livestock production (1 or 3).

- Women's income from other activities (1 or 3).
- Women's freedom of movement (3).
- Respect for women and their activities in the household (3).
- Respect for women and their activities in the community (3).
- Women producers who are regularly attended by service-providing institutions (e.g. percentage of women participating in agricultural extension groups and training events, percentage of credit disbursed to women) (1 or 3).

c. Such monitoring data should provide a reliable image not only of the project's impacts on men and women producers but also of its effects at *institutional or policy level*. In order to see the effects of the project on mainstreaming gender in the participating organizations, one needs to include indicators like the last one mentioned above. Note that this indicator is different from indicators monitoring the participation of men and women in the project activities. Have these organizations or institutions adopted a gender policy? Has their attention to strategic gender issues increased? Have their services to women producers improved in their regular programme and other projects?

Examples of *rating scales* that can be used when measuring these effects:

(1) Proportion (percentage) of women in total:
   a. 0–20%
   b. 20–40%
   c. 40–60%
   d. 60–80%
   e. 80–100%

(2) Women influence / control / decide:
   a. Not at all or only a little
   b. About the same as men
   c. It is mostly women who do so

(3) Increase compared with before the project began
   a. None or hardly none
   b. Some
   c. Very great

Many projects monitor the participation of men and women in the various project activities (e.g. in training and extension activities, distribution of inputs, distribution of loans, etc.). Such 'gender-balanced' data allow one to monitor during implementation the extent to which the project is attending to men and women. However, such data do not say much about the impacts of the project on gender relations and women's empowerment, for which one requires indicators such as those mentioned above.

Many of the tools that were used in the diagnostic phase of the project can also be used for monitoring and evaluation purposes. For example, the tools to analyse men's and women's role in decision making or their degree of control over household resources and the benefits of productive activities can also be used to monitor changes effected in these key issues by comparing the scores before and after the project or – in projects with longer duration – to do so at certain moments during project implementation also. This allows one to monitor what changes are effected and – if needed – to adapt the project strategies.

It should be ensured that all monitoring tools applied are gender-sensitive and allow for the collection of gender-disaggregated data. Also the reporting formats should be such that monitoring data can be reported easily. In the Nakuru case study (Chapter 11), for example, project staff were involved in collecting gender-disaggregated monitoring information (e.g. number of men and women trained, number of beneficiaries by gender per household for both projects, members' involvement in farm activities by gender, time, and date), while records were maintained by households on the performance of dairy goats belonging to men and women (on feeding, weight gain/loss and health), as well as amount of vegetables, milk, and income produced per (male- or female-led) household.

A *pre-test* of the monitoring methods and indicators is often very effective in pointing out certain practical problems with the chosen monitoring methods (or operationalization of indicators) so that improvements can be made before the monitoring and evaluation system is made operational. Making changes in an on-going monitoring system often turns out to be complicated

When planning the monitoring and evaluation activities, one should not only organize the data collection and reporting but also specify at which moments (at a minimum annually) the project partners (including male and female representatives of the beneficiaries) will come together to reflect on the monitoring results and on the facilitating and hampering factors that influenced the results. Gender should be made an important item on the agenda of such meetings. The periodic reflection meetings on the basis of the monitoring results will help to identify changes that have been brought about and how the project activities are differentially affecting men and women. Especially in the field of gender, such meetings are of crucial importance. Gender awareness and gender analysis and planning skills are not created overnight, and such periodic meetings are an important means to further strengthen these. Where such periodic reflection does not take place, gender considerations often tend to fade away during implementation of the project. The reflection on the monitoring data provides an opportunity to reflect on the factors that limited or facilitated attention to women's priorities and may lead to adaptations in the working methodologies and inclusion of new ones. For example, it may be decided to take actions to enable women to overcome specific constraints that limit their possibilities to participate actively and reap the benefits of the project, to shift budget allocations, to enhance gender balance in project actions, to strengthen alliances with women's organizations, etc.

The frequency of the internal evaluations will be determined according to the needs of the stakeholders (including donors) and the situation of the project (more frequent evaluations might be needed in case of communication difficulties between participating organizations, low level of trust among beneficiary groups, etc.).

In the later stages of project implementation, attention will also be given to the *'systematization' of the experiences gained* and *drawing 'lessons learned'* regarding – in this context – gender mainstreaming in the project cycle, with a view to the dissemination and replication of project results and their uptake into local or national policies on urban agriculture. The case studies featured in Part I include sections on 'lessons learned' whereby such insights are used to formulate further research, interventions, and actions. For example, the Carapongo case study (Chapter 13) highlights the need to support social capacity building through provision of training and development of gender-sensitive producer networks based on project findings that reveal low participation and representation of women in such organizations.

For an externally mandated formal mid-term or end-of-project evaluation, additional information on the project's results and impacts might be collected, in addition to the monitoring data. In that case the team that will implement this has to be familiar with gender-sensitive data collection and analysis. Terms of reference for mid-term or end-of-project evaluations should clearly specify the questions to be addressed in the evaluation regarding the project's expected results for men and women respectively and its impact on gender equality in the project area. The evaluation team should have sufficient expertise in recognizing the gendered differences of the project's impacts (GWA, 2006).

Box 15.9 Recommended tool for the monitoring and evaluation phase

• Monitoring and evaluation checklist

Box 15.10 Gender-mainstreaming questions for the monitoring and evaluation phase

• Are monitoring data gender-disaggregated?
• Does the monitoring process capture feedback and information from both men and women beneficiaries and participants?
• Are monitoring data analysed with direct participation of both male and female representatives of the beneficiaries?
• Does the monitoring information show how the project is benefiting respectively women and men?
• Are remedial actions taken if the monitoring data show that women and men do not benefit to the same extent from the project activities?
• Do the monitoring and evaluation teams include members who are gender-sensitive and have gender expertise?

Phase 6: going to scale

In this phase, project results are used to plan follow-up actions, are disseminated to other actors and local policy makers in the same city or similar areas (out-scaling), or are disseminated to policy makers and programme managers at national level (up-scaling). The list of potential users of research findings or systematized experiences gained in a development project is extensive: urban producers in other areas, urban planners, agricultural research and extension organizations, NGOs, local authorities, national policy makers, etc.

The Nakuru case study (Chapter 7) provides details of strategies and tools used to incorporate gender in policy influencing, including holding a regional policy-intervention workshop that brought together municipal and national government representatives, farmers and livestock keepers, civic groups, community organizations, international bodies, and donors. The workshop led to the establishment of City Focal Points in each of the participating cities, as well as a Forum through which meetings are held to discuss policy-related issues regarding urban agriculture among all stakeholders.

Such efforts can ensure that the learning experience and outcomes of a gender-sensitive project are integrated into the planning of new research or action projects and the development of new (or revision of existing) policies on urban agriculture. It should be noted, however, that policy development can cut across the entire project cycle, as was noted in the section on monitoring and evaluation above and is further discussed in the section on 'early involvement' below.

It is also important to be aware of the fact that in most instances going to scale with gender findings and potential actions can be challenging and is often the stage of the project cycle that is left unrealized. The gender element tends to go missing here, because of the broader spectrum of actors that become involved in out- or up-scaling, and more specifically because many of those actors are not as focused on or in tuned to gender-sensitive research or development planning. Further, the capacity of project team members to network with policy makers and planners, or key stakeholders beyond their locality, may not be fully developed at this stage of the project cycle, thus inhibiting the uptake of gender issues.

Project teams should consider how to build capacity for gender mainstreaming beyond their own community, project, or organization, to ensure that differences and inequalities between men and women remain in the forefront of interventions when going to scale. Efforts may begin with incorporating farmer networks beyond the immediate community, as was done in the case studies from Harare (Chapter 6), Nairobi (Chapter 9), Rosario (Chapter 10), and Mexico (Chapter 12); these larger-scale organizations may facilitate opportunities for policy impact, as was the case in Villa María del Triunfo (Chapter 8), which may otherwise not be realized. More attention to gender issues when going to scale may be obtained by making reference to UN Resolutions regarding women's rights and equality (e.g. the Beijing declaration and action programme for equality, development, and peace, 1995; see www. un.org/womenwatch), and by ensuring that grassroots women's groups and gender-sensitive sectors of municipal or national governments are always directly involved at this stage of the project cycle.

Gender mainstreaming throughout the project cycle means that urban governance and policy/planning mechanisms must necessarily pay adequate attention to gender dynamics and issues. To that end, it is important to define potential users of project results early on in the project cycle, getting to know their information needs and assessing the best ways to present information so that channels and formats for the dissemination are appropriate. The results of the project have to be 'packaged' and transferred in the right way, producing gendered materials that have high use value for the targeted audience (e.g. practical manuals, guidelines for professionals, planning models and software, policy briefs, videos, etc.) and using effective channels to convey the key messages and the produced materials to these publics (e.g. study visits, workshops/seminars, radio broadcasts, popular media, formal publications, e-mail lists, websites, etc.). For example, the Harare case study (Chapter 6) identifies common 'myths' about urban agriculture that need to be dispelled and can be done so through municipal television and radio programming; to this end, project results need to be pulled together in a way that helps the public to understand the important role that urban farmers play in enhancing local food security and income raising.

Ensuring sufficient attention to systematization and dissemination of experiences gained, as well as facilitating uptake of gendered project results in local or national policies and programmes, means that a specific project objective should be formulated on achieving certain dissemination results and improvements in the actual policies and programmes on urban agriculture. This will spur the design of specific project activities regarding engagement with policy makers and other institutional stakeholders and facilitating the use of the project results in policy and programme design.

In both RUAF and Urban Harvest, special attention is given to engagement with local and national authorities and most relevant government and non-government organizations that play (or should play) a role in the development of safe and sustainable urban agriculture ('institutional stakeholders'). This facilitates adequate policy uptake and institutionalization of the project results, thereby multiplying and sustaining the impacts of the project. This includes gender aspects.

Few policy makers and senior officers read final project documents or research reports. It is impossible for them to have up-to-date knowledge on all issues relevant to public-policy making and programme development. Yet they frequently make vital policy decisions that have widespread implications for urban agriculture (city development plans, land-use regulations, enhancing or reducing budget allocation to agricultural support programmes, etc.). So, the decision makers need to be adequately briefed with selected information regarding key issues in urban agriculture to enable them to make informed decisions.

This requires a well-planned strategy to engage with policy makers and institutional decision makers in order to achieve the following results:

- Raise their awareness on the actual and potential role of urban agriculture for sustainable urban development (and especially poverty alleviation, enhancing food security, local economic development, recycling, and social inclusion of disadvantaged categories of the urban population, including e.g. female-headed households with young children), since such awareness is generally low.
- Brief these decision makers on the main results of the implemented project in their city/country – the findings of a research project or the systematized experiences of a development project – and the relevance of these results for policy development and design of future programmes and projects on urban agriculture.
- Whenever required and possible, to contribute to processes of policy review and design of programmes on urban agriculture.

Early involvement

As indicated earlier, gender-sensitive policy influencing and development needs to be incorporated within the entire project cycle. This offers the best way to engage with policy makers and senior institutional decision makers. Such participation throughout the project process will provide the opportunity to feed them bit by bit with information regarding urban agriculture and related policy issues, to include their questions and information needs in the project design and implementation, and to gradually enhance their commitment to urban agriculture in general and gender issues in particular. This encourages active use of the project results in policy and programme design or adaptation. During this process, it is important not to let policy makers get the upper hand and 'hijack' the project; rather the project should retain its independence and innovative approach for setting out new avenues for gendered development.

The Urban Harvest research project on Health and Urban Agriculture in Kampala established an advisory 'Health Co-ordination Committee' which included the various actors that co-operated with the project in one way or another, including university departments, NGOs, Kampala County Council (KCC) Agricultural Extension Officers, the Ministry of Agriculture, Animal Industry and Fisheries, and the National Agricultural Research Organisation. Perceived policy concerns, long hampering the formal adoption of a number of ordinances on urban agriculture that were drafted in 1999, were incorporated in the project design. The presence of KCC as a major partner in the Health Co-ordination Committee ensured that policy-relevant research findings could directly feed into the policy-making process. Project results were used in a number of consultations that were held to discuss the draft ordinances, to inform the participants on the real health risks associated with urban agriculture and effective ways to reduce and manage such risks, which led to several changes in the draft ordinances. The project also contributed by field testing the new ordinances to identify challenges faced by producers in observing the new ordinances and assessment of the impacts of these ordinances. This included a study of the gender impacts.

In the pilot cities of the RUAF programme, local authorities and various other stakeholders in urban agriculture are engaged in a process of Multi-stakeholder Policy Formulation and Action Planning (MPAP) for sustainable urban agriculture development. The MPAP involves citizens, producers, civil organizations, private-sector companies, and government organizations in the formulation of a municipal policy on urban agriculture and the joint design of action plans and the establishment of a Multi-stakeholder Forum on Urban Agriculture. This collaboration goes beyond mere consultation, where stakeholders are asked for their feedback on an already defined line of action. Instead, in MPAP, the various stakeholders participate actively in the situation analysis, the definition of problems/constraints and potentials/opportunities, the development of possible solutions and development strategies, as well as the assignment of roles and expected contributions to each actor involved in the implementation of these strategies. For sustainable urban agriculture development, such multi-stakeholder participation is particularly important, since it is a cross-sectoral issue and its development requires the participation of different disciplines and a variety of actors (Dubbeling and De Zeeuw, 2007).

The MPAP process (Ibid., 2007) not only greatly improves dialogue between local authorities and urban producer groups and other stakeholders on existing problems and the required measures to achieve the development of safe and sustainable urban agriculture, but also improves their knowledge of urban agriculture and their capacities to jointly plan, implement, and monitor urban agriculture projects and their commitment to contribute actively to the realization of such projects.

End of project engagement

If early involvement of senior decision makers is not possible, towards the end of the project various communication and lobbying strategies can be used to engage with and influence politicians and other institutional decision makers. First it is important to analyse existing policies regarding urban agriculture and the actual views of the various institutions on urban agriculture (its problems, its potentials, and how they see its future development) and to identify on what policy issues the project can provide relevant insights and recommendations. Second it is important to attempt to understand what triggers the interest/ attention of specific policy makers or other decision makers. The degree to which actual policies and programmes adequately address gender issues in urban agriculture, or may have differential impacts on men and women, should be a key part of the review.

Once one has defined what relevant insights the project may provide to senior decision makers, what the key messages are, and how they relate to the existing policies and programmes and specific interests of the decision makers, including a gender analysis, adequate communication and lobbying strategies can be identified and prepared.

Such strategies may include the following (see for more details Dubbeling, 2004):

- Creating opportunities for direct dialogues with policy makers (oral briefings during a personal visit or meeting, a policy seminar or a site visit, etc).
- Preparing a 'fact sheet' or 'policy brief' on urban agriculture and the results of the project. This is a short and concise document that synthesizes relevant information on selected key issues of a certain policy area, and lessons learned from action and research projects, and suggests how a municipal or national government or another institution may (further) develop their policy and programmes in this area, often illustrated by concrete city cases ('building theory based on practice'). Briefing papers are less likely to be misquoted and can more easily be shared with third persons than oral presentations, but they offer fewer opportunities for interaction and dialogue.
- Mobilizing others to communicate the message (briefing of accessible policy advisers and like-minded staff inside those institutions, linking up with the media and briefing of interested journalists, letting influential people present the story in public debates, etc.).

Often a combination of the above-mentioned strategies needs to be implemented. Because senior staff are often replaced after elections when political parties change, it is wise to diversify relations and policy approaches and to recognize that building relationships is an iterative process, with reversals as well as advances.

When preparing policy briefs and engaging in policy dialogue, the gender dimensions of urban agriculture should be given proper attention, stressing the following:

- Recognition of *women as independent actors and beneficiaries* in/of urban agriculture public policies and projects.
- Acknowledgement of the real *value of women's contribution* to the development of urban agriculture: production, food security, income, etc.
- Recognition that the *needs of men and women are different* and that women's access to and control over resources and participation in decision making are restricted by socio-cultural and institutional traditions.

- Recognition that public policies and projects, as well as economic and technological trends, can have *differential effects for men and women.*
- Recognition that *affirmative actions are needed* to ensure that women (and men) can reap equal benefits from urban agriculture policies and projects.

Gender should be included as one of the 'key issues' for the development of sustainable urban agriculture discussed in policy briefs or fact sheets on urban agriculture. Gender dimensions of other key issues should be outlined carefully (e.g. women's access to land and credit, technical and entrepreneurial training, etc.). Suggestions have to be made for measures to be addressed in a new policy or programme so that the gender dimension is included. The consequences of not including such measures might also be sketched.

Finally, project team members at this stage of the project cycle should consider the potential unintended consequences of suggesting and even accomplishing up- or out-scaling of particular gendered actions. It is possible that what is considered a positive policy or planning initiative, as advocated by the project team based on their research and development planning, may have negative implications for men or women in the community. For example, the formalization of urban agriculture, which has been urged by lobby and producer groups globally, may not necessarily be a positive change for women, given that their access to this activity and its benefits often depends on its informal nature. In another instance, gendered measures may actually end up reproducing particular ideas about women that are not ideal (e.g. enhancing women's capacity to do urban agriculture for subsistence purposes reinforces their place or position as food producers for the household only). It is important then to consider consistently and regularly the positive and negative implications of going to scale with gender issues.

Box 15.11 Recommended tools for out-scaling and up-scaling gender-responsive interventions

- Scoping of gender issues for policy and planning
- Policy brief
- Policy-action matrix

Box 15.12 Questions for gender mainstreaming in out-scaling and up-scaling interventions

- Have the gender dimensions of urban agriculture been given proper attention when preparing publications, fact sheets, manuals, and policy briefs?
- Have gender-research findings been incorporated in policy discussions and dissemination activities?
- Have the project impacts on men and women urban producers been presented to policy makers?

References

De Zeeuw, H. and Wilbers, J. (2004) *PRA tools for studying urban agriculture and gender*, ETC–Urban Agriculture, Leusden. Available from http://www.ruaf.org/node/97

Dubbeling, M. (2004) 'Senior Stakeholder Engagement and Dissemination for Policy Change, ETC-Urban Agriculture', paper prepared for the PAPUSSA project.

Dubbeling, M. and De Zeeuw, H. (2007) *Multi-stakeholder Policy Formulation and Action Planning for Sustainable Urban Agriculture Development*, RUAF Working Paper # 1, Leusden, The Netherlands.

FAO (2001) *Field Level Handbook,* Socio-Economic and Gender Analysis Programme (SEAGA), FAO, Rome. Available from http://www.fao.org/sd/seaga/downloads/En/FieldEn.pdf

Gender and Water Alliance (GWA) (2003) 'Mainstreaming gender in the project cycle', in the *Training of Trainers Package on Gender Mainstreaming in Integrated Water Resources Management*, GWA, Delft, The Netherlands.

Gender and Water Alliance (GWA) (2006) *Resource Guide Mainstreaming Gender in Water Management*, version 2.1, GWA and United Nations Development Programme (UNDP).

Martin, A., Oudewater, N., and Gündel, S. (2002) *Methodologies for Situation Analysis in Urban Agriculture*, Topic Paper 1: RUAF–SIUPA E-conference on Appropriate Methods for Urban Agriculture.

Prain, G. and De Zeeuw, H. (2007) 'Enhancing technical, organizational and institutional innovation in urban agriculture', *Urban Agriculture Magazine 19, Stimulating Innovation in Urban Agriculture*: 9–15.

Pretty, J. (1994) *Gender Differences and PRA, Making a Difference: Integrating Gender into PRA Training*, PRA Notes no 19, IIED, London, UK.

Pretty, J. (1995) *Regenerating Agriculture: Policies and Practices for Sustainability and Self-reliance*, Earthscan, London, UK.

CHAPTER 16

Beyond the project cycle: institutionalizing gender mainstreaming

Abstract

In this chapter the emphasis shifts from integrating gender in the project cycle to mainstreaming gender in organizations and institutions. Different ways to promote the integration of gender into organizations (their organizational culture, their policies, etc.) will be discussed. Thereafter, by way of example, we present the process that was applied by the RUAF Foundation and Urban Harvest to promote gender mainstreaming in the organizations taking part in these networks.

Institutionalizing gender mainstreaming

Up to this point, we have focused on the identification and analysis of critical gender issues in urban agriculture and the measures that can be taken in every phase of the project cycle to ensure that gender differences are taken into account in research and development projects. However, merely integrating gender into the project cycle is not enough. Gender responsiveness should also be integrated in the organizations that execute such projects, which will greatly facilitate gender mainstreaming at the project and activity levels and will also enhance its effectiveness. The current chapter focuses on how gender can be institutionalized and mainstreamed at the organizational level.

Possible measures to stimulate gender mainstreaming in urban agriculture organizations will be discussed, followed by examples of the process applied in the organizations participating in the RUAF network and the Urban Harvest programme when attempting institutional gender mainstreaming.

Why is gender mainstreaming in organizations important?

There are three main reasons for giving attention to gender mainstreaming in organizations (e.g. our own and our partner organizations). Firstly, focusing on projects to improve gender equality at local level may have little impact on changing the wider social position of women, youth, and other disadvantaged groups. Secondly, it is recognized that gender inequality is continuously recreated by the 'mainstream', with existing dominant views and stereotyped roles for men and women informing gender-unbalanced policies and biased institutional practices. Hence, thirdly, we need to influence policies and institutions to achieve greater impact.

We also should seek to bring about changes in the institutional policies, programmes, legislation, and resource allocations of our own organization and our partner organizations in order to enhance the impact on gender equality of our and their development programmes.

The most important changes we seek to bring about are the following:

- Recognition by decision makers that gender equality is a strategic objective of development rather than a 'women's issue'. It should be pursued by leading institutions and reflected in major policies, programmes, and institutional practices.
- Women as well as men are enabled to influence the policy agenda, decision making, and resource allocation, and are fully recognized as agents and beneficiaries of change.
- Institutions develop analytical, planning, and management skills to identify and respond to gender issues relevant to their mandate.
- Design of policies and programmes is based on data regarding existing gender differences and focused on enhancing gender equality.

What can be done to promote gender mainstreaming in organizations?

When gender mainstreaming an organization, institutional commitment and political will are indispensable. The first stage will be to ensure that all stakeholders are familiar with the basic concept of 'gender' and what 'gender-sensitive planning' means. Conducting a gender audit to assess staff knowledge and awareness of gender responsiveness can be helpful at this stage. (A guideline for such a gender audit is provided by the Commission on the Advancement of Women, 2003.) Commitment at the management level needs to be sought, and a simple action/time plan for development of a gender policy made.

Gender expertise should also be made available in the organization, such as by including a gender specialist in an advisory role. However, it should be ensured that a project team itself or an organization's staff in general are responsible for integrating gender throughout the project cycle. The gender specialist should just assist and support staff in their work and should not be made the only one responsible for 'integrating gender' in project work.

Equally important is the organization of staff gender training and the diffusion of gender guidelines. Enhanced capacity of staff on gender responsiveness in an organization will enable the formation of gender committees or task forces with effective representation across divisions, themes, units, and other organs of the organization through which institutional accountability on gender mainstreaming can be achieved. Sensitivity to gender issues and gender balance have to be taken into account when selecting the staff members who will be involved in project planning and management, and they have to be trained in gender analysis and gender-sensitive project planning and management.

Other important measures are the following:

- Facilitate development of a gender statement and ensure that promotion of gender equality is made an explicit aim of the organization.
- Help to ensure that gender integration becomes a routine concern of all units and all staff members.
- Facilitate changes in the organizational culture.
- Promote incorporation of gender analysis and gender planning in standard procedures for project design and facilitate gender specification of institutional reporting and monitoring systems.
- Promote the engendering of staff recruitment and enumeration policies.
- Evaluate whether budget allocations reflect the gender policy of the organization.
- Promote the inclusion of indicators for impacts on gender equality in the institutional monitoring system as well as periodic discussion of the results (self-learning).
- Stimulate exchange with other organizations regarding gender issues in urban agriculture and related subjects.

Gender mainstreaming in the RUAF network

About RUAF

The International Network of Resource Centres on Urban Agriculture and Food Security (RUAF Foundation) consists of eight partner organizations spread across the globe. The RUAF partners are of two kinds:

a. Regional RUAF partners:
   - IAGU, Dakar – regional co-ordination for Francophone West Africa
   - IWMI–Ghana, Accra – regional co-ordination for Anglophone West Africa
   - MDP, Harare – regional co-ordination for Eastern and Southern Africa
   - IWMI–India, Hyderabad – regional co-ordination for South and South-East Asia
   - IPES, Lima – regional co-ordination for Latin America
   - IGSNRR, Beijing – regional co-ordination for China
   - AUB, Beirut – regional co-ordination for the Middle East and Northern Africa
b. International partner:
   - ETC–Urban Agriculture – programme co-ordination, strategy development, and international linkages

The RUAF Foundation has the following mission statement:

The RUAF Foundation contributes to reduction of urban poverty and food insecurity and stimulates participatory city governance and improved urban environmental management, by creating enabling conditions for empowerment of male and female urban and peri-urban producers and by facilitating the integration of urban agriculture in policies and action programmes of local governments, NGO's and private enterprises.

The RUAF Foundation currently is implementing the 'Cities Farming for the Future' (RUAF–CFF) programme (2005–8) in 20 cities in 15 countries. The regional RUAF partners closely co-operate with multi-sectoral teams of local partners, including municipalities, NGOs, producer organizations, research and training institutions, and government organizations.

The main strategies of the RUAF–CFF programme are the following:

- **Capacity development on urban agriculture**: training of trainers, development of training modules, organization of policy workshops and training courses, organization of study and exchange visits, design of an internet-based self-learning package
- **Policy development and action planning on urban agriculture**: establishment of multi-stakeholder forums and interdisciplinary working group in each city, joint situation diagnosis, policy formulation and action planning, implementation, and monitoring of pilot projects
- **Knowledge management on urban agriculture**: preparation and publication of policy papers, production of guidelines and manuals, publication of the Urban Agriculture Magazine in four languages; dissemination of experiences to another 30 cities
- **Learning from urban agriculture experiences**: introduction of participatory impact-monitoring systems at all levels, systematization of experiences, participatory evaluation of pilot projects, regional partner consultations
- **Gender mainstreaming in urban agriculture**: (the activities undertaken in this strategy are elaborated below).

Gender mainstreaming efforts undertaken by the RUAF network

At the global level, gender mainstreaming in the RUAF network was stimulated by:

a. Integrating a gender statement (see Box 16.1) in the statutes of the RUAF Foundation which stipulates how the RUAF partners view gender in relation to their work in urban agriculture, why they think it is important to take gender into account, and how this will be made concrete in the work. The gender statement is included in the co-operation agreements with the regional and local organizations that are participating in the programmes of the RUAF Foundation.

b. Establishment of an Advisory Group on Gender and Urban Agriculture, consisting of a small group of people who have expertise and hands-on experience in gender issues and agriculture in an urban setting, and who advise and guide the RUAF partners regarding the integration of gender issues in their activities and how to 'gender mainstream' urban agriculture.

c. Training of all international and regional RUAF staff by the following means:

- The organization of specific training on gender in Johannesburg (2003) for the international team and regional RUAF co-ordinators regarding 'Gender Issues in Urban Agriculture Research and Development Projects', in order to enhance their capacity to take gender issues into account when planning, implementing, or monitoring their RUAF and other urban agriculture activities.

- The development of six case studies on gender and urban agriculture in the RUAF pilot cities.

- The organization of a 'Workshop on Gender and Urban Agriculture' in Accra in September 2004, together with Urban Harvest and with financial support of the International Development and Research Centre (IDRC) and the CGIAR System-wide Programme on Participatory Research and Gender Analysis (PRGA). The workshop resulted in important lessons and recommendations for dealing with gender issues in urban agriculture projects (differentiated for capacity development, research, policy development, action planning, implementation, and monitoring).

- Production of three RUAF working papers on gender and urban agriculture: (a) on main gender issues in urban agriculture, (b) on gender mainstreaming, and (c) on gender-sensitive tools for participatory situation analysis.

d. Making gender equality a separate objective in the CFF programme, which also ensures that the topic is discussed in every progress report and annual work plan and during the yearly co-ordination meetings. During these meetings, the partners apply the Checklist for the Gender Mainstreaming Activities in the CFF Programme, which indicates for each objective of the RUAF–CFF programme how gender is to be integrated (see also Table 17.15 of the Tool box).

At the regional level, the following efforts to integrate gender in RUAF–CFF and to mainstream gender in the organization and that of the local partners, are developed:

- A gender adviser was appointed in each region to provide support to the regional RUAF team regarding the integration of gender when drafting Terms of Reference of sub-contracts, developing training materials, preparing guidelines, etc.

Box 16.1 The RUAF gender statement

RUAF applies the following definition of gender, taken from IDRC–Cities Feeding People, which reads as follows:

The term gender refers to the culturally-specific roles, rights and obligations of women and men, and the relationships between women and men. These roles, obligations and rights are not fixed; they are contested and vary widely within and between cultures. They also vary according to other social factors such as class, race, ethnicity, age and marital status. Women and men continually participate in defining gender relations. It is therefore essential to explore the multiple meanings given to gender roles, obligations and relationships in historically and spatially-specific contexts.

The partners in the RUAF network promote the exchange of experiences and the generation of knowledge regarding gender issues in urban agriculture. We continue to educate ourselves on this topic and are committed to keeping our knowledge up-to-date about new developments. We share this knowledge with our local partners so that they are enabled to adequately address gender issues in their activities.

The RUAF partners work towards the goal of gender equity in partner countries by both specific initiatives and by ensuring that all of their activities support gender equity objectives. This means that gender differences relevant to urban agriculture projects and policies will be identified in order to improve the relevance and impacts of such projects and policies for both women and men. RUAF partners acknowledge that in all their initiatives, the participation of both women and men – both quantitatively and qualitatively – needs to be ensured. They also promote equal access to and control over productive resources for urban agriculture for men and women and the development of participatory mechanisms that enable women, as well as men, to participate in decision making processes and to influence the policy agenda and the priority setting for development projects in the areas where they live.

In concrete terms, this means that the RUAF partners are committed to integrating gender into their strategies and their methods; showing the importance of gender differences by developing case studies; using gender analysis in their research activities; integrating gender in their training and communications activities; applying gender-sensitive project and policy planning and implementation; using gender-specific monitoring and evaluation methods; and building the capacity of local partner organizations in gender analysis and planning.

- The appointment of gender-sensitive staff in the regional RUAF teams and the creation of a gender balance in these teams.
- Promoting the development of a gender policy in their own organizations.
- Networking with gender-sensitive organizations and interest groups to reinforce their own gender capacities.

At the local level the following strategies were applied:

- Training of the local (multi-stakeholder) teams in major gender issues in urban agriculture and gender-sensitive diagnosis, policy formulation, action planning, and monitoring.
- Inclusion of the RUAF gender statement in all co-operation agreements with local partners and stimulating them to adopt a similar gender statement in their organization.

- Promoting gender balance in the composition of the local teams and among the participants in the Multi-stakeholder Forums in each pilot city.
- Implementation of specific gender case studies in the RUAF pilot studies in order to strengthen the gender perspective.
- Promoting gender-balanced participation in all events organized by RUAF at local level, such as the training sessions, policy seminars, and meetings of the Multi Stakeholder Forum, inclusion of gender as a topic on its agenda and the use of gender-sensitive discussion methods.
- Assisting the staff of local partner organizations involved in RUAF–CFF in the integration of gender issues when drafting a Municipal Policy or Strategic Action Plan on urban agriculture and designing urban agriculture projects and inclusion of gender-affirmative actions.
- Promoting the use of gender-sensitive tools and gender-disaggregated indicators for the monitoring of project results.
- Production and/or distribution of specific materials like the case studies and RUAF working papers on gender and urban agriculture mentioned above, a specific issue of the Urban Agriculture Magazine on this topic, a chapter on gender mainstreaming in the RUAF publication 'Cities Farming for the Future: Urban Agriculture for Green and Productive Cities', and creating a specific page on gender and urban agriculture on the RUAF website that is regularly updated with newly published materials.
- In each region a regional gender workshop will be conducted to facilitate sharing and analysis of the experiences gained in each of the cities regarding gender mainstreaming of urban agriculture.

Results and experiences

The experiences gained and results achieved with the above vary from region to region, reflecting the prior exposure and experience of the regional and local staff involved in RUAF, the degree of gender sensitivity of the organizations involved, as well as the socio-cultural context in which these operate.

One lesson learned is that applying a gender approach (rather than a 'Women in Development' approach) has yielded positive results. Equally, placing the discussion on gender issues in the broader framework of 'social inclusion' also enhances acceptability and commitment.

When reviewing the results of the efforts made to mainstream gender into the RUAF–CFF programme, it is observed that gender is well integrated in the training activities at all levels and is having positive effects on the participants. The training has changed their attitudes to gender, and most of them report that they know how to apply the gender concepts and tools in the planning and implementation of urban agriculture projects. Some of the participants also feel able to facilitate the integration of gender into their own institutions and their activities.

However, when reviewing the various activities undertaken by the trained participants, we observe that the integration of gender in diagnostic research activities advances relatively well but that its integration in subsequent policy formulation and action planning is much less strong and needs more attention.

Factors that seem to play a role here are that the diagnosis stage is still much more controlled by the core partners in each RUAF pilot city, whereas in the action planning and implementation stages many other actors also take part, including large sectoral institutions that may have less gender sensitivity than the core partners. It is also observed that equal participation of men and women in RUAF-organized training events and workshops and in the Multi-Stakeholder Forum is close to equal when it concerns field staff and producers, but when it comes to higher-level officers and politicians, men are still strongly over-represented.

The application of the gender checklist has proved to be very useful to facilitate the integration of gender in the different components of the RUAF–CFF programme and at different levels and helps to quickly identify areas that need to be strengthened.

Local partners indicate that a further refinement of the methods and tools used for gender-sensitive diagnosis, planning, and monitoring and evaluation is needed as well as looking for ways to reduce the costs involved in collecting gender-disaggregated data.

In some regions the functionality of the regional gender adviser has been low, due to difficulties in really involving the adviser in the planning and realization of the RUAF activities in this region (often the role of the gender adviser focused on participation in regional training events and development of training materials and much less so on advice regarding integration of gender into (draft) policies and action plans on urban agriculture).

The efforts made to mainstream gender into the organization of the regional RUAF partners are progressing well. All regional RUAF partners have developed a gender statement and policy if they did not have one yet. Stimulated by the experiences gained in the RUAF project (among other influences), most regional RUAF partners now also are applying a gender-sensitive approach in their other projects.

The gender-mainstreaming effect in the local partner organizations in the pilot cities is less clear. Often the RUAF gender statement is adopted by the local partner organizations (or they had such a statement or policy already). But *living up* to such a statement and its active implementation still requires in most organizations (especially the larger institutions) an internal process of awareness raising, changes in the organizational culture, and further training that might be lengthy (and beyond the possibilities of the RUAF programme).

However, many local partner organizations in the RUAF programme have found that the application of gender-sensitive tools in the diagnosis phase generates valuable information and a better understanding of the actual situation, its problems and potentials when taking women's and men's specific roles and interests into account, and that such information is vital for the design of effective development strategies.

In conclusion: gender mainstreaming in RUAF staff training and diagnostic research activities is progressing well. But effective use of the insights gained on gender and urban agriculture in the policy formulation and strategic action planning in the Multi Stakeholder Forum activities needs to be further strengthened. The regional and local RUAF partners have an important role to play here, as they can provide technical assistance on gender during formulation of the policies and action plans, and provide examples of gender specific strategies and gender-affirmative actions.

Gender mainstreaming in CIP–Urban Harvest

About Urban Harvest

Urban Harvest is a system-wide initiative on urban and peri-urban agriculture of the Consultative Group on International Agricultural Research (CGIAR, the network of 15 international agricultural centres; see www.cgiar.org for an overview of participating institutions and their activities). Urban Harvest aims to mainstream urban-agriculture issues in all centres. Since its inception, it has developed activities with five of the centres so far (CIP, CIAT, ICRAF, ILRI, IITA, and IWMI).

Hosted by International Potato Centre (CIP), Urban Harvest was formally launched in late 1999 to focus the CGIAR issues relevant to urban and peri-urban agriculture. Working through CIP, it has regional co-ordination offices in Kenya, Vietnam, Philippines, and Peru.

Urban Harvest aims to enhance food and nutrition security, increase incomes, and reduce negative environmental and health effects among urban populations via agriculture. These three goals are closely related to the differences in the role of agriculture rural, peri-urban, and urban conditions. Evidence of micronutrient deficiencies among poor urban children underlines the need for a sustainable means to increase the availability of micronutrient-rich foods for this group. The highly unstable employment conditions of developing world cities highlight the need for flexible, alternative employment opportunities that can provide access to supplementary income. Finally, while agriculture offers opportunities to make a positive impact on urban ecosystems under tremendous strain from high populations and poor infrastructure, it can also have negative repercussions through the unsafe use of urban organic wastes and the indiscriminate use of pesticides. Because agriculture in urban and peri-urban areas has both great potential and also certain risks, Urban Harvest recognizes a further goal: through engagement with the particular policy and planning institutions of urban areas to integrate agriculture into urban systems as a safe, accepted component of sustainable cities.

The main research themes of Urban Harvest are:

- the livelihoods context (urban livelihoods and markets)
- the environmental context (urban ecosystems health)
- the urban political–institutional context (stakeholder and policy analysis and dialogue)
- knowledge networking (knowledge exchange, sensitization, capacity development).

Urban Harvest undertakes collaborative research for development projects with various international research centres, government ministries, local municipal departments and authorities, civil-society organizations, and private-sector companies active in urban agriculture in nine cities, as well as national research institutions and universities and government parastatals such as national environment-protection agencies, and water and sewerage companies in six countries.

Gender-mainstreaming efforts undertaken by Urban Harvest

Since its inception in 2000, Urban Harvest has undertaken the following gender-mainstreaming activities targeted at the programme and project activities level and among boundary partners:

- Urban Harvest Sub-Saharan Africa participated in the 2003 RUAF training on gender to upgrade the knowledge, skills, and commitment of staff regarding 'Gender Issues in Urban Agriculture Research and Development Projects', and to enhance its capacity to take gender issues into account when planning, implementing, monitoring, and evaluating urban agriculture research and development activities.
- Urban Harvest (in partnership with IDRC–Cities Feeding People, the RUAF partners ETC–Urban Agriculture in the Netherlands, Municipal Development Partnership in Zimbabwe and International Water Management Institute in Ghana, the Urban Management Programme (UMP) and the Research Development Department of Kenya Government) organized the 'Anglophone Africa Training Course on Urban Agriculture; Concepts and Methods for Research and Management' held in Nairobi Kenya from 8 to 26 March 2004 (the training materials of this course are available on-line as part of a web-based course: http://etraining. cip.cgiar.org). The participants were three-member City Teams, a researcher, a local government official, and a third member involved in project implementation – usually someone from an NGO or CBO. A section on gender issues related to the subject matter of each module was incorporated in each of the seven modules that were prepared and delivered by teams from different countries. In addition, a half-day session specifically on gender was organized involving role-plays, presentations and debate, facilitated by a gender specialist.

- Urban Harvest participated in the organization of the workshop on 'Women Feeding Cities' in Accra in September 2004. Urban Harvest and its partners prepared some of the case studies, based on previous research, and identified a strategy for gender mainstreaming in the Urban Harvest programme, project activities, and boundary partners.
- As a follow-up to the 2004 Ghana workshop, two projects in Kenya (namely 'Positive Selection of Potato Seeds' and 'Assessment of Benefits and Risks in Wastewater Reuse for Agriculture in Urban and Peri-urban Areas of Nairobi') were chosen for the implementation of a project on gender mainstreaming in the research process of the International Potato Centre (CIP) between March 2006 and March 2007, supported by the Participatory Research and Gender Analysis (PRGA) programme (which aims to mainstream gender analysis and equitable participatory research in CGIAR centres and national agricultural research systems). The PRGA programme assesses, develops, and promotes methods and organizational innovations for gender-sensitive participatory research, and works to mainstream their use in plant breeding and in crop and natural-resource management. A staff member of Urban Harvest sits on the Advisory Board of the Participatory Research and Gender Analysis (PRGA) Programme of the CGIAR. The gender mainstreaming in the research process of the CIP project was led by Urban Harvest and implemented in collaboration with CIP and the University of Nairobi, Jomo Kenyatta University of Agriculture and Technology, Kenya Agricultural Research Institute, Ministry of Agriculture, and Kenya Green Towns Partnership Association. The project involved selection of appropriate gender-analysis tools from existing literature (such as the RUAF working paper on gender-sensitive participatory appraisal tools for studying gender in urban agriculture), capacity building of all partners in participatory research and gender-analysis tools and approaches, field application of the tools in the collection of qualitative and quantitative gender-disaggregated data and development of draft guidelines for gender-responsive research in CIP.
- Urban Harvest has been engendering its projects and those of CIP through deliberate incorporation of gender-mainstreaming activities into the projects' background information, objectives, methodology, activities, monitoring and evaluation, budgets, feedback/publications, and community capacity building. Some of the results of these research projects are to be published in *Urban Agriculture in Sub-Saharan Africa – 20 Years On: Case Studies and Perspectives*, edited by Gordon Prain, Nancy Karanja, and Diana Lee-Smith (forthcoming). In the project 'Understanding how to Achieve Impact-at-scale through Nutrition-focused Marketing of African Indigenous Vegetables and Orange-Fleshed Sweet Potatoes', led by CIP between 2007 and 2009, Urban Harvest worked with Farm Concern International in engendering the former's community capacity-building modules.

- Promoting the development of a gender statement (see Box 16.2) was another strategy for enhancing gender mainstreaming in organizations that was applied by Urban Harvest for its programme partners and its parent research centre.

Results and experiences

The efforts of Urban Harvest to mainstream gender analysis in its research for development process have resulted in capacity building of its boundary partners in the use of gender tools and approaches. What these experiences revealed was that if gender mainstreaming in research for development projects was to be effective, the gender-analysis skills of the research teams themselves needed to be enhanced. Sometimes this can be done through identifying existing gender-analysis skills among the partners. The aim is to ensure gender mainstreaming in the whole project cycle.

The experiences with gender mainstreaming in urban agriculture gained by Urban Harvest have indicated the need to adapt some of the available gender-analysis tools to the urban situation. For example, in the urban setting the education status of male and female participants in a participatory appraisal may be highly variable. The bias created by differences in literacy levels of men and women respectively can be limited by the use of visual methods e.g.

Box 16.2 Urban Harvest–International Potato Center (CIP) gender statement

Despite increased attention to gender in the international development arena since the rise of feminism during the 1970s, few international or national agricultural research organisations have yet integrated gender as a central element of problem diagnosis and technology development. The stakeholder needs have not always been taken into account in research, technology development and transfer processes. This may contribute to the prevailing low adoption of innovations.

The concept of gender analysis and diversity is not new in CIP, and there exists a Gender and Diversity Committee which works very closely with Gender and Diversity Programme of the CGIAR and advises the Director General on human resource issues. However, engendering the research agenda still remains a challenge. For this to happen there needs to be political will, technical capacity, organisational culture and accountability for gender integration in CIP's research in all divisions.

Recognising this, Urban Harvest led the drafting of guidelines for ensuring gender responsiveness in CIP's research process, with the support of the CGIAR system-wide programme on Participatory Research and Gender Analysis (PRGA).

Urban Harvest and CIP are committed to integrating gender responsiveness in their research and development activities with partners. Urban Harvest and CIP ensure that the experiences, aspirations, knowledge, opportunities, needs, concerns and constraints of females and males of all ages are integrated in project planning, budgeting, implementation, monitoring and evaluation, reporting and publications. This approach ensures that gender disaggregated data is available for pro-poor policy formulation processes. To achieve gender sensitive research teams and consequently gender responsiveness in the research process, CIP has set gender and diversity staffing, practice and policy goals targeting human resources.

the use of different counters for men and women (like different types of grain or pebbles of different colours) when doing a ranking exercise to prioritize certain issues.

Another lesson learned is that inclusion of gender issues in research proposals is necessary but not sufficient. Achieving this does not avert the danger that gender issues are being completely ignored during implementation and monitoring and evaluation of the project. There is a strong need for aggressive championing of gender issues in all steps of the project cycle, and opposition by some team members (e.g. by asking for evidence of the added value of gender mainstreaming) should be expected. We also learned that budget allocation is necessary to achieve gender-mainstreaming activities in projects.

Some lessons learned from the organization of the Anglophone Africa training course on urban agriculture are the following:

a. The special session by the gender specialist generated great controversy among participants and some heated debate and even conflict. There was likewise difference of opinion in the evaluation process about whether this indicated success or failure, and disagreements about the causes of the conflict. Some participants stated that the gender material should have been presented in a more 'diplomatic' way, given the cultural leanings of the participants. However, discussion among the evaluator and members of the management team suggested that the problem lay in not ascertaining whether all participants had had any previous exposure to gender concepts, as first exposures to the subject tend to generate strong emotions in men and women alike who have not had to examine gender issues previously.
b. The integration of gender in all modules in direct relation to the subject matter of each module went well. However, since the learning modules were delivered by teams with members from various countries, the integration of gender in each module presented a co-ordination challenge for each team, and for the course management.

Also the experiences gained in engendering training materials for community capacity building indicate that incorporating gender perspectives in all modules is more effective than having a separate module on gender. Enhancing gender perspectives in already existing training materials encounters the challenge of property rights and ownership that needs to be addressed through proper acknowledgement of the contribution of each party.

Urban Harvest, through its pioneering role in gender mainstreaming in the research process of the two above-mentioned CIP projects in Nairobi, is currently taking the lead in enhancing the gender responsiveness of research and development activities in CIP as a whole. The CIP project on gender mainstreaming in the research process implemented by Urban Harvest has resulted in the development of guidelines for gender-responsive research that target the institutional set up, the organizational culture, and the research project cycle; these guidelines are being applied now in the Center as a whole and can be found on the PRGA website http://www.pragaprogram.org. This could be another way of getting gender analysis on the research agenda.

There are challenges involved in getting gender mainstreaming on to the agenda of research and development organizations. These challenges could be addressed by forming gender and diversity committees with a wide representation across divisions, themes, programmes, regions, and men and women in an organization. The CIP Gender and Diversity Committee (G&D) provides services and resources to the CGIAR centres in order to raise the gender-and-diversity awareness of all CIP research staff. The committee hopes that this will lead to improved attitudes among staff and greater incorporation of gender analysis into research projects.

The experiences of the CIP Gender and Diversity Committee indicate that in order to influence institutional policy and culture it is important to win the support of the management. Awareness among all staff can be created through e-debate and round-table meetings. Equal importance should be given to presentation of gender-mainstreaming work during organizations' general meetings where other scientific findings are also discussed, in order to enhance the importance which people attach to the former as a cross-cutting issue in research and development.

Urban Harvest has benefited from the system-wide gender programmes of the CGIAR and has also contributed substantively to both of them through its active participation. In this it has been assisted by its partnership with RUAF, both in the form of training and in terms of collaboration on the substantive development of the field of gender in urban agriculture, which has become the focus of starting gender mainstreaming in its parent research centre.

References

Commission on the Advancement of Women (2003) *The Gender Audit: Questionnaire Handbook*, InterAction, the American Council for Voluntary International Action, Massachusetts.

Prain, G., Karanja, N., and Lee-Smith, D. (eds.) (forthcoming) *Urban Agriculture in Sub-Saharan Africa – 20 Years On: Case Study Results from Cameroun, Kenya and Uganda in a Regional Perspective*, IDRC, Ottawa and CIP, Lima.

CHAPTER 17

Tool box for gender-sensitive urban agriculture projects

Abstract

In this chapter the tools suggested in Chapter 15 will be discussed in more detail. First, we consider how some general diagnostic research methods can be used in a gender-sensitive way. Subsequently, we will present more specific tools that are of special value to integrate gender in a particular phase of the project cycle, and we will provide some suggestions how to use each tool.

Introduction

In this section we present a range of tools that have been suggested for each phase of the project cycle in Chapter 15 of the publication. We begin by providing an overview of gender-sensitive main diagnostic research methods, followed by details of specific tools that can be applied as part of a main method (e.g. a problem tree or ranking exercise during a focus-group meeting; an activity-analysis checklist during a semi-structured household interview, etc.) or on their own.

Specific tools, after certain adaptations, can often be used in project phases other than the one for which they are mentioned in Chapter 15. For example, diagnosis tools may be appropriate and useful in condition monitoring and evaluation tasks later in the project cycle. It is recommended that main or specific tools be adjusted according to local circumstances, interests, and needs; in some cases flexibility and creativity can lead to innovative 'new' tools that are a result of mixing and matching the ones presented here. To this end, the tool box is a starting point from which to begin gender mainstreaming in urban agriculture research and development projects; practitioners should feel enabled rather than constrained by the tools and approaches outlined below.

Main diagnostic research methods

Review of secondary data

Reviewing secondary data involves the collection and review of existing published or unpublished data and information relevant to the area or topic, including reports, census data, research findings, municipal and hospital statistics, and aerial photographs (for example on land use patterns). This method is useful to get an initial picture of the situation of the target group and socio-economic and institutional context, as well as to determine gaps and possible contradictions in the available data. This will help to formulate alternative working hypotheses for the field study and to design the fieldwork. With regards to gender, it is important to critically review secondary data for gender bias, seek out gender-disaggregated data, and ensure that data are sourced from as wide a range of research or development areas as possible, especially including those, like household food security, nutrition, and child health, which are sometimes negatively characterized as 'home economics' or 'household studies'. The data review is usually done by team members visiting libraries, government offices, universities, research centres, marketing bodies, etc. Secondary information can be processed in two stages: first, the identification and compilation of the material, and second, the analysis of the collected information. Analysis often begins by grouping the information gathered according to the main themes of the study, followed by the analysis, and finally formulation of the working hypotheses for the field study during which collected secondary data will be cross-checked and gaps will be filled in (Groverman, 1992; Lingen, 1997).

Direct observation

Direct observation involves collecting information by noting down the things one sees happening at the time that they happen (e.g. objects, conditions, events, processes, relationships of people). In most cases a list is used of key items or indicators related to the issue under investigation (e.g. how women participate in community discussions, the activities that men and women perform, decision making on use of the resources, etc.). If it is difficult to take notes at the time of observation, they can be written down later. Observations are analysed afterwards for patterns and trends. This research tool has been adapted from anthropology for problem diagnosis in projects. The hypotheses arising from direct observation should be cross-checked (e.g. with key informants or group interviews), and direct observation can be used to check information gathered from secondary data or interviews/focus groups. Direct observation is useful for gaining a better understanding of a situation, specifically those things that are difficult to verbalize or often go unnoticed because they are grounded in internalized social norms. Observation can enrich insight into various gender aspects of urban agriculture (e.g. the activities that women and men perform, the daily workload of women and men, women's role in decision making in the household, women's participation in community meetings, the self-confidence of women, the behaviour of men vis-à-vis women, etc.). This tool can be used in any phase of the project cycle (Groverman, 1992; Lingen, 1997).

Semi-structured interviews

Semi-structured interviews are informal discussions and conversations structured by using a list of key issues prepared in advance. They can be useful to obtain information in general or about a specific topic, to analyse problems and opportunities, to discuss plans, or elicit perceptions (e.g. on gender relations). It is advisable to take not more than one hour for an individual interview, and not more than two hours for a focus-group interview; timing, venues, and language should be considered in terms of the circumstances of the respondents (e.g. women who work on agricultural plots during the day may not be able to meet for interview during these hours; women may be more comfortable in a particular local language). Interviews are often conducted by a team of two or three people of different backgrounds. Semi-structured interviews can be used in combination with other specific tools (e.g. mapping, ranking, making calendars or timelines, etc.).

How to conduct semi-structured interviews

- Prepare an interview guideline in advance. This is not a questionnaire but a list of topics to discuss with respondents (grouped in such a way that the sequence of the discussion will be easy to manage for those involved). For each topic, prepare initial questions (to introduce the topic and make the respondent think and talk about it) and probing questions (to dig deeper, to get more details: what, why, who, when, how, how do you mean, anything else, why, etc.).
- Select one person to lead the interview. A second person records the questions, answers, and discussion, making notes discreetly to avoid distracting the respondents.
- Deal with topics one by one. Begin questioning by referring to something or someone visible. Ask questions in an open-ended and probing way. Intersperse with probing questions and discussions; ask for concrete information or examples. Ask new questions arising from the answers given. Allow the interviewed person also to raise her/his questions and discuss these too. Involve other people in the discussion, if present. Pay attention to group dynamics.
- With regard to gender, it is important to include women as respondents, to encourage women to participate, and to look for situations and places where women can express themselves freely. Depending on the purpose of the interview and the degree to which women can express themselves freely in the presence of men, group interviews may or may not be carried out with mixed groups. Interviews with small homogeneous groups of women can provide valuable information on their position in that society or on sensitive issues.

Semi-structured interviews can be conducted at any stage of the project cycle (Groverman, 1992; Lingen, 1997) and can be conducted at the individual, household, or community level. A brief description of the content and the use of these specific interview forms is given below; which interview format to select depends on the issues studied and the level on which these issues take place.

Key-informant interviews

Key-informant interviews are conducted with specially selected individuals who have long experience in a certain community or specialized knowledge or skills in a certain topic. Key informants should be carefully selected, on the basis of the issues and themes relevant to the project. One should be aware of possible biases of the persons interviewed, including gender bias; it is important in gender diagnosis to interview not only those key informants who are gender sensitive, but also those who are not, in order to gain understanding of a full range of perspectives. Information should be cross-checked with information from other sources. Key informants may include members of the target group, for example, local leaders or staff of support organizations and development programmes in the sector concerned (male and female), or they may be external to the project (e.g. municipal officials, NGO staff, local historians, academic researchers, etc.) (Groverman, 1992; Lingen,1997).

Semi-structured household interviews

These are interviews with specially selected households to get a view on the differences between different types of household (e.g. various socio-economic categories or farming-system types) in the community regarding the gender-differentiated management of resources, division of labour, specific problems and potentials, etc.

One technique of identifying the criteria to select the households to be interviewed is as follows:

- Bring together a small group of locally well-informed people (e.g. the local health officer, a school teacher, and the leader of a local women's group).
- Ask them to draw a picture of a poor farming household, an intermediate farming household, and a wealthier farming household. The drawing should reflect the main characteristics of such a family: composition and size, resources available (e.g. land, water, animals), farming activities, their non-farming activities, location, position in the community, origin, etc.
- Discuss the drawings with the respondents in order to detect the main differences between the three types of household (e.g. recent migrant/ not recent migrant, land-holding/not land-holding, with cattle/without cattle, family with older children/family with young children, male-

headed/female-headed, ethnicity/class, type of job in addition to farming).
• Select households based on factors identified through this process.

When interviewing a household, one will preferably interview adult male and female members separately, eventually followed by a discussion on certain issues with the whole household. Sometimes it is also interesting to interview older and younger women separately, since this might indicate on-going changes in the position of women. If possible, let a male member of the team interview the male member of the household and the female team member the female in the household (De Zeeuw and Dubbeling, 2004).

(Focus) group interviews

These are interviews with a specially selected (focus) group of six to ten people who have certain factors in common. In such sessions, specific topics are discussed under the guidance of a moderator in order to get their views and perspectives on a certain issue, to gain insights on the position and problems of this specific (sub-)category of the population, or to tap their specialist knowledge of a certain topic or problem. This might be specific gender or age groups, households with similar wealth status, owners of specific resources, people with a specific problem or disadvantage, people involved in a specific role or activity (e.g. women leaders, people involved in marketing of a certain product), etc. (Groverman, 1992).

Focus groups are often a good follow-up to household interviews in order to get more insight in certain topics and to check whether patterns found in the households are validated in the whole group. Focus-group meetings are also very suitable to analyse a certain situation or problem in more detail and to identify and evaluate potential solutions to these problems. Organize group interviews (four to six persons), preferably with men and women in separate sub-groups (each with a team member as facilitator; a female interviewing the women, a male interviewing the men) (Lingen, 1997: De Zeeuw and Dubbeling, 2004). It is important to note that when single-sex focus groups are brought together in a mixed plenary session, voices may be silenced, or people may not be so willing to share sensitive views or insights. It may be useful to ask a representative from each single-sex group to present a summary of the discussion to the mixed-sex group in a way that ensures privacy and at the same time ensures that all voices are heard.

Preferably, focus-group discussions should be followed by quantification of the results of the group discussions for each issue discussed. This can be done with the help of a ranking technique. For example, when discussing the effects of a project on food availability in the household, one first invites some people to indicate by what percentage food availability has increased in their household and then use the variation in the answers to establish a scale (see the example in Table 17.1) and invite all participants to score (men

Table 17.1 Example of quantification of one topic in a focus-group discussion

Availability of home-grown food for consumption	No increase	Less than 10% increase	10-25% increase	25-50% increase	Over 50% increase	Total
Male respondents	0	1	1	3	1	6
Female respondents	0	3	2	1	0	6
Total score	0	4	3	4	1	12

and women separately, for example, by giving women and men stickers of different colours or sticks of different lengths).

Questionnaires

Questionnaires (or surveys) are extensively used in gathering quantitative data on various topics related to urban agriculture. They are applied mainly to collect baseline data and quantitative data on production levels, economic costs and benefits, certain characteristics of the producers, etc. Questionnaires can be costly and time-consuming (collection and processing of the data) so it may be necessary to reduce the number of questions as much as possible by focusing on those questions for which quantification is essential and which cannot be answered in another way (one page 'sondeos').

Questionnaires are applied to a representative of the household, who is often the head of household or at least someone who is old enough to know the answers to the questions. Researchers may decide that they can ask either the man or woman of the household, and may prefer one or the other depending on who is around and easier to get hold of, or who knows the most about urban farming. A weakness of formal surveys is that they assume that the interviewee is knowledgeable about all aspects of agriculture and household livelihoods, whereas in many households there is gender specialization, so that surveys need often to be complemented with more detailed group or household documentation methods. A question on who is the head of the household should always be asked, and then the answer used to classify the data as coming from a man- or woman-headed household for calculations of the answers to all the other questions.

It is good to include questions about who is responsible for specific agricultural tasks in the household, or whether hired help is used. Following the procedures of structured interviews, the answers can be classified as male or female (or adult man, adult woman, girl, boy, or hired labourer). When calculating the results, this will then give tables with numbers and percentages engaging in different tasks. Useful questions to differentiate by gender include those about exposure to training, membership of groups and participation in other livelihood activities, where there may well be significant differences in gender patterns, and thus an opportunity to provide good statistics.

Not all questions need to be gender-disaggregated in a structured survey. Those asked about age and education levels can be made questions about the head of the household, with the answers classified under the sex of the head of household. Similarly, questions about other livelihood activities or sources of income can be asked about the household as a totality. Where these questions are asked about both men and women (usually the husband and wife) in a household, calculating and presenting the results gets more complicated, because of having to process the data from women-headed households differently from those from men-headed households. Most such surveys do not do this, although some ask for the information about all household members and then present these as separate sets of tables, grouped by males and females. It is not advised in most structured surveys, as various complications arise such as the survey time becoming very long and data from children distorting the results (e.g. high proportions of people may be reported as not attending school – and thus classified as lacking education – when in fact a large percentage of the population is under school age).

The interviewers (often junior students are recruited to carry out the interviews) should be properly trained in interview techniques and gender awareness, and the interviews must be implemented under the supervision of experienced research staff. It is important that interviewers take thorough notes, keep a research notebook, and are aware of any qualitative information that may supplement numerical or statistical data gathered through the formal survey method.

The numerical information derived from questionnaires tells us about behaviours of the people we are interested in (namely, men and women urban producers in a particular place) and it may be accurate for that location, yet the numbers may be of limited value if we want to know what proportion of people in the city's population are farming or what role they play in the city's economy. Wider statistical surveys interview both producers and non-producers, mostly in order to determine the proportion of producers in the general population and to allow generalization of findings at city level. In this case, the proper drawing of samples of the population being analysed is essential.

Although survey questions may focus on the behaviour of household members in relation to urban farming, questions about men's and women's interests and preferences are not appropriate for this kind of data-collection method. Focus groups or semi-structured interviews or other participatory techniques are better suited for that. The latter are better suited also for building understanding of main constraints and opportunities, and exploring the causes and effects of certain problems.

Diagnostic tools

Activity-analysis chart

The activity-analysis chart (adapted from Hovorka, 1998: 17-19) is a tool that helps to explore how tasks and responsibilities in urban agriculture are distributed according to gender at the household and local levels.

It is important to accurately understand and document the activities of male and female urban producers in order to assess why an urban agriculture system functions as it does, and the implications of the division of labour for men's and women's roles, responsibilities, and obligations in a particular context. Eventually, gender analysis of activities can inform potential technology and policy options that will benefit both men and women urban producers.

The gender activity-analysis chart may be used to draft interview questions for individuals or focus-group discussions where participants can identify those activities they are actively or partially involved in (indicating the percentage of time involvement of women/girls and men/boys in each activity). Additionally, one may use this tool as a checklist for participant observation in a local neighbourhood or community, as well as when accompanying a producer on his/her 'typical' daily routine.

This tool may also be used in the following participatory exercise: line up three large drawings of (a) a man, (b) a woman, (c) a man and woman. Below these drawings are scattered smaller cards depicting various types of urban agriculture activity. Include some blank cards so that participants can add activities. Ask participants to sort the cards by categorizing them under the three drawings in columns, according to whether the task is generally performed by a man, or a woman, or both. Facilitate the discussion among participants about why they made the choices they did, and why particular tasks are delineated in this way.

Regardless of whether the activity analysis-checklist (Table 17.2) is used in interviews, direct observation, or a participatory exercise, it is important to consider whether respondents are focusing on gender stereotypes (the ways in which activities are divided by gender in society at large) or on the reality of men's and women's lives (the ways in which activities are actually divided up on a daily basis). Further, as noted earlier in this chapter, it may be useful to separate men and women to gather this information, and then bring respondents together to discuss people's activities in a plenary group session.

Seasonal calendar

The gender analysis chart discussed is more apt for getting an overview of gender division of labour in a household, but might be more difficult to answer by the respondents for the cropping activities, since division of labour might differ from product to product, or at different times of the year. This might also apply to certain types of livestock keeping (e.g. division of labour in dairy might be quite different from that in poultry keeping). In such cases the production of a seasonal calendar for the main crops or animals produced might yield a better understanding.

The seasonal calendar helps to explore issues including the following:

- division of urban agriculture tasks/labour among men and women (especially cropping activities) undertaken by the household (and related problems and possibilities);

Table 17.2 Gender-activity analysis checklist

	Activity	Women/girls	Men/boys
Horticulture (this section may be further disaggregated by crop-type)	Finding plot Securing plot (e.g. building fence) Clearing/levelling plot Guarding plot Land preparation (ploughing, hoeing, etc.) Finding/buying seeds Preparing seeds Sowing Transplanting Weeding Finding water source Irrigating Fertilizing Pest control Harvesting Threshing Cleaning Other processing Storing Packing Transporting produce to market Selling produce from home/at market Maintenance (e.g. irrigation system, shed) Composting Interaction with extension workers/municipal officers Sharing information with other producers		
Animal husbandry (this section may be further disaggregated by livestock-type)	Finding plot Clearing plot Building/maintenance of enclosure/stable/fishpond Cleaning of stable, etc. Finding/buying feed Feeding Watering Milking Vaccinations Other animal health care Slaughtering Selling/purchasing animals Processing animal products Selling processed animal products Disposal/reuse of manure Interaction with veterinary services/municipal officers Sharing information with other producers		
Other	Collecting firewood Collecting water Collecting wild fruits/vegetables, nuts, herbs Processing fibres, dyes (arts and crafts) Cooking Cleaning, sweeping Washing, laundry Disposing of household water Child care, care for elderly and infirm Paying land rent House repair Household garbage disposal Assisting neighbours Attending meetings of producers' organization(s) Attending other community meetings		

Source: Hovorka, 1998: 17–19

- which activities are undertaken in what periods of the year (and in what periods there are labour shortages or surpluses);
- in what periods of the year the households receive more benefits from the garden (food, income) and in what periods they experience food and income shortages.

This tool may be used as follows: first, participants list the main crops grown and animals kept by the household (or community group) interviewed. Second, they are asked which are the most important ones (for food and income); these are then selected for further analysis with the help of the seasonal calendar. A matrix is established, putting along the horizontal line the months of the year and along the vertical line the main crops / animals that were selected for further analysis. Respondents are asked to identify when the growing season for product X starts, and which activity they normally will undertake first, how long that activity will take, and who normally will perform it. Indicate the answers by drawing a line in the weeks concerned and give codes for the type of activity (e.g. s.p. = soil preparation, m= manuring, etc.) and the persons implementing it (for example o= female adult, < = male adult; * is girl and + is boy). One can include the estimated amount of time involved in each activity by using blocks /week for each activity instead of drawing a line. Once you have finished with product X, continue with product Y.

Thereafter, discussions take place on other vital issues, for example in which months they (specifying men and women where relevant) experience food shortages or lack of income, labour peaks, periods with water shortages or high incidence of pests/diseases, high market demand, etc. and what might be done to reduce such problems, or how to make better use of periods with labour surpluses, high market demand, etc. The answers are included in the calendar.

In Figure 17.1 a (rural) example of a seasonal calendar is presented.

Key questions to consider

- How exactly are men/women/children involved in specific urban agriculture activities?
- How do men's/boy's workloads compare with women's/girl's workloads?
- How does the task distribution of urban agriculture differ in men-/ women-headed households? Which are the most burdensome tasks? Who is responsible for these tasks?
- Who is responsible for reproductive tasks in the household? And for productive tasks?
- What urban agriculture activities face particular financial or time constraints? Who is responsible for these activities?
- How much flexibility is there in changing the workloads of men/women? Sharing tasks?

Climate Pattern / Month \ Activity	Hot/dry				Warm/wet				Cool/dry			
	Sept	Oct	Nov	Dec	Jan	Feb	Mar	Apr	May	June	July	Aug
Food shortages					xxxxxxxxxxxxxxxxxxxx							
Stumping					□	SP		□				
Hybrid maize	⊡ ◉ LP ◉ ⊡		⊡◉ P ◉⊡		⊡◉	W	⸱F⸱⸱⸱⸱⸱⸱⸱⸱◉⊡		⊡ ◉ ST □	FH PK	H	◉⊡ □	□ TR □
Traditional maize	□ LP P □				□	H □						
Sorghum		□ LP BP □			◉⊡	BS	⊡◉					
			O TW O			O ST	H	O				
Beer brewing	O											O
Finger millet		O	PS	O		O BS	O	H	O			
Beans	O⸱⸱⸱⸱⸱R⸱⸱⸱⸱⸱⸱O⸱⸱⸱⸱⸱⸱⸱⸱⸱⸱⸱R⸱⸱⸱⸱⸱⸱⸱⸱O				O	H	O					
Groundnuts	O R O		P	O	O	L	H		O			
Sweet potatoes	O⸱⸱⸱⸱⸱⸱R⸱⸱⸱⸱⸱O		P	O⸱⸱⸱⸱⸱⸱CH⸱⸱⸱⸱⸱⸱O								
Collecting Firewood	◉⊡											⊡◉
Carrying water	◉⊡											⊡◉
Feeding small Livestock	◉⊡											⊡◉
Coaching	O											O
Childcare	⊡											⊡
Fence and house Construction and repair	⊡⸱⸱⸱⸱⸱⸱⸱⸱⸱⸱⸱⸱⸱⸱⸱⸱⸱⸱⸱⸱⸱⊡								⊡			⊡
Cattle herding	⊡											⊡

Legend

O	Female adult	SP	Stumping (pulling/digging of trees out of fields)	BS	Bird scaring
◉	Female child	LP	Lane preparation	ST	Staking (cutting and staking teepee style
□	Male adult	R	Ridging	L	Lifting
⊡	Male child	P	Planting	H	Harvesting
–	Continuous activity	PB	Planting by broadcast	CH	Continuous Harvesting
...	Intermittent activity	F	Fertilizing	S	Shelling
T	Transplanting	W	Welding	PK	Packing
TR	Transporting				

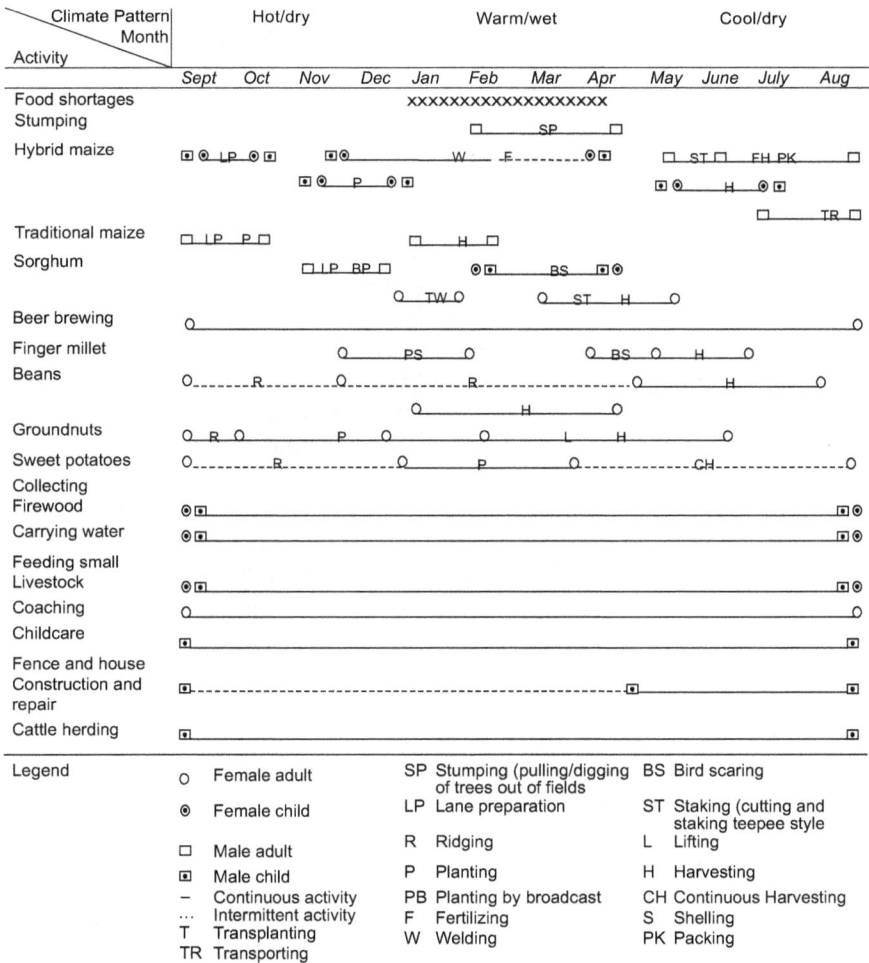

Figure 17.1 Example of a seasonal calendar
Source: Sims Feldstein and Poats (1989)

- Do men/women engage in or rely on social networks to share burdens and workloads (e.g. sharing of equipment/tools, sharing of child-rearing responsibilities)?
- What are the busiest periods of the year for men/women? Boys/girls?
- What is the relationship between surplus/shortages of resources and men's/women's responsibilities?

The following scenario illustrates how activity analysis may be applied in a particular context.

Urban Agriculture Scenario: household production of urban agriculture crops

An initial survey of production of urban agriculture crops reveals that households are producing cassava, corn, watermelons, peppers, and okra. Gender-disaggregated data further reveal that men produce okra, while women are responsible for the production of all other crops. Interpretation and analysis of this division of labour reveals that women produce cassava, corn, watermelons, and peppers largely for home consumption. Women cultivate peppers, for example, because they require little maintenance and produce over a long period. The household division of urban agriculture crops is such that in being responsible for lower-maintenance crops, women have more time to spend on their household tasks, such as cooking and cleaning, in addition to their urban agriculture tasks. Women are not likely to become involved with okra cultivation, due to the greater time constraints and activity conflicts that this high-maintenance crop generates. Men, because they are not responsible for household tasks, grow okra, which requires more work than some of the women's crops but pays better on the urban market. In this household urban agriculture system, men are responsible for urban agriculture crops for income generation and, in turn, the money generated through this activity. Women contribute to household food security and nutrition due to the variety of crops they produce. Technology and/or policy interventions stemming from this research should consider both profitability and nutrition as measures of food security. Interventions that encourage both men and women to participate in okra cultivation may, in fact, lead to a decrease in nutritional well-being among household members.

Resources-analysis chart

The resources-analysis chart (adapted from Hovorka, 1998: 20-23) may be used to draft interview questions for individuals or focus-group discussions where participants can identify those activities in which they are actively or partially involved. Participants can identify those resources they are knowledgeable about and have access to, when these resources are available, who controls or owns the resource, where it is located, the quality of the resource, how often the resource is used, and the amount of time spent on each resource.

This tool may also be used in the following participatory exercise: place the three large drawings of (a) a man, (b) a woman, (c) a man and woman in a row. Underneath these drawings scatter the smaller cards, each picturing a different urban agriculture resource. Include some blank cards so that participants can add resources. Ask the participants to sort the cards by placing them under the three large drawings, depending on *who uses* the resource (has access to this resource). Facilitate the discussion among the participants about why they made the choices they did, and why resources are delineated this way.

Table 17.3 Example of a resources-analysis chart

Resources *(distinguish between individual, household, community, organizational, etc. resources)*	ACCESS *(who normally are using this resource)*		CONTROL *(who decides on who can use the resource)*	
	Women/ Girls	Men/ Boys	Women/ Girls	Men/ Boys
Land o arable land 　– of good fertility/nearby/flat/ well-drained 　– of lower fertility/farther away/slope/water 　　logging o urban forest (fuel wood, grazing) o livestock grazing areas				
Water o river, stream, drain o piped (municipal) water o well/pump o irrigation and/or drainage systems o aquaculture ponds o household wastewater o trickling filters				
Inputs o seeds o livestock o bought fertilizers o composted farm and household wastes o institutional credit o informal loans (friends, neighbours) o tools (shovels, hoes, etc.) o building materials (for animal pens, fences, etc.)				
Infrastructure o on farm 　• shed 　• stable, pen 　• threshing & drying floors 　• storage facilities 　• latrine, septic tank, waste-treatment pond o off farm 　• mill 　• packing station 　• store/shop 　• co-operative 　• trading areas 　• composting site 　• sites of in-kind exchange				
Transportation o public buses o hired transport o bicycle o feet				
Information, training, and knowledge o government extension services o private enterprises o veterinary services o informal social networks; community groups o co-operatives o information centres o non-government organizations o social services (municipal)				

Source: Hovorka, 1998: 20–23

Then put the second set of drawings and cards on the ground, close to the first set. Repeat the exercise but this time focus on *who has control*, ownership, or decision-making power concerning each resource. Again, facilitate the discussion among the participants about why they made the choices they did, and why resources are delineated this way. Ask the participants to compare the ways in which they have arranged the two sets of Resources Picture Cards.

One may also use this tool as a checklist for participant observation in a local neighbourhood or community, as well as when accompanying a producer on his/her 'typical' daily routine. Transect walks can complement the resources-analysis chart, during which one visually notes resource availability and location, as well as people's access to, use of, and control over (where possible) this resource across the landscape. This information can be discussed with local people to gain in-depth understanding of what one sees during the walk.

Resources mapping

The resources-mapping tool can be used to explore what resources the household are using in their urban agriculture activities and who in the household has access to each of these resources and who is controlling them (owning, deciding on who can use it and for what purpose).

Start by asking the participant(s) to identify the main resources they use in their agricultural activities, starting with the land and water resources and on-farm infrastructure available to them, and note this on a map. A good way of doing this is to go around the garden or farm with the household members and make a sketch of the farm/garden layout, indicating:

- boundaries of the plot and approx. length/width; in some cases the household or group may use several plots, for each of which a sketch will be made;
- on-plot and access roads, with fences and gates;
- water sources, pump, irrigation equipment;
- tool sheds, storage facilities, plastic tunnels, greenhouses, shading nets, nurseries, paddocks, pens, etc.;
- locations of raised beds, manure heaps, grazing areas.

Then ask the participant(s) to draw on the map other resources that they use in their particular urban agriculture system. The resources chart might be used to give them some hints of types of resources they may think about (e.g. what inputs do you use and where do you get them?).

Once all resources used have been identified and drawn on the resource map, one will ask the participants to identify for each resource included in the map the individuals who are using this resource (have access to it); the answer will be indicated on the map with the help of simple codes (e.g.: m, f, m/f).

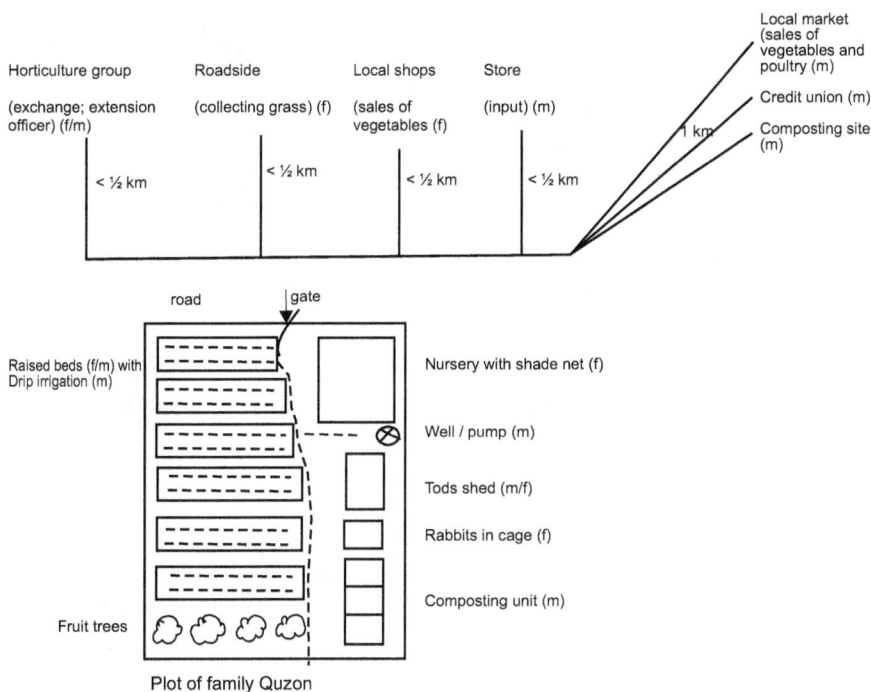

Figure 17.2 Example of a resource-mapping tool
Source: RUAF workshop, Situation Analysis Urban Agriculture, Cape Town, October 2007

This may be repeated for the question about who controls these resources (use the same symbols but in another colour to put this on the map). The flow of resources is a fundamental concept within an urban agriculture system.

Key questions to consider

- What are the main land and water sources? Who makes decisions about who can use these resources? Who owns the land? Who controls water access?
- Which other resources do men and women respectively have access to? Are the rights of access to resources different for men/women or for people from different ethnic or socio-economic groups? For male-/female-headed households?
- What is the quality of these resources? How does this influence the labour time they have to spend and the benefits they derive from it?
- What resources are underutilized? Which of these are degrading or improving?
- What are the linkages between men's/women's labour input and their use and control of resources?

- How does increased/decreased availability of one resource alter men's/women's use of and control over the other resources? Is men's/women's workload increased/decreased as a result?
- Is it men/women who use credit? Who makes the decisions on credit use? What are men's/women's experiences with credit?
- What is the resource-use and decision-making pattern in female-headed households compared with male headed-households?

The following urban agriculture scenarios illustrate how resource analysis may be applied in various contexts.

Urban Agriculture Scenario: the effects of environmental degradation on urban agriculture practitioners (adapted from Haile, 1991: summary)

Researchers are interested in exploring the effects of environmental degradation on urban agriculture practitioners. A survey of men and women reveals that there is a decreasing amount of fuel wood, which is becoming an obstacle for food processing. Women are identified as those persons responsible for fuel-wood collection: women and girls carry loads of branch wood and leaves from surrounding areas into the city. Analysis reveals that while drought is a significant factor in the decreasing supply of fuel wood, women's use of this resource is exacerbating the problem of depletion. This has created specific problems for women, in that they now must travel farther distances in order to gather adequate fuel-wood supplies. In turn, women have less time to spend in actually processing food for household consumption. Not only is the household receiving less nutritious food, the household budget is used to supplement foodstuff bought from street vendors. As a coping strategy, women have attempted to use cheaper, lower-quality, quick-burning biomass fuels. Unfortunately, the fuel generates more indoor pollution than fuel wood; thus it compromises women's health, and is more time-consuming because it must be continually tended due to its quick-burning nature. Another coping strategy involves the removal of girl children from school to help their mothers with fuel-wood collection. This impedes the ability of young women to gain a solid education. By identifying the primary users of a particular resource, and the implications of this, researchers can better assess what solutions may be appropriate and sustainable.

Urban Agriculture Scenario: dry-season and wet-season urban farming (adapted from Ofei-Aboagye, 1997: 5)

During fieldwork, it is observed that mainly men are involved in dry-season farming. Women operating as individual urban agriculture producers engage in wet-season farming. Those women working with husbands in dry-season farming do mainly in terms of weeding and harvesting. Further investigation reveals that in addition to women having less physical strength to clear the dry-season farmland, their access to hired labour or a tractor is hindered because it is too expensive (most women heads-of-household are in lower income brackets). Fertilizers and irrigation pumps are also not affordable for them. Because fewer producers (mainly men) engage in dry-season farming, there is more money made due to relatively lower supply of foodstuffs and unchanged levels of demand in this period of the year. Women producers have recently started to organize a dry-season farming co-operative and intend to share resources and farm a single plot of land for income-generating purposes.

Tools for the analysis of decision making and distribution of benefits

The decision-making and benefits-analysis tools presented below help to examine how decision-making powers about urban agriculture activities and resources and who benefits from them are distributed among the various members of a household and among group members.

Understanding the gender division of decision making and the distribution of benefits (including cash and goods) generated through the urban agriculture activities is important, since it provides insight into power relations between male and female household members or in producer groups, and thus the effects that may be expected of certain projects if such relations remain unchanged.

Decision-making matrix

This tool can be used as a participatory exercise to assess decision-making power either in individual household interviews or in focus-group discussions (through which participants can further expand on their views and perceptions of household and external dynamics).

A decision-making matrix is created by listing vertically the various issues on which decisions have to be taken, and by listing the decision makers horizontally. Table 17.4 is an example of this kind of matrix.

One may go through the decision issues in the matrix one by one with the households or group members. Alternatively, one brings together community members and asks the people to divide into sub-groups by gender and to reflect on the following key questions:

- On what aspects of the households' agricultural activities do you make the decisions (as a result of your gender and your particular responsibilities)?
- What decisions are you not allowed to make?
- Are there aspects of the urban agriculture activities that you are responsible for but on which you do not make decisions?
- What decisions are taken jointly?

The groups are brought together, and their answers to each of the questions might be compared and discussed (and filled into the matrix; either the consensus reached or both male and female perspectives).

Table 17.4 Example of a decision-making matrix

Decisions	Man only	Man and woman jointly			Woman only	Comments/ Explanation
		Man dominates	Equal influence	Woman dominates		
Labour Who decide(s) how the available family labour will be used? Who decide(s) whether to hire additional labour? Etc.						
Inputs Who decide(s) what seeds to buy? Who decide(s) what fertilizers to buy? Etc.						
Production Who decide(s) which food crops to grow? Who decide(s) which cash crops to grow? Who decide(s) where to plant what? Who decide(s) when to harvest? Who decide(s) whether certain products will be processed or stored?						
Marketing Who decide(s) what part of the harvest is sold and how? Who decide(s) what animals or animal products are sold and how? Etc.						
Investments Who decide(s) to buy equipment and tools? Who decide(s) to take a loan? Who decide(s) to buy or rent additional land? Who decide(s) to buy more animals? Etc.						
Reproduction Who decide(s) whether a child goes to school or not? Who decide(s) on going to a doctor? Who decide(s) whether or not to apply birth control? Etc.						

Source: De Zeeuw and Wilbers (2004)

Benefits chart

This technique is applied to analyse the distribution of benefits derived from the products and by-products produced by a household, and the making of decisions on the use of the income raised. The exercise can be done with the members of selected households or in a focus group. One can work with a mixed group, which may lead to lively discussion on household decision making. But if women do not speak up in a mixed group, one may choose to conduct separate focus groups for male and female household members.

The procedure for the benefits analysis is as follows:

- Products and by-products derived from the resources of the household are written down on index cards (one card for each product or by-product). In the example given in Table 17.5, the following tree by-products were identified: fruit, fodder, fuel wood, lumber, bark, and poles. Written words can be replaced by pictures or drawings of the (by)products, to accommodate the participation of illiterate members of the household.
- Cards are discussed one by one, looking into the following questions: how the product is used, who uses the product, who decides how it should be used, and who controls the money if sold. Additional input is sought from other household members. The information obtained in this way is summarized in Table 17.5.

Table 17.5 Example of a benefits chart: palm-tree products

(By-) Product	How used?	Who decides on its use (f/m)?	Who carries out the activity (f/m)?	How is cash used if sold?	Who decides on cash use? (f/m)
Leaf	Leaf veins made into brooms	f	f		
	Leaves wrapped around boiled rice	f	f		
	Sticks	anybody	anybody		
Fruit	Eaten at home	f + m	f + m		
	Sold at market	f	f + children	To buy food + other basic necessities	
	Dried and sold for production of coconut oil	f	f + children		
Husk	Made into charcoal for home use or sale	m	m		
	Fibre used to stuff pillows and mattresses	m	m		
Trunk	Used as fuel wood	m	f + m		
Tree	Shade	f + m	f + m		
	Ornamental use	f + m	f + m		

Source: Lingen (1997)

The following urban agriculture scenario illustrates how gender-benefit analysis may be applied in various contexts:

Urban Agriculture Scenario: decision making and benefits of urban vegetable production (adapted from Ofei-Aboagye, 1997: 8 and Mianda, 1996: 99)

Benefits from vegetable production in a local urban agriculture system, as defined by urban agriculture practitioners, are considered to be earnings and decision-making power over these earnings. Researchers found that on the whole men earn more from vegetable production because of their larger holdings and ability to do two farming seasons, both wet and dry. Men who farm with their wives save on labour costs, produce greater crop yields, and have complete control over how profits are allocated. Those crops produced by men fetch more per acre at the local market, as compared with women's crops that are more fragile and perishable. In those households where only the wives produced vegetables, women have more decision-making control over income generated than women who farmed with their husbands. Interviews with the wives revealed that, in some cases, husbands not involved in vegetable production may impinge on women's benefits (i.e. cash) from this activity. Rather than give the profits to their husbands, women vegetable producers use strategies based on their social role as mothers to keep cash to purchase household needs. Researchers found that women producers who are not landowners demand their share of revenue derived from production, because they are the ones who are primarily responsible for the care of children. However, when they are not successful in convincing their husbands of the need to share earnings, women retain part of the money from their vegetable-produce sales without the knowledge or consent of their husbands. While still in the marketplace, women entrust their earnings to money managers (e.g. floating banks), women friends, or their own children, or will simply deposit the cash in a hiding place where men are not likely to search (e.g. in a kitchen pot or pan, culturally associated with women's domain). They do so to establish decision-making power in their own right.

Identification and ranking of main problems and opportunities

For a gender-sensitive identification of major problems and opportunities, one invites the participants in two separate groups: one of women and another of men. Eventually, these groups may be further divided along the lines of age, ethnicity, class, and so on.

Ask the participants first to think about the main *problems or constraints* that they encounter in the implementation and development of their agricultural activities (which might be technical, financial, organizational, political, or other problems).

Subsequently prioritize the problems according to their importance (in the view of men and women respectively) with help of a *ranking* method. Set up a matrix (on a black board or sheet of wall paper or on the floor) listing all problems identified (by men and women) on the vertical axis (see column 1 in Table 17.6) and add a symbol to each of them (for any illiterate members of the group). Subsequently all participants are asked to vote on which problem(s) they consider the most important ones. Each participant may be given several votes, which can be distributed over the various items in the list according to individual preference.

The voting might be done with help of small stickers (e.g. in two colours for men and women respectively) or pebbles or sticks (if the matrix is on a table or the floor). In the example, six male participants and six female participants each received five stones, which they distributed among the five impacts mentioned according to their own preferences. The example clearly indicates the importance of gender differentiation when doing the listing and ranking. If we had not done so, we would not have discovered that the key problems and priorities of female producers are (partly) different from those of the male producers.

One may also conduct an inventory what of *local potentials, opportunities, recent innovations, plans, ideas, and challenges* the urban producers are identifying. Starting from opportunities normally creates a positive climate and is very stimulating for the participants. In this case the participants may be asked to think about questions such as the following:

- What things have you been trying out/experimenting with recently? What innovations have you observed in other urban farms/gardens that might be of interest for other urban farmers/gardeners too? (Think of technology as well as innovations in the way they process, store, and market their products and in the way they organize themselves.)

Table 17.6 Example of a ranking exercise

Problems encountered	Male producers	Female producers	Total score	Ranking		
				M	F	T
Pests	8	3	11	3-4	6	5-7
Lack of irrigation equipment	15	1	16	1	9	1
Theft/no fence	4	7	11	6-7	4	5-7
Poor sales of surpluses	8	2	10	3-4	7-8	8
Insecure land tenure	4	4	8	6-7	5	9
Harassment by police officers/ illegal status of UA	6	9	15	5	2	2
Lack of technical support	10	2	12	2	7-8	4
Plots are too far away; difficult to combine with tasks in the home	2	12	14	9	1	3
Low quality of irrigation water (contaminated)	3	8	11	8	3	5-7
Total score	20 x 3 votes = 60	16 x 3 = 48	36 x 3 = 108			

- What market opportunities/niches exist that we may utilize? Are some products in high demand during specific periods of the year? Are consumers asking for new products or specific qualities (e.g. organic or certified products) or other ways of presenting produce (small units, dried, etc.)?
- What ideas/plans do you have for the development of your garden/farm and your group/organization?
- What resources are available in the city that are under-utilized and might be accessed and applied by you?

The potentials and opportunities are listed and prioritized by men and women participants respectively by applying the ranking method explained above.

Design tools

Problem and opportunity analysis

The problem and opportunity analysis tool helps to explore the causes of priority problems identified by male and female urban producers, to identify possible solutions to such problems, to further analyse the prioritized (most promising) opportunities for further development of that urban system, and to identify the steps needed to realize such an opportunity.

Applying gender analysis to the identified problems and opportunities is a fundamental step in making recommendations for appropriate, sustainable, and equitable interventions. A solution for a problem or an opportunity initially may be identified by one sex but may have implications for other household members or other households in a neighbourhood.

Problem and opportunity analysis charts

To conduct participatory analysis of the *priority problems* identified in the diagnosis stage, a problem-analysis chart may be used. See Table 17.7. In the first column you list the priority problems identified, making sure that both men's and women's key priorities are included. Then you discuss for each of the *priority problems*:

- What are the causes of this problem? The answers are entered in column 2.
- What do men and women respectively do currently to cope with this problem? List the coping strategies in column 3.
- How is this problem (negatively) affecting men and women? List the consequences for men and women in column 4.

Similarly, an analysis chart (see Table 17.8) can be made for the *priority opportunities* that have been identified, by discussing for each the following:

Table 17.7 Example of a problem-analysis chart

Priority problem	Causes	Actual coping strategies	Consequences of the problem and related coping strategies for		
			All	Men	Women
1	a. b. c. n. etc.	a. b. c. n.			
2					
3					
4					

- What steps/activities should be undertaken in order to realize this opportunity or potential or to further develop this innovation?
- What would be the positive and/or negative effects on men and women when this opportunity was realized? How would these potentials, plans, and opportunities change existing division of men's and women's activities, access to and control over resources, distribution of benefits, etc.?

Table 17.8 Opportunity-analysis chart

Priority opportunities identified	Steps needed to realize the potential	Expected effects of realizing this potential for		
		All	Especially men	Especially women

Urban Agriculture Scenario: gender-sensitive problem analysis (adapted from Slocum et al., 1995: 172–180)

By analysing the problems identified by men and women of a particular urban area, researchers find gender-differentiated impacts of problems, coping strategies, and potential solutions. (See Table 17.9.) For example, the lack of land poses a specific problem for women producers. While both men and women face constraints due to the illegal nature of urban agriculture in this city, women are further disadvantaged because they traditionally have few rights to land tenure, as such men tend to have first choice of any available vacant land suitable for urban agriculture. This often leaves women with low-quality plots of land that are located at a considerable distance from the homestead.

Table 17.9 A combined problem-and-opportunity analysis chart

Problem	Causes	Coping strategies	Resulting problems	Solutions and opportunities
Not enough land for UA	Urban agriculture is considered illegal. Women have no rights to land tenure. Rapid urban expansion; vacant land used for construction of buildings etc.	Men and women using land in the peri-urban areas	Long distances to vacant plots for women. Only marginal land available for women (resulting in lower yields). Men and women still subject to harassment by municipal authorities	Lobby municipality to legalize and support urban agriculture.
Inadequate water supply	Drought (rationing of water). Local pump with poor water quality. Staggered supply of water by municipality	Women buying water from market vendors	More household cash is spent on water; women in charge of budget. Not enough water for crop irrigation; men's activity.	Increased use of grey water for irrigation (technology to be managed by women and applied by men)
Cattle too expensive	Shortage of cattle on urban market	Men travel to rural areas to purchase cows. Goats used for milk production by women	Men away from household for long period; thus less income brought in by men. Livestock activities (once done by men) now women's responsibility (in charge of small animals)	Men to form co-operative in order to share available cattle and resulting milk production for household consumption and sale

The solution for increasing access, as identified by the group, may improve access to land for urban agriculture in general. However, it will not address the issue of inequitable access to land between male and female urban producers. The problem of expensive livestock may initially be considered a problem for men. Yet women are also burdened by a lack of cattle, for when goats are substitutes as milk producers it is women who now tend to this activity. An appropriate solution identified by the women in this group calls for the men to form a co-operative and share in urban dairy-farming activities.

Problem (or opportunity) tree

The problem tree is a useful tool to analyse the causes and consequences of a central problem that has been identified and – by doing so – to identify possible solutions to the problem (tackling the problems) and potential effects of solving the problem (taking away the consequences).

If you have implemented the analysis exercise discussed above, then making a problem tree is quite easy since the causes and consequences have already been identified for each core problem and can be inserted in the problem tree after discussing their causal linkages (see the explanation below).

One may develop a problem tree directly after having identified the priority problems (and without making a chart analysis first). In that case it is recommended to subdivide the list of problems into 'clouds' or 'clusters' of problems that are strongly interrelated. One may do this by writing each problem on a card and then making the clouds, while discussing their relationships. Finally one chooses the core problem in each cloud (the problem that creates many other problems and has the most important effects on the lives of the participants).

Take one core problem and place in the middle of a sheet of paper or blackboard (the trunk of the problem tree). Discuss now for each other problem within the same 'cloud' how it is related to the core problem: a cause (a root of the problem tree) or an effect (or consequence) of the core problem (branch of the problem tree). Also the relation between the various causes and respective effects are discussed (is this one causing the other?). Additional causes or effects may be identified during the discussion and added by writing additional cards. The exercise will result in a diagram such as the one featured in Figure 17.3.

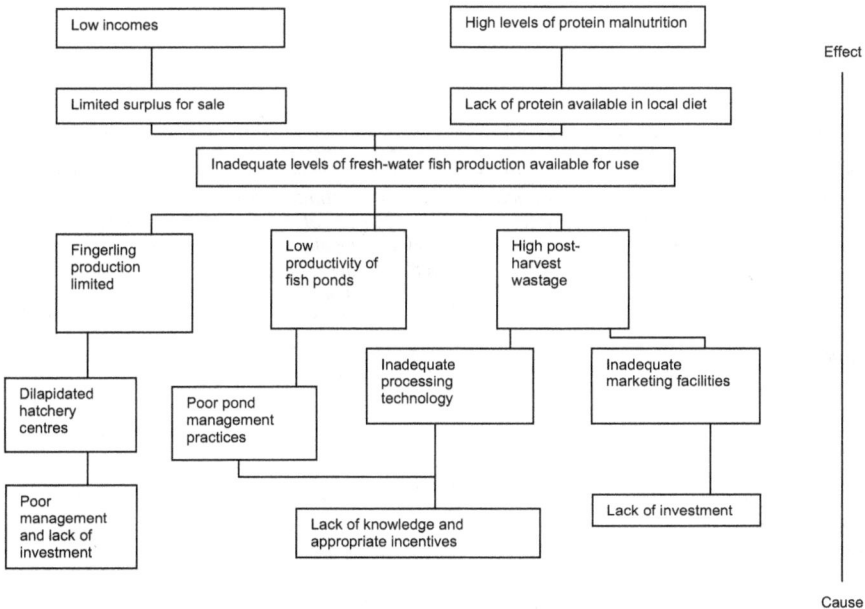

Figure 17.3 Example of a problem tree
Source: AusAid (2000)

This procedure is repeated for all 'problem clouds' (with a maximum of three or four).

When the causes and consequences of each of the core problems have been identified, one proceeds with the identification of possible ways to solve the core problem, by systematically analysing the different possibilities to influence each of the problem causes included in the problem tree. Some causes might be more difficult to change at the local level than others.

Once you have identified potential solutions to each core problem, you will enter into an assessment of these alternative problem-solving strategies. To do so you first need to establish the criteria needed to assess and choose certain solutions/problem-solving strategies. Ask participants which criteria they want to use to assess and select certain solutions. Provide (additional) suggestions where needed. For example, if participants have not considered how equitable a particular solution is for both men and women, encourage them to reflect on this issue: how might this solution affect gender division of labour, their role in decision making, distribution of benefits, etc? Also encourage participants to consider how much will it cost compared with the benefit it will bring? What human and financial resources are needed, and which of these are locally available and which are not? Will this solution have negative ecological or social effects? How long will the solution last?

Once the criteria have been established, ask participants to rank each solution according to these key criteria (see the example in Table 17.10). If not enough information is available to assess certain options, ask participants how they plan to obtain more information.

Table 17.10 Example of an assessment of proposed solutions

Proposed solutions for the irrigation problem	Suggested evaluation criteria						Votes	Rank
	Low investment cost	Low price of water (incl. maintenance)	Easy to manage (esp. by women)	Optimal use of locally available means	Providing enough water	Low health risks		
Collection of storm water	10	10	4	12	–	2	38	2/3
Use of household waste water	14	14	10	12	–	–	50	1
Digging a shallow well plus hand pump	6	6	6	6	6	6	36	4
Digging a deep well plus electrical pump	–	–	–	–	10	8	18	5
Use of municipal piped water	–	–	10	–	14	14	38	2/3
Total	30	30	30	30	30	30	180	

Similarly one can draw a tree to analyse the main opportunities (under-utilized resources, local innovations that can be further developed, market demand/niches) that have been identified. The highest-ranking opportunity is selected for further analysis (the trunk). In this case the roots will represent the activities that will have to be implemented in order to realize that opportunity, and the branches represent the effects or results that one expects to obtain by realization of that opportunity. These opportunities can be assessed in the same way as indicated above for the assessment of problem solutions.

Formulation of results-based objectives

Once key problems and opportunities have been analysed, it becomes possible to formulate the project objectives. But before doing so it might be the right moment to consider whether key interests in the group are homogeneous enough to have one group/one project, or whether it might better to have more groups with more homogeneous interests, each with a separate project, or to have a number of (co-ordinated) sub-projects (e.g. a specific women's sub-project next to other sub-projects for both sexes). See also the group-definition tool.

The formulation of project objectives starts by reviewing the results of the problem-and-opportunity analysis: which problems and opportunities will be tackled? Include these in the first column of the chart in Table 17.11.

Subsequently discuss for each of these key problems/opportunities:

- How can the actual situation be described in measurable terms? (column 2);
- What results can realistically be expected from the selected problem-solving or opportunities- realizing strategies? (column 3).

Finally, formulate the results-based project objectives (column 4), which should:

- clearly reflect the changes in the situation that are expected to be realized as a result of the project;
- be formulated in measurable terms so that the results can be monitored (both the magnitude of the effects that will be realized as well as the number of beneficiaries that will be affected);
- be achievable in the given local conditions, with the people and means available (or that realistically can be expected to be generated) and within the project period.

Group definition (adapted from Slocum et al. 1995: 120–123)

The group-definition tool helps a project group or producers' group to clarify the limits and resources of their own group. Urban agriculture projects may fail either because they excluded certain members of the community, or

Table 17.11 Example of a chart with results-based objectives

Selected key problems or opportunities	Actual situation	Realistic achievable results at the end of the project	Results-based objectives
Severe water shortages in dry season	Only 20% of the producing households have access to irrigation water during the dry season. Water costs per acre are on average $ X	60% of the households will have access to water for irrigation during the dry season. Water costs involved will have been reduced by 30% (excl. inflation).	Access of the urban producers to irrigation water during the dry season will have been enhanced from 20 to 60% of the producing households. Costs of irrigation water during dry season will have been reduced from $ X to $ (X-30%).
High incidence of pests in dry season	Average yield per acre in dry season is only about 70% of the average yield in the wet season	Average yield per acre in dry season is about 90% of average wet-season yield	By reducing losses due to pests, the average yields during dry season have risen from 70 to 90% of average wet-season yields
Weak producers' organization	All decisions are taken by the group leader and an informal clique of friends. No clear development plans. Functionality of the organization is low, due to poor internal structuring and low involvement of the members. No maintenance of equipment, due to lack of funds.	Decision making is taking place in a management committee. A strategic five-year development plan has been made which orients planning and decision making. Annual action plans and budgets are being made. Functional sub-committees are operating (buying/ selling inputs, marketing, savings scheme).	At the end of the project the producer organization is led by a management committee according to five-year strategic and annual plans; functional sub-committees and a savings scheme are operational

because they were too heterogeneous in terms of the priority interests of the group members, or because they were too small or too big to realize the set objectives.

Discussion of the following questions might help to clarify the group's definition, and eventually in rearrangements (inclusion or exclusion of new members, creation of more groups or sub-groups with more homogeneous key interests, etc.):

- Who is in this group as presently formed? How can the membership be characterized (e.g. main interests, socio-economic status, position in local community, gender, ethnicity, age)?
- Who are at present excluded from or barely included in this group as presently formed in terms of gender, ethnicity, class, age, etc.? Are there others with the same or similar key interests that should be included?

Would any of those currently excluded from the group have resources to offer in terms of valuable knowledge or skills, tools or materials? Will we recruit others to join our group or limit participation? If we want to include others, what should be done to realize that?

- What would be the optimal group size for the things we want to do; can all voices be heard? If the group is large, how can we use small groups to do some of the work?
- How are decisions made at the present time? Who can make what kinds of decision? What decisions can best be made by the group as a whole? What other decisions can be made by sub-groups or functionaries?
- What other groups and organizations should we interact with? Why? What interests do they represent? How we should interact with them? Report to those outside? Hide certain information from those outside? Seek regular or occasional support and inputs from those outside? Form alliances or networks?

It is recommended to summarize the outcomes of the discussions in a clear group definition: who we are, how we make decisions, our potential allies, etc.

Mapping institutional linkages

The institutional networking tool helps a project group or producers' group to identify key actors, explore their roles, and establish the perceptions that people have about these institutions. It clarifies which institutions are the most important, which have the respect and confidence of men and women, and who participates in and is represented by which ones. This tool is also a visioning exercise whereby participants can identify key institutions that they would like to engage in an urban agriculture project, and the ways in which linkages will be forged.

A (non-exhaustive) listing of institutions that might be of significance in a particular urban agriculture context is as follows:

- Co-operative
- Producers' association
- Marketing association
- Garden club
- Composting group
- Community kitchen
- Other community-based organizations (e.g. youth league, parents' association, churches, community centres, etc.)
- Schools, local clinics, local churches
- NGOs
- Municipal authorities (e.g. Department of Housing, Land Affairs, Health, Local Economic Development, Social Affairs and Community Development, Parks and Forests and/or Agriculture, etc.)

- Government agricultural extension and veterinary services
- Credit services (e.g. bank, lending centre)
- Private enterprises (e.g. inputs, processing, marketing, services)

As a participatory exercise, participants in a group meeting are divided into two sub-groups: one representing women, one representing men. Each group is given four sets of different-sized circles made of coloured paper (at least six each of very large, large, medium, and small sizes). Participants of each group list the groups, organizations, or institutions with which they co-operate in their urban agriculture activities in one way or another. Subsequently they write the name of each organization in a circle (size of circle corresponding to the importance they give to the said organization for the development of their agricultural activities). Finally, the participants lay the circles in a configuration which indicates the relationship between and among the different groups and organizations, with their own group in the middle, groups/organizations based in their own community close to their own group, and external institutions further to the outside.

Participants then discuss:

- Why is this organization more or less important for our urban agriculture activities? What is its actual role? What services does it provides?
- What might be the potential value of this organization for the realization of our plans and solving our problems? What limits the actual value of their services for us? How they might improve, and what can we do to promote that?
- What relations do we want to maintain with each of them? How will we do that?

The sub-groups then share and compare their results and jointly draw their conclusions. Participants may decide to also invite to a meeting a representative of one or more organizations that have been identified as important, to establish the appropriateness and potential of collaborating with such institutions. Next to questions about their mandate, programmes they implement, services they provide, how to apply for support or come to an agreement to collaborate, it will be important to look into their degree of gender sensitivity, the availability of female staff, the degree of attention to female producers, and integration of gender issues in their activities.

The following urban agriculture scenario illustrates the type of information garnered from this exercise:

Urban Agriculture Scenario: assessing the effectiveness of extension services for urban agriculture practitioners (adapted from Hovorka, 1998: 24-25 and Slocum, 1995: 127-131)

The Venn diagram (see Figure 17.4) shows a relatively small circle, representing local extension service, compared with a larger circle indicating the greater

Men's perceptions

Women's perceptions

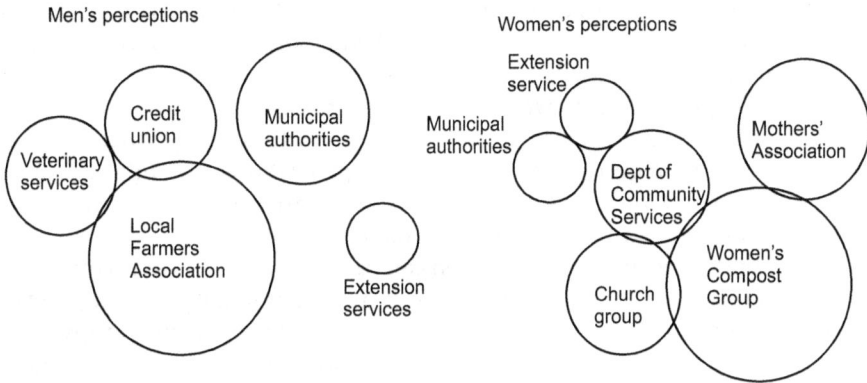

Figure 17.4 Men's and women's assessment of the importance of various organizations

importance of the local farmers' association (men's diagram) and the compost group (women's diagram). During a focus-group discussion men and women participants voice their concerns regarding female extension workers' unwillingness to do field visits in a 'dirty environment', and their inability to handle larger livestock such as cattle. Further gender analysis, through individual interviews with men and women, reveals that while all persons share the former concern, women producers are satisfied with female extension workers' support in gardening and poultry activities, for which women are largely responsible in this urban agriculture system. It is the men who are frustrated with extension services related to larger livestock. This distinction is a key element in improving extension services so that it is of benefit to both men and women producers.

Activity-planning tools

Activity-planning matrix

The activity-planning matrix is a tool that helps to establish concrete and realistic plans for implementation of priority development activities. To produce the activity-planning matrix, all urban producers who share certain development priorities are brought together. Interested support organizations and other potential partners might also be invited to participate in the planning activities.

The activity-planning matrix (see the example in Table 17.12) starts with entering the results of the design phase. In the first column the group's 'priority problems or opportunities' are stated; in the second column, are entered 'Solutions/Steps', the most effective solutions to solve a problem (or steps to realize the opportunity) that have been previously identified. Subsequently the participants are asked to review all the concrete activities necessary to achieve each of the solutions (or steps), which are filled in the third column,

'Activities'. For the fourth column, 'Actors' (Who will do it?), ask participants to review what they can do themselves and what kinds of support they need from other organizations. For the fifth column, 'Resources needed', ask the participants first to identify what they can contribute themselves or what can be obtained from other local sources, and second to identify what external resources may be required. To identify the organizations that might be able and willing to provide such support and resources, the participants will review the results of the institutional linkages exercise (see above).

It is important to ensure that women can participate equally in this exercise. In case there is a risk that men will dominate the planning, it might be good to do all steps of the planning exercise with two homogeneous subgroups, each of which gives its input to each step. One should also ensure that the gender-

Table 17.12 Example of a project-planning matrix

Priority problems or opportunity	Selected solutions (steps to realize the opportunity)	Activities to be undertaken	Actors (Who will do it?)	Resources needed (Who will contribute what?)
Water shortage	1.Collection of storm water	1.1 Acquire and deliver the required materials. 1.2 Instruct group members in (a) how to connect collection pipes to tin roofs and link to water tank and (b) how to maintain the system. 1.3 Install the system.	1.1 Project committee 1.2 Water Dept 1.3 Producers guided by Water Dept staff	1.1/1.2 200m pipes, 40 water tanks, transport, and 2 instructors x 1 month by Water Dept. 1.3 Labour, nails, wood for platforms, hammers, etc. by producers
	2. Re-use of household wastewater	2.1 Instruct the group members on safe reuse of household waste water	To be completed	To be completed
	3. Link community gardens to municipal piped water system (back-up system for dry periods)	3.1 Apply to the Municipal Water Department. 3.2 Organize a savings account in order to be able to pay the water charges in dry season		
High pest incidence in dry period	1. Capacity development on identification and integrated pest management	1.1 Prepare an Urban Farmer Field School on IPM. 1.2 Implement the UFFS		
	2. Change to resistant varieties in dry season	2.1 Identify potential resistant varieties. 2.2 Participatory design implementation and evaluation of local trials		

differentiated results gained from previous diagnosis and design activities are fully used and that both men's and women's priority problems and preferred solutions are taken into account.

Participatory budgeting

The objective of this tool is to facilitate preparation of a budget estimate for projects. The exercise can be prepared by listing basic budget components and by preparing a template to guide group work (see Table 17.13).

First, ask participants if they have ever participated in preparing a budget for a project. Ask them to identify the types of thing that they included in the budget. As participants respond, sort items into the major categories such as:

- **Investments** (purchase of animals, equipment, and construction materials, and costs of construction – including local labour – of infrastructure like cowsheds, poultry pens, composting units, etc.).
- **Costs of training, research, and extension activities** (other than personnel): production of educational and instructional materials and audio-visuals; costs of workshops, training courses, exchange visits, demonstrations, extension meetings (including related travel costs and meals of participants, allowances, costs of trainers/instructors and required training materials).
- **Costs of co-ordination, planning, monitoring, and evaluation activities** (other than personnel): costs of planning and evaluation meetings, data gathering and processing, etc..
- **Personnel costs** (salaries, fees, wages (plus related costs like social security and taxes) of persons from the producers' organizations who will play a role in the project (e.g. co-operative manager, store keeper, secretary) and others who will be specially hired for this project either on a more continuous basis (e.g. project co-ordinator) or for the realization of specific activities (e.g. trainers, specialists, etc.).
- **Travel expenses** incurred by such personnel (transportation, lodging/ meals or DSA).
- **Related office and communication costs** (office supplies, photocopying, fax/telephone/mailing costs, office expenses (rent/water/electricity/etc.), computer supplies, banking costs, etc. as far as directly related to the project; if that is difficult to define, one might also opt for including a certain percentage (e.g. 10%) of the personnel costs as 'overhead costs'.
- **Unforeseen:** (5%).

Explain that in most projects, costs can be broken down into these categories. Subsequently, one will use the information included in the 'Resources needed' column of the activity-planning matrix that was developed earlier to formulate the budget, translating person months and materials into monetary terms, as is illustrated in Table 17.13.

It is good custom to add a number of columns to the right of the column '(total) amount', to specify who will contribute what (e.g. own contributions of the producers, Water Department, Ministry of Agriculture, an NGO) and to indicate what will be contributed in kind and what will be made available in cash.

The 'participatory' in the title of this tool refers to (a) the participation of male and female producers in preparing the budget so that it is fully 'owned' by them; and (b) the involvement of other actors/stakeholders who commit part of their institutional resources to this local project in accordance with their own mandate and programmes.

Finally check the following:

- Have certain budget items been gender-specified where required?
- Are both men's and women's priority activities funded?
- Have gender-affirmative actions been funded?
- Are all means that are required to implement the proposed activities included in the budget? Are cost estimates realistic (take into account inflation and salary increases over time)?

Table 17.13 Example of participatory budgeting

Activity set	Personnel	Unit price	# of units	Amount
1. Activity #1	Personnel	Rupees		Rupees
	1. Project Manager	2,000	11	22,000
Example. Establish	2. Research Officer	1,500	20	30,000
and support gender	3. Short-term Adviser	1,500	26	39,000
committees to work	4. Committee Clerical Support	500	50	25,000
with local government	Total personnel activity #1		46	116,000
officials to bring more				
women into the local	Travel			
planning and decision-	1. Trips between A and B via C	200	10	2,000
making process.	2. Accommodation		Nil	
	3. Meals while travelling	200	10	2,000
	Other (list specifics)			
	Meeting expenses	5000	5	25,000
	Supplies			3,000
	Communication			3,000
	Total activity #1			151,000
Activity #2 (specify)	As above			
Activity #3 (specify)	As above			
Etc.				
Activity #X: Project Management	As above			

Source: CNGO, 2003: 31; 58

- Could the same activities be implemented more efficiently if other methodologies were applied or the project was organized in another way?
- Are resources that will be contributed by the producers themselves well specified? Can such contributions realistically be expected?
- Can the planned contributions of other actors realistically be expected?

It should be pointed out to the participants that different funding organizations have different formats for preparing their budgets. So if the budget is prepared to apply for funding to a certain organization, they should first enquire what budget format is preferred.

Scheduling the work

The objective of this tool is to teach the participants how to formulate a work plan to implement the project and how to schedule planned activities. A work-plan template can be prepared to guide group work (see Table 17.14 for an example). The format of the work plan can be changed according to the needs of the project, so facilitators should give examples of how and when this can be done.

First, discuss and highlight the importance of a schedule that specifies what will be done, when, and by whom. Then explain the format of the work plan you have prepared. Following this, form specific groups for sub-projects or specific activities and have participants begin to prepare a schedule for the activities they have identified in their activity-planning matrix: estimate for each activity and sub-activity the time that will be required to realize that activity, the sequence required, and the best moment to start/end it. Allow about one hour for group work, then assemble in plenary for presentation and discussion.

When scheduling the work, it is important to take into account the seasonal calendar and division of labour, for example by asking the following:

- What are very busy times for men and women (what might prevent them from participating)?
- What are the best periods of the year to discuss and tackle certain issues (when men and women can be involved in planning or implementing that activity or that problem can be observed in the field)?

Implementation tools

Implementation checklist

During preparation for the implementation, a checklist can be developed which includes the various ways in which gender will be integrated into project activities. This checklist can easily be shared with others and is a quick reminder of key things to consider when implementing a specific activity. The

Table 17.14 Example of scheduling the work

Name of Project	Gender in Local Planning and Decision-Making													
Implementing Agency	Nepal NGO													
Time Period (Year)	January 2004 to December 2004, One Year													
Intended result	Main and sub-activities	Person responsible	Months of the Year 2004												Remarks
			J	F	M	A	M	J	J	A	S	O	N	D	T	
Increased and more effective participation of women in local planning and decision-making	Main:															
	Conduct baseline study and gender analysis	Research officer/ monitor	15												15	
	Network with local officials and agree on strategy for creating committees	Project Manager		3										2	5	
	Create joint committees of local government, NGOs and other community organizations	Project Manager			3										3	
	Support and assist committees to develop practical strategies and methods for increasing the involvement of women	Short-term technical adviser				10	2	2	2	2	2	2	2	2	2	26
		Clerical support for committees				10	5	5	5	5	5	5	5	5	5	50
	Monitor extent to which participation increases	Research officer/ monitor							2						3	5
	Adjust strategies and plans, as required	Project manager													3	3

Source: Adapted from CNGO, 2003: 31; 58

same checklist also can help in monitoring the implementation of gender-mainstreaming initiatives. To that effect a column might be added in which one can indicate the percentage of completion of each planned gender-mainstreaming activity (progress thermometer).

The checklist that was developed by partners for use in the RUAF–CFF programme is shown in Table 17.15 as an example. The checklist can be reviewed by the project staff when preparing new activities and during meetings with other team members and/or with producers and other stakeholders in the project.

The Urban Producer Field School (UPFS)

The Urban Producer Field School (UPFS) method applies adult-education thinking and experience and group-based learning processes to agricultural capacity building. UPFS originated as Farmer Field Schools (FFS), initially developed to facilitate farmer understanding and application of integrated pest-management principles in rice farming, for which conventional technology-transfer training approaches were found to be inadequate (Röling and Van de Fliert, 1998). The approach has subsequently been expanded for use with other crops and to address a wider range of issues than integrated pest management (Röling, 2003; Züger, 2005).

FFSs provide the setting and the materials for producers to explore and discover for themselves new knowledge about agricultural production. They are based on the presumption that knowledge actively and repeatedly obtained in this way will be more easily internalized, retained, and applied after completion of the training. Repetition is important for retention, which is one reason why FFSs are repeated, usually on a weekly or fortnightly basis, with the same structure throughout the growing season. FFS was developed in rural settings, and the urban situation has required some adjustments in the tool, especially in the major time commitment required by FFS participants, which can be problematic in urban conditions where agriculture may be only one of several livelihoods activities.

The participation of both men and women in urban field schools is extremely important, in order to capture fully the different experiences and expertise present in the local population, and also the different needs that men and women may have. For example, evidence from field schools in Lima, Peru indicates that women are more time-constrained than men, because of domestic obligations in addition to income-earning activities like petty trading or the provision of services. This limited their participation in horticulture-based field schools, in which both men and women have an interest. It is advisable to consider ways to limit the total duration of the field school so as to better accommodate women. Since women are almost exclusively involved in raising small animals, it is especially important that livestock field schools adjust themselves to women's schedules.

Table 17.15 RUAF–CFF checklist: implementation of gender-mainstreaming activities

---

o **Integration of gender in the situation analysis**
- local team trained in gender issues of urban agriculture
- gender-sensitive and disaggregated data collection; use of gender-sensitive PRA tools
- external gender adviser assists the team in preparing the activities and comments on draft results
- policy review to be done with attention to gender
- implementation of gender case studies on all aspects of the gender framework
- gender-sensitive analysis of results
- attention to gender in report on the situation analysis (facts and recommendations)

o **Integration of gender in all training activities**
- gender-sensitive trainers
- specific module on gender and UA
- gender aspects in every other module highlighted
- gender balance in participants
- gender adviser advises on contents training modules

o **Integration of gender in policy seminars**
- gender balance in participants
- gender-sensitive facilitator; use of gender-sensitive discussion methods
- gender issues are included as a separate topic on the agenda of the meeting
- gender aspects are highlighted when dealing with other topics

o **Integration of gender in knowledge materials**
- specific working material and/or Urban Agriculture Magazine issue on gender and UA
- gender aspects highlighted in each of the other knowledge materials

o **Integration of gender in the City Strategic Action Plan or Municipal Policy on Urban Agriculture**
- members of the working group(s) are trained in gender-sensitive planning
- gender adviser comments on the draft proposals of the working groups
- gender-sensitive facilitator of the Multi-Stakeholder Forum on urban agriculture (MSF)
- gender balance among participants of working groups and forum participants
- gender capacity of each forum member is enhanced
- gender issues on the agenda of the MSF
- gender-positive actions are taken and special funds earmarked
- MSF statement on gender

o **Integration of gender in M&E of RUAF**
- use of gender-sensitive indicators
- use of gender-sensitive tools and methods for data collection
- gender-sensitive analysis of results
- gender-aware researchers and facilitator

o **Gender mainstreaming in own organization**
- promote adoption of a gender statement
- promote development of a gender policy
- ensure gender balance among the employees of the organization (starting with the RUAF team)
- announce that the RUAF gender adviser will be appointed to advise the whole organization

---

Each field school session, both in its 'classical', rural version and in the adapted urban field school, is structured into different parts, which – with their facilitation – are described in different methodological sources (see for example Arce et al., 2006a; 2006b) and are not discussed in detail here. It is important to incorporate a gender dimension in all parts of the session structure, and not simply add a special topic on gender in one field school session. The typical structure for an urban producer's field school is included in Table 17.16.

Table 17.16 Session structure of an (*engendered*) Urban Producers Field School

1. Welcome (*during first meeting: self-introductions of men and women participants*).
2. Review of last week's session (*gender-related issues that arose are highlighted*); clarification and distribution of the tasks for this session (*equal participation of men and women is ensured*)
3. Sub-groups: diagnosis of a specific problem and identification of possible solutions
4. Presentation and discussion of results (*gender-linked aspects of the identified problems and solutions are identified*)
5. Key theme of the session (*gender is incorporated into each key theme*)
6. Agreements on next session (*the suggestions of both men and women are taken into account*)
7. Evaluation of group process; strengthening group dynamics (*attention is given to the contributions of men and women to the session and related gender dimensions*)

Participatory innovation development (PID)

A local or farmer innovation can be described as a 'new and better way of managing resources' that is discovered through learning from own experiences, local experimentation, sharing with other farmers, and creative use of their local knowledge (including knowledge originally stemming from external sources that is fully internalized within local ways of thinking and doing) (Waters Bayer and Van Veldhuizen, 2004). A local innovator then is a person or group who – on their own initiative – is developing a new technology or a new way of doing things, using basically their own knowledge and experience and locally available resources.

Participatory innovation development (PID) is a process that seeks to stimulate and strengthen local innovation by taking farmers' own experiences, knowledge, and capacities as a starting point, enhancing local innovation capacity and facilitating further development of local innovations by linking local innovators to other experiences, information, and ideas coming from other actors (e.g. farmers, researchers, development agents). Inspiring experiences with the PID approach are presented by Critchley (1999), Waters-Bayer and Van Veldhuizen (2004), and Wettasinha (2007). See also www. prolinnova.net.

Participatory innovation development includes participatory technology development (PTD) but is more comprehensive, since it goes beyond technological innovation to include also innovations in the way people are gaining access to – or regulating the use of – natural resources, organizational innovations, innovations in marketing the products, and in institutional linkages and policies, etc. See Van Veenhuizen, Waters-Bayer and De Zeeuw (1997) for detailed methodological explanations of PTD.

The PID process is usually developed through the following six stages: partnership building, identifying local innovation and innovators, documentation, local action planning, implementing local innovation activities, and institutionalization.

Partnership building; multi-stakeholder steering group for PID

This involves bringing together different stakeholders who have an interest in starting a PID process in a certain locality or region (e.g. farmer organizations, NGOs and government agencies involved in agricultural and natural-resource management research and development, and educational institutes) and/or are already promoting local innovation.

It also involves the implementation of training-cum-planning workshops to familiarize the staff of these organizations with the PID approach and tools and to enhance their capacities regarding:

- identification of local innovators and innovations
- documentation and sharing of local innovations and innovation processes
- provision of support to local design and implementation of PID processes
- participatory monitoring and evaluation of joint activities, outcomes, and impacts
- facilitation of learning from on-the-ground experiences and dissemination of results
- raising institutional awareness on PID and engaging in policy dialogue.

Identifying local innovations and innovators

For stimulating local innovation, a positive approach is needed. Focusing on problems and weaknesses will increase the negative image of the situation and will increase the feeling that outside intervention is necessary and consequently farmers will likely remain passive.

The first step in this process is therefore to start from the strengths and capacities of the producers, to identify and recognize local innovations and to regard farmers as equal partners in the process of the further development and dissemination of these innovations. Local innovations and innovators can be identified by the following means:

- Through observation (what efforts are people undertaking to solve existing problems or grasp opportunities they have already identified,

who are the people that do things differently, that do 'strange' things, that are experimenting with something on their own initiative?).
- Through identification by key informants (ask extension agents, farmer leaders, teachers to identify local innovations and innovators with the help of the questions above).
- Through chain or 'snowball' interviews: ask farmers to identify innovative farmers (with the help of questions like the ones above); when interviewing these farmers, ask them to identify other innovators.
- Reconstructing innovation: ask a group of farmers to list one or more agricultural innovations that have been made in the last ten years and are relevant for most of the families in the area; ask them to identify the farmers who played an important role in introducing, adapting, or developing these innovations, and go and talk with these farmers (Wettasinha et al., 2007: 10).

There is often a male bias in identifying innovation or innovators. The reason for this is often a gender bias in the interviewers (male interviewers tend to focus on male respondents), and among the respondents (male producers who are used to talking on behalf of the female members of the household, women who lack confidence or have not been brought up to think of themselves as innovators).

Some ways to identify women innovators are listed below:

- Analyse the gender division of responsibilities and labour and focus in interviews (when asking for innovators and innovations) on the activities that are mainly the responsibility of women (since those will be the areas they know most about and will probably seek to improve themselves).
- Let the female team members interview women, to make them feel more at ease and to encourage them to talk freely about their local innovative activities.
- Identify local innovations by broadcasting a radio programme in which women speak about their innovations and respond to phone-in questions from listeners. This stimulates still more women to phone in about their own innovations or ask their children to write on their behalf to the radio station.

Documentation and sharing of innovations and giving recognition to innovators

Subsequently, the local people involved will be inspired and assisted to document these innovations (and the innovation process followed) so that these can be shared and discussed with others. There are various means of documenting and sharing information between farmers in an informal way, including for example posters, farmers' magazines, community radio, participatory video, exchange visits and study tours, farmer trainers, farmer competitions, etc. It is important to give recognition to local innovations and innovators. One way to do so is to periodically celebrate innovations and reward innovators (through invitations to workshops, media attention, awards, certificates) in ways that stimulate all actors involved.

Group building and local action planning

Once recognized as valuable partners, local innovators will be motivated to collaborate with each other and with outsiders, combining local and outside knowledge. The identified local innovations/innovators become the foci for community groups to examine opportunities, to plan joint experiments, to explore the ideas further, and to evaluate the results together. Special attention should be given to strengthening the capacities of weaker stakeholders (especially women and poorer farmers) to make them equal partners in the PID process, starting with their active involvement in the action planning.

Where the innovation has a technological character, existing manuals on participatory technology development (see for example: Van Veldhuizen et al., 1997) may be of help in the participatory design of the local experiments. The joint design (and implementation and monitoring) of the innovation activities is enhancing the innovative capacity of the local community. In addition their knowledge of specific aspects of the innovation can be enhanced though training activities, demonstrations, visits to other innovative farmers and research centres.

Implementing local innovation activities; capacity development; organizational strengthening; dissemination

In this stage the farmer groups develop their innovations further, with the active support of the supporting organizations, and monitor and evaluate the results. This process, involving concrete joint activities, also helps to strengthen community organization for development. The innovators' groups are places where farmers regularly meet and analyse and discuss existing and new practices, where they learn and share experiences, and enhance collaboration and collective action. Results of the experiments will be documented (as well as the processes applied) and shared with other groups of innovators, as well as with other producers. That can be done in many ways: posters and picture guides, participatory video, farmer-to-farmer training, storytelling in radio programmes, etc. The shared experiences of local innovations will inspire other groups. Phases 3 to 5 will be reiterated.

Institutionalization; policy influencing; up-scaling

As a parallel process, the partners in the process should seek to create better institutional and policy conditions for local PID, by engaging in policy dialogue and facilitating integration of the approach into the programmes of relevant institutions.

Monitoring and evaluation tools

A large number of monitoring and evaluation tools may be applied to assess the effects of a project on practical and strategic gender issues. In fact most of the participatory appraisal tools described in earlier chapters might be used also for monitoring. To prevent overlap, we will not discuss such tools here; rather we will present a checklist (Table 17.17) that provides main points for attention when evaluating the impacts on gender of a particular urban agriculture project.

Table 17.17 Monitoring and evaluation checklist

Capacity development (project staff)
- Was the gender sensitivity of the project staff enhanced and their skills in gender analysis planning and monitoring improved prior to the start of the project? When training them on other topics, were the gender dimensions of that topic reviewed? Was the gender training effective? What gaps were identified later on?
- Have team members responded favourably to the inclusion or expansion of gender analysis within the project? What difficulties and/or successes have there been?
- Was a gender resource person consulted throughout the process? In what form(s) did this take place? How effective was their involvement?

Project design and planning
- Was the project design based on a gender-sensitive situation analysis yielding gender-disaggregated data on all issues investigated, and insight into the roles and responsibilities of men and women in urban agriculture, access to and control over available productive resources, decision making and distribution of benefits, etc? What proved to be very important? What improvements are needed?
- What were the main specific gender issues coming out of the analysis? Have these been properly incorporated in the project design? Which of these issues have been properly addressed by the project and which less so?
- Were women involved in decision making on the priority issues to be attended by the project and in the selection of the activities to be undertaken? How is this reflected in the project objectives and strategies and related budget? What could have been done to further strengthen women's participation in the project design?

Project implementation
- Were representatives of men and women producers actively involved in periodic project management/co-ordination meetings and decision making on the course of the project? Were gender issues regularly discussed during those meetings? How did they influence the decision-making?
- Was there equal and active participation of both male and female staff and producers in the implementation of the project activities (training, experimentation, improvement of infrastructure, playing certain roles in the project, etc.) and the distribution of resources provided by the project (information/knowledge, credit, tools, land, equipment, seed, irrigation water, etc.) during project implementation? What hampered the full participation of women producers in such activities and resources? With what effects?
- Which gender-affirmative actions have been undertaken to overcome existing inequalities and barriers for female participation in the project? With what effects? What else could have been done?
- Which activities have been implemented to promote gender mainstreaming in the participating producer organizations? With what effects?

Monitoring and evaluation
- Have the monitoring data been collected in a gender-disaggregated way? What data were more difficult to collect in a gender-disaggregated way; which others were collected in that way but proved to be less useful?
- Have the monitoring data been collected and analysed with direct participation of both male and female representatives of the beneficiaries? How did this influence the quality of the monitoring information and the use that was made of such information?
- What does the monitoring information show regarding the distribution of project benefits (enhanced income, improved nutrition, knowledge and status gained, control over resources, etc.) among respectively women and men? Which are and which not? Due to what factors?
- Which remedial actions were taken when monitoring data showed that women and men were not benefiting equally from the project activities? With what effects?
- What is the combined effect of the project on gender equality and strategic gender issues? Due to what factors mainly?

Going to scale; institutionalization
- What has been undertaken to facilitate gender mainstreaming in the organizations that participate in the project? With what effects? Which mainstreaming strategies worked well and which did not or less so?
- Have project findings regarding gender dimensions of urban agriculture (and how to attend these in urban agriculture research and development projects) been given proper attention when preparing publications and other dissemination activities (website, seminars, etc.)? With whom these findings have been shared (newspapers, NGOs, farmer organizations, municipal departments, etc.)? With what effects?
- Have efforts been made to influence policy makers to apply project findings regarding gender and urban agriculture? How exactly? How? With what effects?

Going to scale (planning follow-up actions, dissemination, policy influencing)

Scoping of gender issues for policy and planning (adapted from FAO, 2003: 186–187)

This tool helps to scope out the possibilities for integrating gender issues into urban agriculture planning and policy realms. Participants should represent both community and institutional actors if at all possible.

In small groups, have a discussion about the implications of what you have learned about gender and urban agriculture during the project, and its relevance and use for policy and planning. The following questions may guide the discussion:

- What has been learned about gender and urban agriculture in this project? e.g. regarding the roles of and contributions of men and women to household and community food security, differential needs and priorities of male and female producers, experiences gained with specific strategies and affirmative actions to develop urban agriculture in a gender-sensitive and equitable way.

- What might be some of the implications for policy and planning of what has been learned about gender and urban agriculture in this project? What policy measures would you like to recommend?
- What needs to be done or can be done to bring about such policy changes? Who should participate?
- What might be the implications of the lessons learned for your own organization (activities, working methods, staffing, training, budget allocation, etc.)?
- What needs to be done to effect such change? With whom would you like to partner to make such changes occur?
- What might be the implications of such policy and planning changes on the practical and/or strategic interests of men and women?

Policy briefing paper

Since policy makers are busy people who have little time to read lengthy project documents or research reports, the preparation of a short and concise document that is specially designed to inform a policy maker on the main policy issues involved is often crucial for impact on policy development and planning.

A policy briefing paper synthesizes data on key issues most relevant for a certain policy area and lessons learned from the recent research or development project(s). It provides recommendations (i.e. suggested policy measures, changes needed in existing regulations, recommended actions) on how a municipal or national government or other institution may (further) develop its policy and programmes regarding the area of gender and urban agriculture, often illustrated by concrete city cases ('building theory based on practice').

Below we present the process as applied in the Urban Management Programme for Latin America and the Caribbean (UNDP–Habitat) to develop a series of seven policy briefs on various aspects of urban agriculture, including one on 'Urban Agriculture and Gender' (UMP–LAC, 2004).

- Scoping exercise with topics discussed in the policy briefs defined by representatives of various partners in the project during a workshop in Quito–Ecuador in 2000, based on review of the results of the research and development activities that they had implemented to date.
- Elaboration of short 'synthesis' papers on each of these topics included in the priority list by one of the project partners, analysing the available information and experiences gained regarding this topic, illustrated with clear cases/examples.
- Design and drafting of a four-page policy brief on each topic with assistance of a professional designer/social communicator.
- Discussion and 'validation' of the draft policy briefs with the project partners and elaboration of final versions of the Briefs.

- Presentation of the set of policy briefs to each of the municipal authorities participating in the programme, plus dissemination to other cities and actors.
- Monitoring of the use of policy briefs by these municipalities (in development of policies, norms and regulations, action plans, project development, training, etc.) and results of such activities.

When preparing a policy brief and engaging in policy dialogue, special attention should be given to the following elements.

- Recognition of *women as independent actors and beneficiaries* in/of urban agriculture public policies and projects
- Acknowledgement of the real *value of women's contribution* to the development of urban agriculture: production, food security, income, etc.
- Recognition that the *needs of men and women are different* and that women's access to and control over resources and participation in decision making is restricted by socio-cultural and institutional traditions
- Recognition that public policies and projects, as well as economic and technological trends, can have *differential effects for men and women*
- Recognition that *affirmative actions are needed* to ensure that women (and men) can reap equal benefits from urban agriculture policies and projects.

Policy-action matrix (adapted from FAO, 2005: 48)

The policy-action matrix helps to relate policy objectives to specific policy actions, responsibility for execution, costs, and time frames. It works well to establish interactions among national, regional, and local stakeholders to make decisions on policy objectives, actions, costs, and time lines. NGOs, community-based organizations, producers, and private-sector service providers can make significant inputs and decide on their roles in policy planning and implementation. With minimal guidance, all stakeholders can participate effectively in the construction of this action matrix.

The project team starts the process by organizing a stakeholder workshop with the following steps:

- Specify the policy objectives (and related policy measures) that are recommended by the project, and select one objective for further analysis.
- Review the recommended measures to achieve this objective and identify additional or alternative recommendations; select the most important recommendations for possible action.
- Define at what level (local, regional, or national) each recommendation would be implemented.
- Determine who is responsible for deciding on each recommendation, and who is responsible for executing the recommendation (these are normally different people or institutions).
- Determine a strategy to implement each recommendation.

- Cost each recommendation, taking into account staffing, operational, and infrastructure requirements for the next three to five years.
- Suggest the time frame for execution.

Repeat the same procedure for each policy recommendation. Fill in the form shown in Table 17.18 as the proposed plan of action is developed for the policy objective(s):

Table 17.18 Example of a policy-action matrix

Strategic objective 1:			
	Specific objective 1		
	Recommendation 1	Recommendation 2	Recommendation 3
Achievement targeted			
Execution level • Local • Regional • National			
Responsible stakeholder for: • Decision making • Execution			
Strategy to implement recommendations			
Cost • 2006 • 2007 • 2008 +			
Timeframe for execution • Short (1–2 yrs) • Medium (3–5 yrs) • Long (5+ yrs)			

Source: FAO, 2005

References

Arce, B., Prain, G. and Salvo, M. (2006a) 'Toward the integration of Urban Agriculture in Municipal agendas: an experience in the District of Lurigancho-Chosica', *Urban Agriculture Magazine 16: Formulating Effective Policies on Urban Agriculture.*

Arce, B., Prain, G., Valencia, C., Valle, R., and Warnaars, M. (2006b) *The Farmer Field School (FFS) Method in an Urban Setting: A Case Study in Lima, Peru,* Urban Harvest, CIP, Lima, Peru.

AusAid (2000) *AusGuidelines, The Logical Framework Approach,* AusAid, Australia.

Canada–Nepal Gender in Organizations Project (CNGO) (2003) *Gender in Project Planning.* Module 1 of Training Package #3 on Gender Responsive Community Development, CNGO/CIDA, Nepal.

Critchley, W. (ed.) (1999) *Promoting Farmer Innovation. Harnessing Local Environmental Knowledge in East Africa*, Workshop report no 2, UNDP, SIDA.

De Zeeuw, H. and Dubbeling, M. (2004) 'Minutes of the RUAF Partners Meeting 2004 in Hyderabad', (unpublished).

De Zeeuw, H. and Wilbers, J. (2004) *PRA tools for studying urban agriculture and gender*, ETC–Urban Agriculture, Leusden. Available from: http://www.ruaf. org/node/97

FAO (2003) *Gender-Disaggregated Data for Agriculture and Rural Development*, SEAGA Guide, FAO, Rome.

FAO (2005) *Participatory Policy Development for Sustainable Agriculture and Rural Development*, Guidelines from the Sustainable Agriculture and Rural Development, p. 48, Farming Systems Evolution Project, Rural Development Division, Sustainable Development Department, FAO, Rome.

Groverman, V. (1992) *Rapid Rural Appraisal/Participatory Rural Appraisal – A Tool for Gender Impact Study*.

Haile, F. 1991 'Women Fuel Wood Carriers in Addis Ababa and the Peri-urban forest', Report to the International Development Research Centre (IDRC) and National Urban Planning Institute (NUPI), Geneva: International Labour Organization.

Hovorka A. (1998) 'Gender resources for urban agriculture research: methodology, directory and annotated bibliography', CFP report 26, IDRC, Ottawa.

Lingen, A. (1997) *Gender Assessment Studies: A Manual for Gender Consultants*, ISSAS & Ministry of Foreign Affairs, The Hague, The Netherlands.

Mianda, G. (1996) 'Women and garden produce of Kinshasa: the difficult quest for autonomy', in P. Ghorayshi and C. Belanger (eds.) *Women, Work, and Gender Relations in Developing Countries*, Greenwood Press, Westport, Connecticut.

Ofei-Aboagye, E. (1997) 'Memo on Gender Analysis of Agriculture in Ghana', Report for IDRC Project No. 96-0013 003149. IDRC, Ottawa, Canada.

Röling, N.G. (2003) 'From causes to reasons: the human dimension of agricultural sustainability', *International Journal of Agricultural Sustainability* 1: 73–88.

Röling, N. G. and Van de Fliert, E. (1998) 'Introducing integrated pest management in rice in Indonesia: a pioneering attempt to facilitate large-scale change', in N.G. Röling and M. A. E. Wagemakers (eds.) *Facilitating Sustainable Agriculture*, Cambridge University Press, Cambridge, UK.

Sims Feldstein, H. and Poats, S. V. (eds.) (1989. *Working Together: Gender Analysis in Agriculture*, vol. 2, Kumarian Press, West Hartford, Connecticut.

Slocum, R., Wichhart, L., Rocheleau, D., and Thomas-Slayter, B. (eds.) (1995) *Power, Process & Participation: Tools for Change*, Intermediate Technology Publications, London.

UMP–LAC (2005) *Guidelines for Municipal Policymaking on Urban Agriculture No 7, 'Urban Agriculture: Fostering Equity between Men and Women'*, UMP-LAC; Available from http://www.idrc.ca/uploads/user-S/10530124730E7.pdf.

Van Veldhuizen, L., Waters-Bayer, A., and De Zeeuw, H. (1997) *Developing Technology with Farmers; A Trainer's Guide to Participatory Learning*, Zed Books, London/New York.

Waters-Bayer, A and Van Veldhuizen, L. (2004) 'Promoting Local Innovation: Enhancing IK Dynamic and Links with Scientific Knowledge', *IK Notes* no 76 (World Bank).

Wettasinha, C. et al. (eds.) (2007) *Recognising Local Innovation: Experiences of PROLINNOVA Partners*, IIRR, The Philippines.

Züger, R. (2005) 'Participatory Development Projects in the Andes – Looking for Empowerment with Q-Methodology', paper presented at PRGA Impact Assessment Workshop, October 19–21, 2005, CIMMYT Headquarters, Mexico. Available from http://www.prgaprogram.org/index.php/module= htmlpages&func=display&pid=65

CHAPTER 18
Resources

Abstract

This chapter offers a variety of selected resources on gender and urban agriculture, grouped as follows:
- *literature dealing with* conceptual frameworks *regarding gender in agriculture*
- field studies *on gender and urban agriculture*
- *literature providing* gender-sensitive methods and tools *for gender mainstreaming*
- *relevant* websites.

Annotated bibliography

Gender and urban agriculture: concepts

Cummings, S., H. Van Dam, and M. Valk (2002) *Natural Resources Management and Gender: A Global Source Book*, **KIT, Amsterdam.**
Various ways in which women's and gender perspectives and considerations have been incorporated in natural-resources management are documented in the case studies from Mesoamerica, India, Pakistan, Uganda, and West Africa. An introduction discusses the history of and contemporary thinking on women, gender, and environment; the country studies illustrate the relationship between women and land rights, gender approaches to the management of water and wetlands, mainstreaming gender in environmental policy, and the need for a gender-differentiated participatory approach. An annotated bibliography of printed and on-line publications, and web links complement the case studies. A co-production with Oxfam GB.

De Zeeuw, H., G. Prain, and J. Wilbers (2004) *Women Feeding cities. Gender Mainstreaming in Urban Food Production and Food Security*, **Proceedings of the workshop: Women Feeding Cities, 20–23 September 2004, Accra, Ghana. RUAF, Leusden, The Netherlands; CIP, Lima. Available from www.ruaf.org.**
During this workshop, participants presented 15 cases, which were critically reviewed in order to identify key issues in gender and urban agriculture. Also a priority agenda was developed with important aspects and actions that will need attention when integrating gender in future urban agriculture research activities, training activities, policy development, and action planning and

implementation. The workshop presented the concept of mainstreaming gender and identified effective strategies for mainstreaming gender in the projects of the workshop's organizers: RUAF and Urban Harvest. Participants recognized that the differentiation of the roles played by urban men and women in urban food production, processing, and marketing, and the documentation of their specific interests, knowledge, constraints, and opportunities, as well as the mechanisms of disadvantage (especially in existing values, policies, and institutional practices) are critical to the design of effective policies and interventions aiming at urban food security (as well as human and socio-economic development).

FAO (2002) 'Gender and access to land', *FAO Land Tenure Studies* **no. 4.**
Women, elderly people, minorities, and other sometimes marginalized groups can be at risk in land reform and land administration projects. Very often, when land values increase as a result of external investments, women become marginalized in the process and risk losing former benefits. Women may be at risk even if it is intended that they share the benefits. For example, improving irrigation on women's fields may have the unintended effect that these newly valuable fields are reclaimed by men in the community. Enhancing housing in a community or peri-urban area may have similar unintended results when the units become more marketable. Children and elderly people may also suffer, although the original intention was to include them in the intervention. The purpose of these guidelines is thus to provide background information for land administrators and other land professionals on why gender issues matter in land projects; and to provide guidelines to assist development specialists and land-administration agencies to ensure that land administration enhances and protects the rights of all stakeholders.

FAO (2005) *Building on Gender, Agro-biodiversity and Local Knowledge: A Training Manual*, **FAO, Rome.**
This training manual focuses specifically on the links between local knowledge systems, gender roles and relationships, the conservation and management of agro-biodiversity, plant and animal genetic resources, and food security. Its aim is to promote a holistic understanding of these components. The training objective is to strengthen the institutional capacity in the agricultural sector and to recognize and foster these links in the relevant programmes and policies. Other manuals may cover these same topics, but there is an obvious lack of integrated training materials that address all three topics. Moreover, FAO's local partner organizations have requested specific training materials that focus on these cross-cutting issues. We strongly believe that a better understanding of the key concepts, and their linkages, will lead to improved project planning and implementation. This manual therefore aims to explore the links between agro-biodiversity, gender, and local knowledge, and to show the relevance of doing so, within the context of research and development. This manual will not equip you with the skills needed to conduct participatory or action research at the field level, or provide guidance for research tools

and methods. However, it is meant to complement existing manuals covering tools, methods, and approaches, such as the FAO/SEAGA handbook material for socio-economic and gender analysis (www.fao.org/sd/seaga).

Hahn, N. (1991) 'Backyard gardening: a food security system managed by women', *Entwicklung + Laendlicher Raum* **1, pp. 24–27.**
This article highlights the important contribution of home gardening to food security, and, particularly, the role that women play in it. The author suggests various directions for further research and innovation. Interestingly, homestead production of spices receives special attention.

Hovorka, A. J. (2006) 'Urban agriculture: addressing practical and strategic gender needs', *Development in Practice,* **(16, no 1).**
This paper considers the role of urban agriculture in addressing the practical and strategic needs of African women and assesses the gender implications of embracing urban agriculture as a development-intervention strategy. Empirical evidence from Botswana and Zimbabwe points to the multi-faceted role of urban agriculture, an activity used by some women to support their households on a daily basis, and by others as an avenue for social and economic empowerment over the longer term. In order to benefit rather than burden women, the promotion and support of urban agriculture must take on an emancipatory agenda which supports individual, practical, and strategic goals, and ultimately challenges the structural conditions that give rise to women's involvement in the activity in the first place.

Hovorka, A. J. (undated) 'Gender and Urban Agriculture: Emerging Trends and Areas for Future Research', Graduate School of Geography, Clark University, Worcester MA, USA.
Over the past decade, literature on women and urban agriculture has emerged, revealing significant insights that arguably can change the future focus of the field at large. This overview presents a synthesis of lessons learned from recent studies that have begun to recognize and assess women as farmers in urban areas. The extent, nature, and role of urban agriculture vary considerably between and within countries, as well as throughout the urban hierarchy. Moreover, evidence tends to be scattered or speculative, with little supportive data to substantiate general statements. As such, it is difficult to formulate a synthesis of trends that hold for every context, or even the majority of contexts.

Nevertheless, there are several broadly identifiable trends in recent literature on women and urban agriculture which warrant recognition and further explanation. First, studies now recognize women as urban farmers. Indeed, women play significant roles in urban food production and contribute to both household and market economies. Second, women benefit from urban agricultural activities that allow them to successfully combine their multiple roles in subsistence, production, and environmental management sectors. Third, researchers document the constraints hindering women's participation in urban agriculture activities. Obstacles exist at both sectoral and household levels. Fourth, studies identify women farmers' survival strategies and social activism in response to structural constraints and urban food issues. Together these trends have enriched the understanding of urban agriculture. Yet gaps persist in the literature, and a discussion of future trends and considerations for urban agriculture research in general is required.

Kabeer, N. and R. Subrahmanian (eds.) (1999) *Institutions, Relations and Outcomes; Framework and Case Studies for Gender-Aware Planning*, Kali for Women, New Delhi; Zed Books, London & New York.
This study develops an analytical framework and a set of tools for planners and trainers to integrate gender systematically into different aspects of their work. It offers an inventory of the kinds of assumption which lead to gender-blind policy. A selection of case studies from the Indian context illustrates the different aspects of the framework and its applications.

Kabeer, N. (2003) *Gender Mainstreaming in Poverty Eradication and the Millennium Development Goals: A Handbook for Policy-makers and Other Stakeholders*, Ashford Colour Press, London.
In this book, Naila Kabeer brings together a set of arguments, findings, and lessons from the development literature which help to explain why gender equality merits specific attention from policy makers, practitioners, researchers, and other stakeholders committed to the pursuit of pro-poor and human-centred development. Neglect of gender inequalities in the distribution of resources, responsibilities, and power in the processes of economic accumulation and social reproduction has a high cost, not only for women themselves but also for their children and other dependants and for the development of society as a whole. This book highlights the interconnections between production and reproduction within different societies, and women's critical role in bridging both, and points to the various synergies, trade-offs, and externalities which these generate. All over the world, women from poor households play a more critical role in the income-earning and expenditure-saving activities of their households than women from better-off households, and they are concentrated in the informal economy. The relationship between household poverty and women's paid activity has, if anything, become stronger over recent decades, partly in response to economic crisis and the 'push' into the labour market and partly in response to new opportunities generated by globalization. Improving women's access to economic opportunities, and enhancing returns to their efforts, will be central to the goal of poverty eradication and the achievement of the MDGs.

Kanji, N. (2003) *Mind the Gap: Mainstreaming Gender and Participation in Development*, International Institute for Environment and Development (IIED), London and Institute for Development Studies (IDS), Brighton.
This publication, the fourth in a series on institutionalizing participation, highlights lessons from gender-mainstreaming work for those who seek to

institutionalize participation. After a discussion of conceptual frameworks and strategies, and the suggestion that there has been a shift from participation to governance (along with the shift from women in development to gender in development), the tensions between gender mainstreaming and participatory development are explored. Suggestions are made to overcome this tension.

Koenraadt, C. (2001) 'Household resource management and urban horticulture in relation to the action plan for women in development', in *Proceedings on the Sub-Regional Expert Consultation on the Use of Low Cost and Simple Technologies for Crop Diversification by Small Scale Farmers in Urban and Peri-Urban Areas of Southern Africa*, **Stellenbosch, South Africa, 15–18 January 2001.**
This presentation discusses the key areas of development efforts in which FAO's Women in Development Service concentrates its efforts and how this is linked to gender and urban agriculture mostly through the Household Resources Management issue. It further introduces the Socio-Economic and Gender Analysis Programme (SEAGA) methodology, being a possible suitable tool to support research in urban agriculture as well as to programme, monitor and evaluate UA programmes where women play a critical role.

Lee-Smith, D. (1994) 'Gender, Urbanization and Environment: A Research and Policy Agenda', prepared for the International Seminar on Gender, Urbanisation and Environment, held in Nairobi, Kenya, 13–16 June 1994. Mazingira Institute, Nairobi, Kenya.
This agenda was produced at the International Seminar on Gender, Urbanisation and Environment, held in Nairobi in June 1994. The purpose of the seminar was to draw together researchers from different regions of the world to discuss issues, theories, and methods, to put the priorities of poor women at community level on to the research agenda and to influence policy to take account of gender issues emerging from research. The agenda is presented according to the four main themes of the seminar, for each of which a working group of scholars and others identified the priorities for research and policy action, but it starts with common themes and priorities which represent the crucial findings of the seminar. Women's property rights and urban agriculture are highlighted.

Lee-Smith, D. (2006) 'Urban agriculture risk perceptions, communication and mitigation: community participation and gender perspectives', in A. Boischio, A. Clegg, and D. Mwagore (eds.), *Health Risks and Benefits of Urban and Peri-urban Agriculture and Livestock (UA) in Sub-Saharan Africa*, **Urban Poverty and Environment Series Report No. 1, IDRC, Ottawa.**
This paper addresses the perception of health risks among poor communities practising urban agriculture, and especially the differences between men and women. It also addresses some aspects of the communication of health risks to communities, and the way in which communities may themselves adopt mitigation strategies. This theme is specifically addressed as a gender issue.

Sometimes women are seen as more vulnerable than men, and the aspect of vulnerability of different groups to different levels and types of risk is briefly touched on. However, the paper explains that the capacity of communities in general and women in particular to respond to risks as active agents should also not be overlooked.

Lingen, A., R. Brouwers, M. Nugteren, D. Plantenga, and L, Zuidberg (1997) *Gender Assessment Studies: A Manual for Gender Consultants, Operational Guide,* **Women and Development Series, Ministry of Foreign Affairs, The Hague, The Netherlands.**
This manual presents the Gender Assessment Study (GAS), an instrument which has been developed for the formulation and appraisal phases of projects. It is a tool for designing project interventions in order to ensure that they will affect the empowerment of women positively. The manual explains the research methodology and its practical implementation in detail and is especially geared to the needs of those who undertake the study in the field. An interesting feature of the instrument is that it combines a gender analysis of the target group and the context with a gender analysis of the future implementing organizations, leading to a gender analysis of the project (proposal).

Mapetla, M., H. Phororo, and G. Prasad (1994) 'Urbanization, Gender and Environment: the Role of Wild Vegetables', paper presented at the International Seminar on Gender, Urbanization and Environment, Nairobi, Kenya, 13–16 June 1994. Mazingira Institute, Nairobi Kenya.
Although many people in Lesotho move to towns in search of better job opportunities, they miss out on access to natural wild-food resources. It is argued in this paper that such resources can be adapted to the urban environment, and wild vegetables could contribute to the nutrition and cash income of urban dwellers. Gathering, preparing, and eating are related to gender and culture. Earlier studies have shown that collecting and preparing wild vegetables is a strategy for rural women to provide a balanced diet for their families. Urbanization in Lesotho has affected women's access to natural resources such as edible wild plants. Settlements now occupy former agricultural land, and wild foods become scarce in peri-urban areas as a result of over-harvesting. Urban women have to spend much more time than previously to find enough wild plants for a meal. The study reveals that only a few people collect wild plants, and children are no longer taught about edible plants due, in part, to schools adopting westernized curricula which do not foster recognition of the value of indigenous plants. For many people in towns, wild foods have low status and they would rather buy cultivated vegetables from the market. Towns have created markets, but only rarely are wild vegetables sold. Women from the countryside sell the wild vegetables through informal networks in town. A promotion strategy for wild vegetables, focusing on knowledge, attitude, value, and nutrition is suggested. Outward rather than inward growth of urban areas is also recommended by the authors. It is argued that gardens

within urban housing sites in Lesotho are agriculturally more productive than fields in rural areas, and agricultural production does not diminish when fields are converted into housing sites with gardens in the present urban extension pattern.

Maxwell, D. G. (1999) *Internal struggles over resources, external struggles for survival: urban women and subsistence household production*, Urban Agriculture Notes [online]. Available from http://www.cityfarmer.org/danmax.html.
Presents data on Kampala in an attempt to examine and understand semi-subsistence urban farming and the way in which the practice has been incorporated into the economic strategies of urban households and individuals. The paper argues that in contemporary Africa farming spans a continuum from a survival strategy for some to a large-scale high-return investment for a few. For the most part it should be considered as a deliberate crisis response on the part of urban women, to provide for themselves and their households a source of food which is not dependent on cash or volatile markets. The paper explores patterns of engagement in farming, the use of food, and division of labour. It also discusses reasons for farming and divisions of household responsibilities before drawing conclusions.

Moser, C. O. N. (1993) *Gender Planning and Development; Theory, Practice & Training*, Routledge, London, UK.
Gender planning is a new tradition whose goal is to ensure that women, through empowering themselves, achieve equality and equity with men in developing societies. *Gender Planning and Development* explores the relationship between gender and development, and provides a comprehensive introduction to Third World gender policy and planning practice. It describes the conceptual rationale for a new planning tradition based on gender roles and needs, and identifies methodological procedures, tools, and techniques to integrate gender into planning processes. It emphasizes the role played by training in creating gender awareness, and highlights the entry points for women's organizations to negotiate for women's needs at household, community, state, and global levels.

Palacios, P. (undated) *Policy Brief: Urban Agriculture, an Opportunity for Gender Equity* (Spanish), PGU-ALC / FEMUM–ALC.
This policy brief focuses on the steps that can be taken at the policy level to promote urban agriculture with a gender-equality perspective, recognizing that urban agriculture can respond to the specific problems faced by different social sub-groups but also that these different groups each have different potentials to confronting the problems. According to the brief, the following intervention strategies are necessary for equal and sustainable urban management: (1) diagnose reality from a gender perspective, (2) articulate the role and contribution of urban agriculture and integrate a gender perspective in the planning process, and (3) install the equal participation of citizens

as a social and economic right. These strategies are each broken down into practical steps. The brief concludes by presenting conceptual, methodological, and operational challenges concerning gender-sensitive urban agriculture, roles and responsibilities of actors involved in urban agriculture, and relevant contacts in the Latin American region.

Ratta, A. (1993) 'City women farm for food and cash', *International Ag-Sieve,* **6 (2): 1–2.**
This brief article outlines women's involvement in urban agriculture and highlights barriers and solutions to such activities. Farming is a viable alternative to wage labour for women and enables them to work close to home. Women's role is not limited to food production but includes processing food for home and market. These activities are rarely reflected in official statistics, nor are they recognized as a contribution to the family budget. Thus women do not fully benefit from research or extension services.

Van Esterik, P. (1999) 'Gender and sustainable food systems: a feminist critique', in M. Koc, R. MacRae, L. J. A. Mougeot, and J. Welsh (eds.), *For Hunger-proof Cities: Sustainable Urban Food Systems,* **pp. 157–161, IDRC, Ottawa.**
This paper explores conceptual and practical links between women and food and suggests how feminist analysis may further our understanding of food security. It argues that women's special relationship with food is culturally constructed and not a product of a natural division of labour. Women's identity and sense of self are often based on their ability to feed their families and others; food insecurity denies them this right. Food socialization and body image are also strongly gendered. The paper concludes with a working definition of feminist principles and a call for the development of a model of feminist food praxis. Food praxis refers to the practical 'mastery' of routines of producing, preparing, and consuming food. The paper proposes 10 points to guide further research and action. These include acknowledging women as gatekeepers of the food system, giving priority to the elimination of hunger, using multiple research methods, recognizing how political forces control people's access to food, emphasizing the temporal complexity of food routines, and providing a critically reflexive guide to advocacy action.

Wilbers, J., A. Hovorka, and R. Van Veenhuizen (eds.) (2004) 'Gender and urban agriculture', *Urban Agriculture Magazine Number 12: Gender and Urban Agriculture.* **Available from www.ruaf.org.**
Urban agriculture can have positive and/or negative consequences for men and women, depending on the situation and conditions. Data gathered on urban agriculture demonstrate that it generally has a positive impact on household food security, and thus will be beneficial to women, who most often are responsible for it. This issue of UA Magazine explores how urban agriculture relates to existing gender dynamics.

Field studies on gender and urban agriculture

Barndt, D. (1999) 'Women workers in the NAFTA food chain', in M. Koc, R. MacRae, L. J. A. Mougeot, and J. Welsh (eds.) *For Hunger-proof Cities: Sustainable Urban Food Systems*, pp. 162–166, IDRC, Ottawa.
Efforts to develop sustainable urban food systems must take into account the role of women in the various stages of production, preparation, and consumption of the food we eat. The Tomasita Project explores women's shifting roles in the restructured global labour force, tracing the journey of a tomato from a Mexican field to a Canadian table. This essay focuses particularly on salaried workers in Mexican agribusiness. The Tomasita project also aims to connect women food workers in Mexico and Canada, in both dominant and alternative food systems, through photo-stories, films, and video letters. Sharing these stories across borders helps women to understand how they are part of a broader global process while they learn from each other's tales of survival and resistance.

Boulianne, M. (2001) 'Urban agriculture and collective gardens in Quebec. Empowerment of women or "domestication of the public space"?' (original in French), *Anthropologie et Sociétés*, 25 (1): 63–80.
In 'developed' countries, self-provisioning urban agriculture practices reappear periodically during phases of adjustment of the capitalist economy. It is actually the case in North America. The most recent experiences encourage collective production. They emerge in Quebec in the context of an increasing recognition of the social economy by the state. Supported by community groups, collective gardens appear as an alternative to food help for the impoverished, and as a tool to alleviate social exclusion. Women are the principal protagonists of these recent initiatives, either as members of the supporting groups or as producers. While certain researchers argue that women's community organizations active in the domain of food security have a strong potential for the individual and collective empowerment of women, others think that in times of decentralization of social programmes, these are rather associated with a 'domestication of public space'. Are the emerging experiences of collective self-provisioning gardening empowering for women, or are they contributing to the domestication of the public space? Relying mainly on fieldwork data collected in Quebec, the present article seeks to bring answers to this question.

Dennery, P. (1995) 'Inside Urban Agriculture: An Exploration of Food Producer Decision Making in a Nairobi Slum' (M.Sc. thesis*)*, Wageningen Agricultural University, The Netherlands.
This study examines urban food producers and their households in Kibera, a large informal settlement of Nairobi. One of the main features of this study is the addition of a qualitative dimension to urban agriculture research in east Africa. Empirical evidence is provided on gender relations, labour relations, and the multiple uses of produce at the individual, household, and

community levels. The traditional division of agricultural labour was noted during fieldwork: men preparing the soil for planting, and women responsible for harvesting food for daily needs. Women decide how much produce to sell and what food to buy, in consultation with the spouse in order to provide a means of preserving marital harmony. Women's decision-making power may be undermined by factors such as size of plot, need for cash, and personal health. Women are also less likely than men to have knowledge about inputs, such as pesticides or use of sewage water, due to their limited exposure to commercially oriented agriculture. The study also reveals that numerous labour issues are directly related to the prevailing gender ideology in Kenya. Female urban producers must carry out most of the care and maintenance of the household, regardless of the time they devote to food production or other livelihood activities. Thus, women tend to stay in the field longer than men and are expected to fetch water and prepare meals upon returning to the house. In other cases, women's ability to control their own agricultural labour time is limited by responsibilities to others.

Dima, S. J. and A. A. Ogunmokun (2001) 'An Overview of Socio-Economics and Gender Aspects in Urban and Peri-Urban Agriculture: The Potential of the City of Windhoek, Namibia.', paper prepared for the Sub-Regional Expert Consultation Meeting on Urban and Peri-Urban Horticulture, University of Stellenbosch, Cape Town, South Africa, January 2001.
Urban and peri-urban agriculture can be defined as the process of producing agricultural commodities within demarcated urban areas and edges of urban areas. Because of colonial and post-colonial local-authority laws, rules, and regulations, urban agriculture has been practised illegally, discreetly, and without technical support by local authorities or the relevant ministries of agriculture. This paper provides an overview of the resources available, and the technologies used for urban and peri-urban horticulture in Namibia. This is followed by a survey of the recent literature on urban and peri-urban agriculture in Africa with a view to assessing its extent and contribution in terms of food production, employment creation, improvement in nutrition status, income generation, innovation, creation, adaptation, and appropriate technologies development. This is followed by a case study of urban and peri-urban horticulture in the city of Windhoek.

Drescher, A. (1997) 'Management Strategies in African Home Gardens and the Need for Extension Approaches', paper presented at the International Conference on Sustainable Urban Food Systems, 22–25 May 1997. Ryerson Polytechnic University, Toronto, Canada.
The relationship between urban food production, food security, and urban environments has been largely neglected. This paper focuses on results from a household garden survey conducted during 1992 and 1993. The main objective of the survey was to clarify the role of household gardens for household food security in Zambia, and to identify differences and problems in management strategies and their effects on production in different areas. The results reveal

that the main actors in urban agriculture are often women. In all compounds studied in Lusaka, women were to a greater extent involved in cropping and gardening than men. Gender analysis is used to reveal differences between men's and women's urban agriculture techniques with respect to alternative methods of plant production, crop species, and use of fertilizer, manure, and compost. The paper argues that gender-specific differences in agricultural activities need to be given more attention by extension services in urban and peri-urban areas.

Egziabher, A.G. (1994) 'Urban farming, cooperatives, and the urban poor in Addis Ababa', in A.G. Egziabher et al. (eds.), *Cities Feeding People: An Examination of Urban Agriculture in East Africa*, pp. 85–104, IDRC, Ottawa.
This paper focuses on urban agriculture in Lusaka and is based on a household garden survey conducted during 1992 and 1993. The main objective of this survey was to explore the role of household gardens in the context of household food security in Zambia. The findings reveal that women are more involved in agriculture and gardening in all compounds of Lusaka than men. In many ways, women play an important role in the food supply of households; through their productive labour, their decisions on production, consumption, and division of food, and through their income. A household-gardening model was developed to enable a better understanding of urban gardening activities in the social and environmental context. The model can assist in highlighting and clarifying some of the factors influencing urban agriculture. The household itself is based in the centre of the model, with various internal and external factors determining the vulnerability of the household. The study reveals that gardening contributes to food security directly by providing food and indirectly by creating income, respectively saving expenditures in the urban environment. Strengthening the role of women is recommended to policy makers seeking to develop the urban agriculture sector to address household food security.

Ethangatta, L. K. (1994) 'Households Headed by Elderly Women in the Slums of Kawangware and Kibagare in the City of Nairobi; Poverty and Environmental Concerns', paper presented at the International Seminar on Gender, Urbanization and Environment, Nairobi, Kenya, 13–16 June 1994. Mazingira Institute, Nairobi, Kenya.
The aim of the study is to determine the social–cultural, health-related factors and economic characteristics of elderly Nairobi women which may affect their nutritional status. The study is based on data collected from 201 elderly women from slums of Kawangware and Kibagare, and the low-income areas of Dagoreti and Waithaka in Nairobi. The findings reveal that women in low-income areas use their land for growing food crops such as beans, kale, cabbage, and bananas. Some of these women also had a dairy cow that produced milk for their own household's consumption or for sale to neighbours. The marketing of fruit and vegetables was also cited by women as an income-

generating activity in both slum and low-income areas. Elderly women in the slum areas turned to growing vegetables on the edges of roads and any other open spaces, in response to economic hardships. This activity created further environmental degradation, due to the uprooting of natural vegetation and consequent soil erosion in these spaces. The author concludes that there is an urgent need to provide basic means of raising incomes and standards of living for families headed by elderly women. In turn, better living conditions will improve the environmental conditions of slums and low-income areas.

Freeman, D. (1993) 'Survival strategies or business training ground? The significance of urban agriculture for the advancement of women in African cities', *African Studies Review,* **36 (3): 1–22.**
This paper suggests that, contrary to findings in other research, urban agriculture is not a stop-gap activity, nor a means to become wealthy. The motivations of the urban cultivators appear not to be influenced by what planners, researchers, or urban administrators feel ought to be the correct attitude to urban farming. This paper analyses interview data collected in Nairobi, Kenya in 1987 as part of a survey of active women cultivators. It first considers the stated motives of women cultivators and then assesses information gathered through detailed case studies of three individual women involved in urban farming. The most common motivation of women cultivators was the need to avert hunger, but also important was the availability of home-grown food so as to free up scarce cash earned by family members. The paper concludes that female cultivators face major impediments to meaningful advancement in Nairobi. Women are, in turn, dependent on the low-wage, formal manufacturing and service sectors for a market for their produce, and their activities are necessarily seasonal in the absence of the means to irrigate crops. The importance of this group as role models for other Third World women, the paper concludes, greatly outweighs their actual numbers.

Freidberg, S. (1999) 'Tradeswomen and businessmen: the social relations of contract gardening in south-western Burkina Faso', *Journal of African Rural and Urban Studies.*
This article examines the prospects for the expansion of export-oriented contract horticulture in south-western Burkina Faso, specifically in the area surrounding the city of Bobo Dioulasso. It sets out the main reasons why any discussion of West African contract farming must take account of the gender roles and moral codes which have historically informed relations between peasants and different members of the urban merchant community. Day-to-day provisioning and commercialization depends on the services of itinerant and local traders, many of whom are, especially in fresh-produce commerce, women. The crucial role of women traders, contrasted with that of male politicians and entrepreneurs, has created a distinctive culture of contract farming. The case study presented demonstrates how women traders have made the most of limited career opportunities by placing the flexibility

and durability of their commercial relations above season-to-season profits. In the interest of developing trading relations that they could pass to their daughters, women wholesalers set standards of trust and commitment that contractors find hard to match. The presence of a well-established, gender-based regional trade network poses potential obstacles to profitable contract horticulture schemes, because this network is essential to the economic security and occupational identity of both women traders and village gardeners in the Bobo Dioulasso 'garden belt'.

Gabel, S. (2004) 'Revealing Social Dimensions of Open Space Cultivation by Older Women in Harare: Advancing a Social Planning Discourse for Urban Agriculture', a thesis submitted in partial fulfilment of the requirements for the degree of Master of Arts (Planning) in the Faculty of Graduate Studies (School of Community and Regional Planning).
This research on urban agriculture in Harare, Zimbabwe, highlights women's ideas, needs, concerns, and agency, contextualizing these findings through an investigation of the institutional and policy environment governing the practice of open-space cultivation in the city. A feminist methodology provides an overall framework, while also incorporating ethno-methodology and participatory research methodologies to highlight the broader social, political, and cultural contexts of urban agriculture. A multi-method approach was adopted which included the use of semi-structured interviewing, focus groups, strategic meetings, participatory methods, visioning interviews, and action methods (such as field trips, creating a stakeholder forum, and organizing income-generating projects). Findings from this research have been used to develop a gender-aware history of women and urban agriculture (UA) in Harare. Key findings show that the forms of organization for open-space cultivation (SOSC) developed by older women have been historically unacknowledged, ignored, and impeded by those with decision-making power, most often male elites. Nine legal channels available for SOSC in Harare are uncovered in the research, dispelling the myth that UA is an illegal activity in the city. This research further elaborates on the impacts of legal ambiguity that have resulted in conflicts between various land-tenure systems and categories, demonstrating the serious governance challenges at the heart of supportive policy development for UA in the City. The voices of women are used to illuminate the dire need for local- and neighbourhood-level leadership, and the importance of addressing the cultural context in which UA is imbedded. A discussion of planning and governance in Harare reveals the exclusionary practices that operate to make the work of women, their UA- and land-based livelihoods invisible in planning practice and city decision making. The research shows the potential for shifting planning practice and discourse towards more people-centred, democratic forms of planning for UA.

Gianotten, V., V. Groverman, E. Van Walsum, and L. Zuidberg (1994) *Assessing the Gender Impact of Development Projects: Case Studies from*

Bolivia, Burkina Faso and India, Royal Tropical Institute, Amsterdam; ETC International, Leusden; Intermediate Technology Publications, London.
This book presents a preliminary methodological framework which was established in accordance with the objectives of a gender-assessment study (which investigates a development project's expected impact on women, compared with its impact on men, and also assesses the extent to which the project responds to the specific interests and needs of different categories of women). It also presents pilot studies in which the framework is tested, which provide information and recommendations for designing projects that will optimally strengthen women's position. The book is intended primarily for readers interested in policy on women and development, and in the methodology of gender-assessment studies. It is also intended for those interested in improvements to women's position, and to gender relations, in the specific context of planning and implementing development projects.

Hasna, M. K. (1998) *NGO Gender Capacity in Urban Agriculture: Case Studies from Harare (Zimbabwe), Kampala (Uganda) and Accra (Ghana)*, CFP series, report # 21, IDRC, Ottawa.
This study examines the marginalization of gender issues within urban-agriculture research agendas, drawing on data collected from field visits to the cities of Accra, Kampala, and Harare. It provides an overview and analysis of the nature and extent of NGO policies and strategies regarding the integration of gender into urban agriculture research. It is found that many NGOs are working with 'women in development' approaches. A proper understanding and scope of 'gender analysis' within these organizations is needed. Participatory learning partnerships should be developed among relevant groups, organizations, and institutions to share gender-sensitive research findings and create provisions for effective gender-focused policy interventions.

Hetterschijt, T. (2001) 'Our Daily Realities: A Feminist Perspective on Agro biodiversity in Urban Organic Home Gardens in Lima, Peru', Wageningen University, Wageningen.
The research for this thesis was conducted in the slums of Chorillos, a district in Metropolitan Lima. The main objective of the research was to study the factors which influence decisions concerning crops and plants in the organic home gardens of the urban farmers. The thesis recognizes that a key role in the establishment, maintenance, and development of the urban organic home gardens is played by the women. It is they who are mostly responsible for all activities involved in cultivating the home gardens and keeping guinea pigs, as the men work outside the house in another kind of job. In addition, the thesis underlines the important role that women play in the dissemination of knowledge, enthusiasm, and use of species, in their roles as female urban farmers and mothers.

Hovorka, A. (2005) 'The (re)production of gendered positionality in Botswana's commercial urban agriculture sector', *Annals of the Association of American Geographers* 95 (2): 294–313.

Urban agriculture studies tend to aggregate data such that they mask differential experiences of men and women farmers, and fail to explain adequately the influence of location and human-environment relations on production systems. People's ability to create productive and sustainable urban agricultural systems is premised on who they are, where they are located, and how they interact with the environment in that location. This article presents an empirical investigation of the effects of gender on commercial urban agriculture in Greater Gaborone, Botswana. It employs a conceptual framework that bridges socio-spatial and human-environment traditions in geography and highlights gendered environments to facilitate this convergence. The investigation reveals that gender clearly influences the quantities and types of foodstuff produced for the urban market. Gender matters because men and women enter into agricultural production and participate within this urban economic sector, on unequal terms based on socio-economic status, location, and interactions with the environment. If urban agriculture is to contribute to food security and economic growth, as well as urban sustainability more generally, gender relations of power, as produced and reproduced through socio-spatial and human-environment relations, must inform our understanding of this phenomenon.

Hovorka, A. (2006) 'The No. 1 Ladies' poultry farm: a feminist political ecology of urban agriculture in Botswana', *Gender, Place and Culture* 13 (3): 207–225.

The research draws on a feminist political-ecology perspective to demonstrate that agrarian restructuring and rural–urban transformation in Botswana offers women opportunities to renegotiate their marginalized positionality within the commercial urban agricultural sector in Greater Gaborone. Men and women participate in equal numbers, and both perceive this sector as offering them new and accessible avenues for economic and social advancement. Although there is continuity of women's social and economic disadvantage relative to men from rural to urban contexts, women are actively making claims on land and capitalizing on their traditional roles and responsibilities associated with poultry production. This negotiation of continuity and change in gendered positionality reflects and indeed suggests positive changes for women in urban Botswana, pointing specifically to the transformatory potential of urban agriculture despite constraints at the sectoral level. The research highlights the ways in which women are (re)defining their constraints and seeking out alternative opportunities for empowerment and action. To this end, gender remains an integral part of and key element to understanding agrarian restructuring and rural–urban transformation in Botswana.

Hynes, P. H. (1996) 'Why so many women?', in P. H. Hynes (ed.), *A Patch of Eden; America's Inner-City Gardeners*, Chelsea Green Publishing Company, Vermont.
In this chapter, the author considers whether there is a larger, broader history of women and gardens which underlies the community garden movement. What meanings – personal, social and political – have gardens held for women of different classes and ethnicities? The practice of gardening has been stratified by wealth and by gender. Millions of subsistence, kitchen, and medicinal gardens planted and tended by women have been central to household economy, village health, and local biodiversity. The garden has been a source of natural beauty for the urban and rural poor. Yet the value of this work is generally not counted in the economy, because it is unpaid and not market-based; nor is it recorded in environmental history, because it is considered insignificant work of many 'ordinary' women. Women's contribution through gardening to the world's food supply is chronically underestimated. The author draws a parallel between inner cities in the United States and the Third World: the urban community garden has the potential to feed households and generate local cottage industry, restore a measure of community life, and recycle organic wastes.

Ishani, Z. and D. Lamba (2001) 'Applications of Methods and Instruments in Urban Agriculture Research: Experiences from Kenya and Tanzania', paper for the workshop 'Appropriate Methodologies for Urban Agriculture', October 2001, Nairobi, Kenya. Available from www.ruaf.org.
The paper deals with methodology applied in two studies of urban agriculture in Kenya and Tanzania, conducted by, and in collaboration with, Mazingira Institute. The first study, entitled 'Urban Food Production and the Cooking Fuel Situation in Urban Kenya', was published by the Institute in 1987. The second study, 'Gender and Urban Agriculture and its Implication for Family Welfare and the Environment in Dar Es Salaam, Tanzania', was completed in 2000 (not published). The Kenyan study comprised six cities and towns, covering the various agro-climatic zones. It analysed the patterns of food and fuel production and consumption by the urban households in Kenya by considering the socio-economic characteristics of the sample population, crop production, livestock production, and fuel. In addition, it raised issues for consideration by policy makers. The Tanzanian study, ' Tanzania – Gender and Urban Agriculture: Cattle Raising and its Implication for Family Welfare and the Environment in Dar es Salaam, Tanzania', was at the city level. It analysed gender roles in cattle raising in the district of Kinondoni in Dar Es Salaam.

Ishani, Z. and D. Lamba (eds.) (2001) *Emerging African Perspectives on Gender in Urbanization; African Research on Gender, Urbanization and Environment*, Kenya Litho Limited, Nairobi, Kenya.
This book contains five studies, from Kenya, Uganda, Ghana, Nigeria, and Egypt. It addresses the themes of women's access to and control of housing, or

their role in housing production, and women's access to and control of other resources in the urbanization process.

Ishani, Z., K. Gathuru, and D. Lamba (2002) *Scoping Study on Interactions Between Gender Relations and Livestock Keeping in Kisumu,* **Natural Resources International, Chatham, UK.**
The focus of the study is the improvement of gender-biased division of labour, inequality between males and females in power and resources, and gender biases in rights and entitlements to increased productivity, remuneration, and development of women livestock keepers. The study was conducted in Kisumu, an urban area located on the shores of Lake Victoria, in Kenya.

Jaiyebo, O. (2001) 'Women and Household Sustenance: Livelihood Impacts from Development Changes and Strategies for Survival in the Peri-Urban Areas', paper prepared for the DPU International Conference: 'Rural–Urban Encounters: Managing the Environment of the Peri-Urban Interface', London, 9–10 November 2001. DPU, London.
In Nigeria, the woman is regarded as the homebuilder, having a pivotal role in the society. This study focuses on women and provides insight into the women's perceptions of their own livelihood status and the impacts of developmental changes and strategies employed for survival in the peri-urban areas. The study area is Ibadan, Nigeria, the choice of which stems from its geographical location, socio-economic heterogeneity, and population size. Ibadan, the capital city of Oyo state, is located in the humid south-western zone of Nigeria and accommodates more than half of the total population of the State. A random sample of 96 women was obtained from two locations, north and south of the main city. The study reveals that the major developmental changes which have had the greatest impact on the women's livelihoods are changes in land use as a result of urban growth, and decreased purchasing power resulting from an ailing national economy. Strategies employed for survival include income diversification and the involvement of children in income generation.

Kiguli, J. (2003) 'Mushroom cultivation in urban Kampala, Uganda', *Urban Agriculture Magazine no. 10 Micro-Technologies for Urban Agriculture,* **pp. 20–21. Available from www.ruaf.org.**
Urban agriculture in Uganda is mainly viewed as a household survival strategy, in the context of rising poverty. Mushroom cultivation is a recent trend in Kampala. This paper explores the rationale for growing mushrooms, with a focus on gender participation, as more women are involved in mushroom cultivation than men, and the necessary conditions for success.

Kreinecker, P. (2000) 'La Paz: urban agriculture in harsh ecological conditions', in N. Bakker et al. (eds.) *Growing Cities, Growing Food: Urban Agriculture on the Policy Agenda,* **pp. 391–411, DSE, Feldafing, Germany.**
The climate in La Paz enhances the development of adapted techniques for urban agriculture. Officials tend not to see urban agriculture, although

in fact it is everywhere. Urban agriculture is a survival strategy for socially marginalized people because it fits well in their economy, which is based on social relations. Several urban farming systems can be found, of which private home gardens and communal gardens are more important. Land titles are unclear, and little capital is used. Farmers are organized in informal and formal groups and networks. Women play a central role in farming, and urban farming contributes to women's independence. Urban agriculture contributes little to food-energy supply but increases the diversity of food consumption. Many factors hamper the development of urban agriculture: among others the ecological conditions, cultural heterogeneity, and land-tenure situation. A future strategy needs to emphasize existing structures and socially accepted Andean varieties to improve the situation of marginalized people.

Kusakabe, K., C. Monnyrath, C. Sopheap, and T. C. Chham (2001) *Social Capital of Women Micro-Vendors in Phnom Penh (Cambodia) Markets: A Study of Vendors' Association*, UMP–Asia Occasional Papers no. 53, United Nations Urban Management Programme, Regional Office for Asia and the Pacific, Bangkok.

In Cambodia, because of the long history of civil strife, it is said that mutual trust has been destroyed, and that because of people's negative experience of 'co-operatives', there is a stigma attached to organizing and working together and sharing information together. Efforts to revive social capital and initiatives to organize people are taking place in Cambodia. One such initiative is the micro-vendors' association in Phnom Penh markets. With the support of a local NGO, Urban Sector Group (USG), and The Asia Foundation, micro-vendors in public marketplaces are forming an association under the Women's Economic and Legal Rights Project (WELR). This study examines how being a member of the vendors' association influenced the members' sense of mutual trust and confidence in making changes in the society, and how such trust in turn influenced their gender norms and ideologies, and how they see their own positions in the households. The study examines this collective process and shows how balance is struck in the micro-vendors' association in Phnom Penh. Through their collective action against the authorities, are they able to overcome their existing economic and social subordination? Or are they encouraging their members to conform to the existing norms through building social capital?

Lee-Smith, D. (1997) 'My House is my Husband – A Kenyan Study of Women's Access to Land and Housing', Thesis 8, Department of Architecture and Development Studies, School of Architecture, Lund University of Technology, Lund University, Sweden.

This thesis explores women's access to property in Kenya. Part 1 gives the Kenyan background and the theoretical and methodological approach. Part 2 presents findings about the social construct of the gender contract, and elaborates these in the context of women's subsistence work, colonization, and women's formation of organizations. Part 3 consists of the conclusions, which include a model of change in gender relations at the micro-level. This is a feminist work in the field of women and housing, using gender-contract theory. Empirical data were gathered through random household sample surveys and in-depth interviews in the peasant, plantation, and urban poor areas during 1990–92. The issue is how women in Kenya get access to property (land and housing), and specifically the social mechanisms that govern men's and women's relationships with each other and with property. The gender contract is identified as the social mechanism, and two forms of the contract are identified in Kenya: the subsistence gender contract and the market gender contract. The Kenyan gender contracts delineate a power relationship in which women's lack of access to property keeps them subordinate to men and requires them to provide subsistence. Women's actions are based on their strategies for improving their lives. Collective action is one such strategy with important implications for housing. A gendered housing policy is needed which recognizes women's access to property as a human right and which builds upon their proven housing-production capability. It should support the values and objectives of women's groups, namely the provision of subsistence to their families.

Levin, C. E., D. G. Maxwell, M. Armar-Klemesu, T. Ruel, S. S. Morris, and C. Ahiadeke (1999) *Working Women in an Urban Setting; Traders, Vendors, and Security in Accra*, **FCND Discussion Paper no 66, IFPRI, Washington.**
Data collected from a 1997 household survey carried out in Accra, Ghana, are used to assess the crucial role that women play as income earners and in securing access to food in urban areas. One third of the households surveyed are headed by women. For all households, women's labour-force participation is high, with 75 per cent of all households having at least one working woman member. The high number of female-headed households and the large proportion of working women in the sample provide a good backdrop for looking at how women earn and spend income differently from men in an urban area. Livelihood strategies for both men and women are predominantly labour-based and dependent on social networks. For all households in the sample, food is still the single most important item in the total budget. Yet there are important and striking differences between men's and women's livelihoods and expenditure patterns. Compared with men, women are less likely to be employed as wage earners, and more likely to work as street food vendors or petty traders. Women earn lower incomes, but tend to allocate more of their budget to basic goods for themselves and their children, while men spend more on entertainment for themselves only. Despite lower incomes and additional demands on their time as housewives and mothers, female-headed households, petty traders, and street food vendors have the largest percentage of food-secure households. Women may be achieving household food security, but at what cost? This paper explores differences in income, expenditure, and consumption patterns in an effort to answer this question and suggests ways

in which urban planners and policymakers can address special concerns of working women in urban areas.

Made, P. (2000) 'A field of her own: women and land rights in Zimbabwe', in J. Mirsky and M. Radlett (eds.), *No Paradise Yet: The World's Women Face the New Century*, pp. 81–100, Panos/Zed Books, London.
Land tenure, a prime issue for women in urban and rural agriculture, is addressed here from the point of view of women farmers in the capital of a southern Africa country. More than 70 per cent of Zimbabwe's agricultural work force is female; an equal share of urban agriculture is managed by women. Women cannot inherit land, either from their fathers or husbands. The author concludes that economic development will be slow, with only half the population empowered by legal access to land.

Maxwell, D. G. (1993) 'Land Access And Household Logic: Urban Farming In Kampala', Makerere Institute of Social Research, Kampala (Uganda).
The objective of this paper is to evaluate the various claims made about urban agriculture in Kampala, Uganda. This includes reviewing the limited literature on the importance of this activity in Kampala; attempting to assess direct evidence on nutritional status; examining the means of access to land; and understanding the logic of various households involved in urban food production. Gender analysis is applied to examinations of land access and household logic. The paper contends that commercial producers may be either men or women, and male and female household members may collaborate in business ventures. In production for food security, it is common for senior women in the household to gain access to land through borrowing, renting, squatting, or purchasing use rights. Urban agriculture contributes to household food security and enables women to use cash income on items other than the purchase of food. Urban agriculture often becomes a survival strategy for low-income female-headed households, widows, and families suddenly abandoned by a primary wage earner.

Maxwell, D. G. (1995) 'Alternative food security strategy: a household analysis of urban agriculture In Kampala', *World Development* 23 (10): 1669–1681.
The author contends that little is understood about the forces behind urban farming or its impact at the household level. Intra-household and gender relations, as well as declining wages and economic informalization, are all important to an understanding of urban farming. The paper presents an overview of the household analysis of urban farming, as based on research carried out in Kampala, Uganda, between November 1992 and October 1993. This includes a discussion of intra-household dynamics, access to land, and a comparison of food security and nutritional status in farming and non-farming households. Underlying the evidence gathered is the fact that urban farming is almost completely under the control of women, who bear responsibility for the provision of food. Discussion also centres on the implications of urban

farming and possible policy alternatives. The author suggests that programmes promoting urban farming should give priority to low-income, female-headed households for equity reasons. Such programmes could be established through women's organizations, such as informal savings and credit groups, and should be closely monitored, both in terms of the direct effect on women's income and in terms of food security and child nutritional status.

Mbiba, B. M. (1995) *Urban Agriculture in Zimbabwe: Implications for Urban Management and Poverty*, Avebury, Aldershot (UK).
This book addresses the phenomenon of urban agriculture in Zimbabwe. While it acknowledges that the activity is a significant source of food and income for the urban poor, the book draws attention to the development conflicts raised by the activity. It attempts to place urban agriculture within the context of urban economy, the environment, institutional concerns, gender, and urban poverty. Evidence presented confirms the role of urban agriculture for employment of women and children. A review of gender dimensions of informal urban cultivation highlights the needs, problems, and experiences of women's double burden of production and reproduction. Men's social and economic motivations for urban cultivation activities are also noted. Issues of 'gate-keeping', female landlordism, and decision making are discussed in terms of gender dynamics. It is noted that women are not a homogeneous group, hence the need to revise generalizations about poor women and extend research issues to high-income groups. The author contends that urban cultivation should form only one part of a strategy designed to improve the position of urban women, for it does not tackle the problems of women's access to education, skills, wage labour, and self-employment. Based on on-going research, the book demonstrates the potential for urban agriculture as part of the urban economy, but argues that the urban poor, including women-headed households, are not major beneficiaries of the activity.

Mianda, G. (1996) 'Women and garden produce of Kinshasa; the difficult quest for autonomy', in P. Ghorayshi and C. Belanger (eds.), *Women, Work and Gender Relations in Developing Countries*, pp. 91–101, Greenwood Press, Westport Connecticut.
This chapter focuses on women and the organization of garden production in Kinshasa, Zaire. Gender relations are viewed as power relations, whereby garden production becomes a power game played between women producers and husbands. Women undertake garden production to acquire economic independence from their husbands, as well as to meet the financial needs of their families. Through various strategies and tactics, women manipulate the sexual division of labour, despite its constraints, for their own benefit. They gain advantage over their husbands for initiating the production. In order to claim total autonomy, women gardeners establish control over the management, marketing, and revenue derived from production. They thus modify, at the level of garden production and at all levels of power related to this production, the traditional image of women.

Mitullah, W. (1991) 'Hawking as a survival strategy for the urban poor In Nairobi: the case of women', *Environment And Urbanization*, pp. 13–22.
This paper considers the role of hawking as a survival strategy of low-income women and their families. Drawing on findings of a study carried out in Nairobi, Kenya in 1987/1988, it presents information on the importance of such activities to household income, offers a brief history of the trade, and discusses the concept and nature of the informal sector.

Mudimu, G. D. (1996) 'Urban agricultural activities and women's strategies in sustaining family livelihoods in Harare, Zimbabwe'. *Singapore Journal of Tropical Geography* 17 (2): 179–194.
Though a widespread practice, urban agriculture is not planned for or supported by urban planners and managers as a legitimate form of urban land use in Harare, Zimbabwe. As women are the main participants in urban agriculture, their activities come into direct conflict with planning provisions for urban space. This study examines the role of women in urban agriculture, and views and perceptions of the use of urban space for agricultural activities in Harare. The large presence of women cultivators is indicative of women's reduced opportunities for formal employment in urban areas, and the perceived notion of women having primary responsibility for providing family sustenance. While women were the predominant 'owners' of the plots, the men in the fields were primarily cultivating land on behalf of their spouses or as hired hands. A significant proportion of female respondents were heads-of-households, and urban agriculture is practised by women of all socio-economic classes. Those women in professional occupations tended to hire contract workers for their plots. Data collected also support the fact that larger households are more likely to be under pressure to supplement their food sources and incomes via urban agriculture as a survival strategy. Urban agriculture offers women the opportunity to enhance their economic power within the household, although not without negotiating with their spouses, and increases their ability to provide food for family consumption. The study identifies two immediate issues requiring the attention of policy makers. First, the potential increased competition for land, as reduced employment opportunities push more men to pursue urban agriculture activities, may pose a threat to women's future access to land. Second, current urban planning concepts must be reviewed so that a clear policy on urban agriculture is formulated, in order to support women's struggle for sustaining family livelihoods in the urban economy.

Ofei-Aboagye, E. (1996) 'Gender Critique on Urban Agriculture: Food Security and Nutritional Status in Greater Accra (Ghana)', Report for IDRC Project No. 96-0013 003149, IDRC, Ottawa.
This proposal review is based in an IDRC study on food security and nutritional status in Greater Accra, Ghana. It highlights key issues in gender considerations for research on urban agriculture. It focuses on the reviewer's expectations regarding objectives of the study and suggestions for improvement along gender lines. The reviewer notes that resources of land, water, credit, information,

and other inputs need to be considered from a gender perspective. Gender proportions of poverty and its influencing factors should inform the design of conceptual framework and proposed methodology (with qualitative methods facilitating particularly rich gender enquiry). The reviewer contends that the participation of a female lead-researcher does not necessarily guarantee incorporation of a gender perspective. Use and involvement of policy makers, the National Council on Women and Development, women's organizations and the media are critical at various stages of the research.

O'Reilly, C. and A. Gordon (1995) *Survival Strategies of Poor Women in Urban Africa: The Case of Zambia*, NRI Socio-economic series 10, Natural Resources Institute, Chatham, UK.
This research was carried out with the aim of providing better definition and targeting of project interventions, recognising the importance of natural resource-related activities (food production, processing, and trade) to poor women in urban areas. The research paper discusses ways in which women's livelihood strategies could be strengthened, emphasizing the need for multi-dimensional initiatives. Credit (particularly for the poorest), confidence building, access to information, and specific income-generating activities (urban agriculture, skills training and food processing, strengthening trading links with rural areas) are highlighted as major requirements.

Rakodi, C. (1988) 'Urban agriculture: research questions and Zambian evidence', *The Journal of Modern African Studies*, 26 (3): 495–515.
This article explores the forgotten or ignored area of food-crop cultivation in urban areas in the 1980s. The author contends that the first stage in studying any neglected area is to review existing evidence and policy, in this case from Zambia, to reveal gaps and suggest avenues for further enquiry, policy formulation, and experimentation. The author situates urban agriculture within a wider framework of the gender division of labour, specifically the economic activities of women. Food production in Zambian cities is predominantly a women's activity, determined by the size of household, income per capita, stability of urban residence, and the availability of land for cultivation around the house and/or within reasonable walking distance. A strategy to increase the household production of fruit and vegetables for consumption and sale must be examined in the context of household decision making, and especially the labour time available to women. Women's response to opportunities to grow more food will depend on the extent to which they make decisions about cultivation, the use or sale of produce, and the distribution of benefits within the household. More detailed evidence from urban agriculture projects and wider implications of such a policy must be assessed before more widespread cultivation is advocated. This includes assessing the benefits to households, and especially to women, compared with alternative economic opportunities which might be made available by other initiatives.

Methodologies and tools for gender mainstreaming in urban agriculture

Ahlers, R. (2005) 'From Increasing Participation to Establishing Rights: Evading the Gender Question in Water Resource Management', EMPOWERS Regional Symposium: End-Users Ownership and Involvement in IWRM, 13–17 November 2005, Cairo, Egypt.

Gender-sensitive policy in water-resources management has shifted from an emphasis on recognizing women as water users and increasing their participation in decision-making bodies, to a focus on securing formal rights to water. Increased involvement of women in male-dominated institutions was perceived to address gender inequality in access to and control over water resources. In recent years, rights-based approaches have gained ground in response to increasing private-sector presence in the water sector and the policy of water supply.

Amaratunga, C. (2005) 'Creating Learning Cultures for Gender Mainstreaming; Strategic Approaches for Impact Assessment of Multi-sectoral Approaches; the Case of HIV/AIDS in Subsistence Agriculture and Artisanal Fisheries', Institute of Population Health, Faculty of Medicine, Dept. of Epidemiology and Community Medicine, University of Ottawa, Ottawa.

The inclusion of gender mainstreaming and multi-sectoral frameworks in agricultural planning can be extremely helpful in understanding and measuring how development interventions affect women and men, girls and boys differently. This case study of gender mainstreaming in the health sector provides a useful example of how a gender analysis can serve as a critical tool for social change. Not only does a gender-mainstreaming approach help to measure empowerment differentials between the sexes, it is also useful for assessing stakeholder assets and needs. This, in itself, provides a practical starting point in the programme-planning process. This paper provides an overview of strategic approaches for impact assessment of multi-sectoral approaches for gender mainstreaming in agriculture. It includes examples and illustrations of gender-audit guidelines, checklists, and programme interventions.

Derbyshire, H. (2002) *Gender Manual: A Practical Guide for Development Policy Makers and Practitioners*, DFID, London.

This gender manual is designed to help non-gender specialists to recognize and address gender issues in their work. The intention is to demystify gender, make the concept and practice of gender 'mainstreaming' accessible to a wide audience, and clarify when to call in specialist help. It focuses on the processes of gender mainstreaming which are similar in all sectoral and regional contexts, and also similar, in some instances, to other processes of social development and organizational change.

De Zeeuw, H. and J. Wilbers (2004) *PRA Tools for studying urban agriculture and gender*, RUAF, Leusden, The Netherlands. Available from www.ruaf.org

This document has been prepared to facilitate the gender case studies to be undertaken by the regional RUAF centres as a training exercise and an input to the gender expert consultation. After giving an introduction to PRA, it presents a range of PRA tools which can be used to investigate various important issues when analysing gender and urban agriculture in local situations, such as the access to and control over resources, decision-making power, division of labour, external factors that influence gender, and constraints, problems, and opportunities.

Feldstein, H. S. and J. Jiggins (eds.) (1994) *Tools For The Field: Methodologies Handbook for Gender Analysis in Agriculture*, Kumarian Press, West Hartford, Connecticut.

From agricultural production to post-harvest activities, this handbook offers a practical set of tools for the novice or experienced professional working on gender analysis in agriculture. It provides real-life examples of how to assemble and use all the research tools you need to collect gender-sensitive data in a timely and cost-effective way. Covering Latin America, Asia, and Africa, the handbook consists of 39 original cases by contributors from the North and the South. The cases illustrate a range of techniques from making gender-sensitive interview guides to ensuring participatory rural appraisal methods that include a gender dimension. Contents include learning about the system and initial diagnosis; research planning; on-farm experimentation and trial assessment; on-going diagnosis and special studies; extension, training and institutionalization.

Gabel, S. (2001) 'Methodological Reflections on Using Participatory and Action Oriented Research with Women Farmers in Harare', paper for the 'Appropriate Methodologies for Urban Agriculture' workshop, 1–5 October 2001, Nairobi, Kenya. Available from www.ruaf.org.

The task of conducting qualitative, participatory and/or action research in a setting that is far removed from one's own can be somewhat daunting, especially for a student organizing her field work independently for the first time. The author, in doing her Master's study with the Municipal Development Programme in Harare, Zimbabwe, experimented with participatory and action-oriented research from many academics and practitioners of participatory development. In this paper she gives the results and reflects upon the use of these methods.

Gender and Water Alliance (GWA) and UNDP (2006) *Resource Guide. Gender in Water Management*, GWA, Dieren, The Netherlands. Available from www.genderandwater.org/page/2414.

This resource guide is intended as a reference document to assist water and gender practitioners and professionals, as well as persons responsible for gender mainstreaming, and anybody else who is interested in the water sector. The Guide was developed in response to an identified need for information on gender mainstreaming in water resource management. While considerable information exists, it is dispersed among different institutions and organizations, making it difficult to know where to get specific resources for particular aspects of gender mainstreaming in the water sector. This Guide supports the efforts of those trying to mainstream gender in their programmes and projects, and those seeking to improve their knowledge and skills in gender and water resource management.

Guijt, I. (1994) 'Making a difference: integrating gender analysis into PRA training', *PLA Notes*, **IIED, London.**
This article argues that only in few instances does PRA training focus on the 'who?', and that it rarely explores the issue of social differences based on gender. It recognizes, however, a slow increase in recent years in the number of attempts to improve the practice of PRA by allowing gender issues to shape both the practical work and the analysis. It argues: 'If gender is to become a concrete and meaningful concept for PRA trainees, then those who are learning to work with PRA will need to become aware of gender before any fieldwork takes place.' The article on the other hand also provides some words of caution, warning against the 'gender average' (assuming harmony and homogeneity among women or men) and possible confrontations between local women and men over issues of power and autonomy.

Guijt, I. (1995) *Questions of Difference: PRA, Gender and Environment, A Training Video*, **IIED, London.**
This two-hour video with provoking images can be used to stimulate discussion and to lead into class-based exercises. It gives a summary of the key elements for using PRA to understand gender and environment. The video is structured in thematic segments of 2–14 minutes, from which users can select those of interest or for specific training. The three case studies show workshop participants using PRA methods to explore issues relating to gender and the environment.

Hill, C. L. M. (2005) *Making the Links: Addressing HIV/AIDS and Gender Equality in Food Security and Rural Livelihoods Programming*, **A toolkit for CIDA staff working on initiatives related to food security and rural livelihoods, Canadian International Development Agency (CIDA).**
This toolkit was designed to enable CIDA staff to address HIV/AIDS and gender equality in food security and rural livelihoods programming. While it is primarily geared towards the CIDA programme cycle, the toolkit is also useful to a wider audience, including NGOs. It includes seven guidance sheets, providing entry points for addressing HIV/AIDS and equality in the programme cycle in a gender-sensitive manner, and suggests how to consider gender-equality results in performance assessment and in results-oriented logical framework analysis.

Hovorka, A. J. (1998) 'Gender Resources for Urban Agriculture Research: Methodology, Directory and Annotated Bibliography, CFP series report # 26, IDRC, Ottawa.
Although there is a growing interest in the mainstreaming of gender in development research, there is also a general lack of understanding of how this type of analysis can be applied. The purpose of the publication is to provide researchers with simple and systematic methodological tools for practical application of gender analysis within urban agriculture. It was developed primarily for the Cities Feeding People team members, but can be applied by anyone doing a similar type of research. The methodology covers all stages of a research project: (1) proposal; (2) data collection, interpretation, and analysis; (3) monitoring and evaluation. Still, it is not intended as a blueprint, but rather gives guidelines. Included are a directory of gender resource persons for urban agriculture research and an annotated bibliography on gender and urban agriculture. It is a very useful publication, both with regard to research and development methodology and for practical information.

Morris, P. T., S. Kindervatter, and A. M. S. Woods (2003) *The Gender Audit Questionnaire Handbook*, American Council for Voluntary International Action (InterAction) Commission on the Advancement of Women (CAW), Washington.
This handbook presents the rationale behind mainstreaming gender in an organization's culture and structure. The gender audit is a tool that is used in organizations to identify staff perceptions of how gender issues are addressed in their programme portfolio and internal organizational process. It has a list of questions and describes strategies for sampling, data collection, analysis, and results presentation. It also illustrates how to use the gender-audit results in action planning.

National Institute of Urban Affairs (NIUA) (undated) *Working With the Urban Poor: A Manual for Trainers*, NIUA, New Delhi.
This manual aims to build the skills of trainers in training programme partners at all levels – community, city, district, and state – to work towards the empowerment of women and the achievement of national social-sector goals.

Njenga, M. E., N. Karanja, C. Kabiru, P, Munyao, G. Kironchi, K. Gathuru, and G. Prain (2007) *Mainstreaming Gender Analysis in the Research Process of the International Potato Centre (CIP)*, Urban Harvest, CIP, and PRGA. Available from www.prga.org
Urban Harvest, the CGIAR system-wide initiative on urban and peri-urban agriculture, which is convened by the International Potato Center (CIP), has incorporated a strong gender component within its research approach. The project reported here carries forward various initiatives, among them a small project to help to mainstream gender in CIP's research programme, whose overall objective was to initiate a process of gender mainstreaming in CIP

through institutional mechanisms, and by piloting the approach in research projects in sub-Saharan Africa. In addition, the report discusses the strengths and challenges encountered in the field application of the PR and GA tools and approaches in both urban and rural farming and how to build on the strengths and address the challenges.

Norwegian Agency for Development Cooperation (NORAD) (1999) *Handbook in Gender and Empowerment Assessment*, **NORAD, Norway.**
As NORAD's role is to assess project proposals presented by partners, and not to participate in project formulation and planning, this manual does not aim to conduct full-scale gender and empowerment analysis of projects but rather assists the user to identify the need for such analysis. It also provides guidelines for NORAD's requests for gender analysis at the various stages of the project cycle: project appraisal, appropriations, progress reporting, project reviews, and evaluations.

Ostergaard, L. (1992) *Gender and Development: A Practical Guide*, **Routledge, London.**
People concerned with development work have awoken in the past two decades to the urgent need to involve women as active participants and beneficiaries of their programmes and projects. But the transition from idea to reality is occurring slowly. As a result, too many development efforts are still being badly designed, failing to achieve their goals, and having a negative impact on women. In this book, eight experts explain the importance of gender awareness and illustrate how gender relations vary from culture to culture. They argue that every development effort should be preceded by a gender analysis and they show in practical terms how this leads to more success in both the long and the short terms.

Oxfam GB (2006) *A Change in Thinking; Gender Budgeting – Now's the Time*, **Oxfam GB, UK Poverty Programme, Oxford, UK.**
This DVD includes experiences in gender budgeting across the United Kingdom. Oxfam and its partners believe that gender budgeting can help local and national governments to improve the lives of both women and men. It includes information on the concept of gender budgeting, the view from a local-government perspective, and a range of case studies.

Palacios, P. (2003) 'Why and how should a gender perspective be included in participatory processes in urban agriculture', in *Research and Intervention in Urban Agriculture: Methodologies and Instruments for Analysis*, **Module 3 of the regional training course on urban agriculture in Quito-Ecuador, IDRC / UMP–LAC / ETC.**
This document analyses the reasons why it is important to include gender equality in participatory processes concerning urban agriculture and explains how this can be achieved, addressing the distinction between strategic and practical needs, as well as the development of specific and affirmative actions.

Quisumbing, A. R. and B. McClafferty (2006) *Gender and Development: Bridging the Gap between Research and Action*, IFPRI, Washington.
This guide offers a non-technical presentation of research findings from IFPRI's multi-country research programme on gender and intra-household issues, along with implications and key questions for integrating gender research findings into project cycle and policy decision-making processes. The volume draws on work undertaken by IFPRI and its collaborators in developing and developed countries since the early 1990s. It presents empirical evidence – based on IFPRI's field research, using both quantitative and qualitative techniques – on the ways in which gender and intra-household issues affect the success of development interventions, and then shows readers how to incorporate the findings effectively into development programmes. The guide – its findings, format, and presentation – has been field-tested in workshops with practitioners in Kathmandu, Nairobi, Guatemala City, and Washington, DC.

Quisumbing, A. R. and B. McClafferty (2006) *Food Security in Practice; Using Gender Research in Development*, IFPRI, Washington.
This guide bridges the gap between research and practice by providing up-to-date, relevant information on why and how gender issues, when taken into account, can improve the design, implementation, and effectiveness of development projects and policies. It presents key research findings from IFPRI's gender and intra-household programme in the framework of project and policy cycles. The authors took the additional step of field-testing the guide among practitioners in Africa, Asia, and Latin America to see whether the findings were relevant outside the study countries. Finally, they conducted a workshop with US-based practitioners and policy makers to see how the findings related to the policy cycle. The guide records the insights, comments, and suggestions of the ultimate users of this research.

Van Dam, H., A. Khadar, and M. Valk (2000) *Institutionalising Gender Equality; Commitment, Policy and Practice, A Global Source Book*, Critical Reviews and Annotated Bibliographies Series: Gender, Society and Development, KIT, Amsterdam.
Encouraged by the growing international women's movement and recent world conferences on women, many governments, NGOs, and other development organizations have made commitments to the goal of promoting equality between women and men. The majority have taken steps towards turning commitments into action by including a concern for women and gender in planning, policy, and, to some degree, practice. As a result, a wide array of measures, approaches, and practical actions intended to introduce a gender perspective and bring about change in organizations is beginning to emerge. However, it has become clear that integrating gender issues is a complex and contested process, apt to encounter resistance from both organizations and individuals.

Wassenaar, N. (2006) *Incorporating gender into your NGO*. Available from www.Networklearning.org
Often NGO/CBOs have difficulties knowing how to incorporate gender into all aspects of their organization and thus ensure gender mainstreaming, balance, and equality within the organization and its activities. This manual assists with this process, starting with an explanation of basic concepts and definitions of gender, followed by 'what to do and how to do it', both within and outside an organization, in order to scan all aspects with a gender-sensitive eye. In the process it explains the need for change and the constraints that must be addressed; it offers an awareness of wrong concepts and influencing factors. Several steps are needed in the process of 'engendering' an organization, and these steps are described one by one, from analysis of the organization, development of an action plan, implementation, and then monitoring and evaluation. Finally, related resources and websites are given; links to tools, such as gender checklists for the project cycle, are provided.

Zwarteveen, M. (2006) *Effective Gender Mainstreaming in Water Management for Sustainable Livelihoods: From Guidelines to Practice*, Both Ends, Amsterdam.
This working paper presents the main findings of a joint project of the Gender and Water Alliance and Both Ends, namely the comprehensive assessment of water management in agriculture. The authors analyse the difficulties of mainstreaming gender in water management. They develop a minimum agenda with practical and realistic recommendations to practitioners, policy makers, researchers, and gender specialists working in the field of water and agriculture, to help them to genuinely and effectively address gender differences and inequities in policy and research in this field.

Websites

www.fao.org/gender
This is a Gender and Food Security site of the Food and Agriculture Organization of the United Nations. It contains articles on projects and programmes, and a thorough set of statistics and other information is also available.

www.genderdiversity.cgiar.org
The purpose of the gender and diversity (G&D) programme of the Consultative Group on International Agricultural Research (CGIAR) is to help the CGIAR centres to make the most of the rich diversity of their staff to increase research and management excellence. It promotes such activities as diversity-positive recruitment, international teamwork, cross-cultural communications, and advancement of women. The website addresses issues of inclusion, opportunity, dignity, and well-being. It also has a database of women scientists and professionals around the globe, resource centres, and newsletters.

www.genderandwater.org

The mission of the Gender and Water Alliance (GWA) is to promote women's and men's equitable access to and management of safe and adequate water, for domestic supply, sanitation, food security, and environmental sustainability. GWA believes that equitable access to and control over water is a basic right for all, as well as a critical factor in promoting poverty eradication and sustainability. The website of GWA serves as a platform for the alliance and network and provides key documents on gender mainstreaming of the water-resources management sector.

http://www.networklearning.org

On this website, many manuals, field guides, and training courses have been made available to groups who need them. The site stimulates the free distribution of information, in an attempt to encourage global learning processes. It also includes a section dedicated to gender issues.

http://topics.developmentgateway.org/gender

This website is part of the dgCommunities, an interactive place where knowledge resources focused on development can be found and where members of the communities can share their own work, participate in discussions, find people with similar interests, etc. The gender community specializes in Gender Mainstreaming links to documents, events, websites, etc. as well as feature articles, special reports, online discussion forums, and periodic online events.

http://www.prgaprogram.org

The CGIAR System-wide Program on Participatory Research and Gender Analysis (PRGA) develops and promotes methods and organizational approaches for gender-sensitive participatory research on plant breeding and on management of crops and natural resources. The programme's website contains links to PRGA publications and other resources such as the Newsletter.

http://www.siyanda.org

Siyanda is an on-line database of gender and development materials from around the world. It is also an interactive space where gender practitioners can share ideas, experiences, and resources.

List of contributors

Editorial Committee

Alice J Hovorka is an Associate Professor in Geography at the University of Guelph, Canada. Her research programme and broad area of expertise concern urbanization in Southern Africa, with a specific focus on gender and urban agriculture. Alice has worked in collaboration with Cities Feeding People at the International Development Research Centre, and she serves as a member of the Gender Advisory Group for the Cities Farming for the Future Programme of the RUAF Foundation. She has an on-going collaborative relationship with the government of Botswana, through which she serves as an adviser on urban agriculture and gender issues. She co-facilitated the launch of the Botswana Policy Initiative on (Peri-) Urban Agriculture in June 2004. Her email address is ahovorka@uoguelph.ca

Henk de Zeeuw is the Director of the RUAF Foundation (International Network of Resource Centres on Urban Agriculture and Food Security), ETC Urban Agriculture, The Netherlands. He is an agricultural sociologist with more than thirty years of experience in agricultural development, specializing in participatory agricultural research and extension. He was among the first in the Netherlands to focus on the role of women in development, and he has conducted training courses on gender in agriculture in the Netherlands and various developing countries. He is the co-ordinator of the Cities Farming for the Future programme and as such also responsible for the gender-mainstreaming component of this programme. His email address is h.dezeeuw@etcnl.nl

Mary Njenga is an agricultural research officer at the International Potato Center (CIP)–Urban Harvest, Kenya. She specializes in participatory research methods as applied to rural and urban farming communities. Since joining Urban Harvest in 2002, she has played a significant role in the mainstreaming of gender in research for development projects initiated by Urban Harvest and the CIP, and in the organizational culture of these two institutions. Mary has written and co-authored a number of publications on gender and urban agriculture, exploring the research, development, and policy issues relating to the needs of women and men, including the young and the old. She sits on the Gender and Diversity Associates

Committee of CIP, through which she contributes to the development of gender-responsive policies. Her email address is m.njenga@cgiar.org

Diana Lee-Smith is an Associate at the Mazingira Institute, Kenya. She has published widely on gender and urban agriculture, including a book entitled *Women Managing Resources: African Research on Gender, Urbanisation and Environment*, published by the Mazingira Institute (Nairobi 1999). She headed the global Women and Shelter Network of Habitat International Coalition from 1988 to 1993 and was responsible for gender mainstreaming in UN-Habitat between 1998 and 2001. From 2002 to 2005 she was African Co-ordinator of Urban Harvest, the system-wide initiative of the Consultative Group on International Agricultural Research (CGIAR), and is the co-editor of two forthcoming books based on the research done in this period. Her email address is Diana.LeeSmith@gmail.com

Gordon Prain is the global co-ordinator of Urban-Harvest, Peru. Urban Harvest is a cross-cutting programme of the Consultative Group on International Agricultural Research (CGIAR) on urban and peri-urban agriculture. He is a social anthropologist whose research programme in Latin America from 1984 to 1991 with the International Potato Center included studies of women's management of root-crop genetic diversity and seed. From 1991 he co-ordinated UPWARD, a participatory research and development network in Southeast Asia, which included research on women's management of vegetable gardens and support for household-run peri-urban agro-enterprises. He has been co-ordinating the Urban Harvest programme since 2000, supporting gender-focused research and development activities in cities in Asia, sub-Saharan Africa, and Latin America. His email address is g.prain@cgiar.org

Joanna Wilbers, who has been succeeded by Femke Hoekstra, is a Junior Consultant with the International Network of Resource Centres on Urban Agriculture and Food Security (RUAF Foundation), ETC Urban Agriculture, The Netherlands. Her email address is joanna.wilbers@gmail.com

Authors of case studies

Case study	Authors
Gender dimensions of urban and peri-urban agriculture in Hyderabad, India	**Gayathri Devi** Researcher, International Water Management Institute, Hyderabad, India mgayathridevi@yahoo.com **Stephanie Buechler** Research Associate, University of Arizona, USA buechler@email.arizona.edu
Gender in jasmine flower-garland livelihoods in peri-urban Metro Manila, Philippines	**Raul Boncodin** Assistant Manager, Intellectual Property Management Unit, IRRI, Los Baños, Philippines. boncodin@cgiar.org **Arma Bertuso** Network Affiliate, Users' Perspectives With Agricultural Research and Development (UPWARD), International Potato Center (CIP), Philippines a.bertuso@cgiar.org **Jaime Gallentes** Research Fellow, CIP–UPWARD, Philippines j_gallentes@cgiar.org **Dindo Campilan** Leader, Southwest and Central Asia Region, International Potato Center, Delhi, India d.campilan@cgiar.org **Rehan Abeyratne** BSc. Economics Student, Brown University, USA Helen Dayo Researcher, University of the Philippines, Philippines helenfd@yahoo.com
Gender and peri-urban vegetable production in Accra, Ghana	**Lesley Hope** Junior Researcher, International Water Management Institute (IWMI), Ghana l.hope@cgiar.org **Olufunke Cofie** Senior Researcher, International Water Management Institute (IWMI), Ghana o.cofie@cgiar.org **Bernard Keraita** Research Officer, International Water Management Institute (IWMI), Ghana b.keraita@cgiar.org **Pay Drechsel** Principal Researcher, International Water Management Institute (IWMI), Ghana p.drechsel@cgiar.org
Gender in urban food production in hazardous areas in Kampala, Uganda	**Grace Nabulo** Lecturer, Makerere University, Kampala, Uganda gracenabulo@hotmail.com **Juliet Kiguli** Lecturer, Makerere University, Uganda jkiguli@musph.ac.ug **Lilian Kiguli** Graduate Student, Rhodes University, South Africa lilykivw@yahoo.co.uk

Gender dynamics in the Musikavanhu urban agriculture movement, Harare, Zimbabwe	**Percy Toriro** Urban Agriculture and Planning Specialist, RUAF–Cities Farming for the Future Programme, Municipal Development Partnership for Eastern and Southern Africa (MDP-ESA), Zimbabwe ptoriro@mdpafrica.org.zw
Key gender issues in urban food production and food security in Kisumu, Kenya	**Zarina Ishani** Programme Officer, Mazingira Institute, Kenya zarinaishani@yahoo.co.uk
Urban agriculture, poverty alleviation, and gender in Villa María del Triunfo, Peru	**Noemí Soto** Adviser, Urban Agriculture Projects, IPES, Peru Noemi@ipes.org.pe **Gunther Merzthal** Regional Co-ordinator, RUAF–Cities Farming for the Future Programme, IPES, Peru gunther@ipes.org.pe **Maribel Ordoñez** Adviser, Urban Agriculture Projects IPES, Peru maribel@ipes.org.pe **Milagros Touzet** Adviser, Urban Agriculture Projects, IPES, Peru milagros@ipes.org.pe
Gender perspectives in organic waste recycling for urban agriculture in Nairobi, Kenya	Kuria Gathuru National Co-ordinator, Kenya Green Towns Partnership Association, Kenya greentowns2002@yahoo.com **Mary Njenga** Research Officer, Urban Harvest, International Potato Center, Nairobi, Kenya m.njenga@cgiar.org **Nancy Karanja** Regional Co-ordinator SSA, Urban Harvest, International Potato Center, Nairobi, Kenya nancy.karanja@cgiar.org **Patrick Munyao** Research Assistant, Urban Harvest, International Potato Center, Nairobi, Kenya p.munyao@cgiar.org
Urban agriculture as a strategy to promote equality of opportunities and rights between men and women in Rosario, Argentina	**Mariana Ponce** Gender Adviser, Centro de Estudios de Producciones Agroecológicas (CEPAR), Argentina lasnornas@hotmail.com **Lucrecia Donoso** Psychologist, Secretaria de Promoción Social Municipalidad de Rosario, Argentina areamujer@rosario.gov.ar
The role of women-led micro-farming activities in combating HIV/AIDS in Nakuru, Kenya	**Mary Njenga** Research Officer, Urban Harvest, International Potato Center, Nairobi, Kenya m.njenga@cgiar.org **Nancy Karanja** Regional Co-ordinator SSA, Urban Harvest, International Potato Center, Nairobi, Kenya nancy.karanja@cgiar.org

Kuria Gathuru
National Co-ordinator, Kenya Green Towns Partnership Association, Kenya
Greentowns2002@yahoo.com
Samwel Mbugua
Nakuru Site Co-ordinator, Urban Harvest, International Potato Center, Nairobi, Kenya and Lecturer, Egerton University Kenya
kasimbax@yahoo.com
Bernard Ngoda
Lecturer, Egerton University, Kenya
ngoda5@yahoo.com
Naomi Fedha
MA Student, Egerton University, Kenya
fedhanaomi@yahoo.com

Gender dynamics of fruit and vegetable production and processing in peri-urban Magdalena, Sonora, Mexico

Stephanie Buechler
Research Associate, Bureau of Applied Research in Anthropology, University Arizona, USA
buechler@email.arizona.edu

Urban agriculture and gender in Carapongo, Lima, Peru

Blanca Arce
Project Leader, Colombian Corporation of Agricultural Research (Corpoica), Colombia
barce@corpoica.org.co
Gordon Prain
Global Co-ordinator, Urban Harvest, International Potato Center, Lima, Peru
g.prain@cgiar.org
Luis Maldonado
Research Assistant, Impact Enhancement Division, International Potato Center (CIP), Lima, Peru
l.maldonado@cgiar.org

Gender and urban agriculture in Pikine, Senegal

Gora Gaye
Project Manager, African Institute for Urban Management IAGU, Senegal
gora@iagu.org
Mamadou Ndong Touré
Geographer/Urbanist, Cabinet d'Architecture et d'Urbanisme du Sénégal (CAUS), Senegal
ndongtoure@yahoo.com

Co-ordinators of field testing of the guidelines and tools

Project where testing was implemented	Implemented by	Co-ordinator of the tests
Community building through waste recycling and agro-enterprise development, Kampala, Uganda	CIP–Urban Harvest, Environmental Alert, Kampala City Council, and Center for Tropical Agriculture (CIAT)	**Buyana Kareem** MA Student, Women and Gender Studies, Makerere University, Uganda buyana@ss.mak.ac.ug.
Community-based briquette production, Nairobi, Kenya	CIP–Urban Harvest and Kenya Green Towns Partnership Association (KGTPA)	**Kuria Gathuru** National Co-ordinator, Kenya Green Towns Partnership Association, Kenya greentowns2002@yahoo.com
Combating HIV/AIDS through women-led livestock and vegetable enterprises, Nakuru, Kenya	Urban Harvest, Ryerson University, University of Toronto, International Livestock Research Institute (ILRI)	**Kuria Gathuru** National Co-ordinator, Kenya Green Towns Partnership Association, Kenya greentowns2002@yahoo.com
Narrowing crop-management knowledge gaps in sampaguita producers through training, Manila, Philippines	CIP–UPWARD and University of Philippines	**Arma Bertuso** Network Affiliate, CIP–UPWARD a.bertuso@cgiar.org
The impact of urban agriculture on food security and nutrition of women and young children (Zona de Nievería, distrito de Huachipa, Lima, Peru)	Instituto de Investigación Nutricional, McGill University	**Rosario Bartolini** Consultant, Instituto de Investigación Nutricional, Peru rosario.bartolini@gmail.com
Establishment of a producers' centre for research, training, and demonstration on arid urban gardening (Villa María del Triunfo, Lima, Peru)	Municipalidad de Villa María del Triunfo, la Red de Energía del Perú–REP, and IPES Promoción del Desarrollo Sostenible.	**Noemí Soto** Adviser, Urban Agriculture Projects, IPES, Peru noemi@ipes.org.pe
Musikavanhu Project, Budiriro, Harare, Zimbabwe	Musikavanhu Project with incidental support of FAO, AREX, Zimbabwe Fertiliser Company, SEEDCO, MDP	**Percy Toriro** Urban Agriculture and Planning Specialist, RUAF–Cities Farming for the Future Programme, Municipal Development Partnership for Eastern and Southern Africa (MDP–ESA), Zimbabwe ptoriro@mdpafrica.org.zw

Improving availability of treated domestic wastewater for poor urban farmers to enhance food security and livelihoods in Bulawayo City, Zimbabwe	Bulawayo City Council, MDP–RUAF, IWSD, SNV, Environment Africa, ZELA, and Zimbabwe Open University	**Nomusa Mhlanga** Lecturer, National University of Science and Technology, Zimbabwe nmhlanga@nust.ac.zw
Projet de sécurisation des agriculteurs et agricultrices urbain (es) et de valorisation des trames vertes de la Commune de Bobo-Dioulasso	IAGU–RUAF, CPAU–B	**Marie Madeleine Bengali** Ingénieur Agronome, Polyvalent Agricultural Centre of Matourkou, Senegal bengalimade@yahoo.fr
Urban agriculture programme, Rosario, Argentina	Municipality de Rosario, CEPAR, la Red de Huerteras y Huerteros de Rosario	**Mariana Ponce** Gender Adviser, Rosario's Centre for Studies in Agroecological Production (CEPAR), Argentina lasnornas@hotmail.com
Safe use of wastewater for agriculture, Hyderabad, India	IWMI India, ICRISAT, BMZ, Municipality Seringampally and Uppal, Andhra Pradesh Urban Services for the Poor and District Rural Development Agency of (DRDA) of Rangareddy district.	**Madhu Murthy** Consultant, Access Livelihoods Consulting Ltd, India nmadhumurthy@gmail.com Venkata Krishnagopal Grandhi Consultant, Access Livelihoods Consulting Ltd, India gvkgopal@gmail.com
Establishment of a sustainable system for the delivery of inputs and equipment of urban producers in Pikine, Senegal	UPROVAN, ANCAR, ENDA, CDH, CFPH	**Gora Gaye** Project Manager, African Institute for Urban Management IAGU, Senegal gora@iagu.org

Index

www.ingramcontent.com/pod-product-compliance
Lightning Source LLC
Chambersburg PA
CBHW072043020426
42334CB00017B/1370